MECHANISM AND MATERIALISM

MECHANISM AND MATERIALISM

British Natural Philosophy in An Age of Reason

ROBERT E. SCHOFIELD

PRINCETON UNIVERSITY PRESS
PRINCETON, NEW JERSEY
1970

L.C. Card: 72-90960
I.S.B.N.: 0-691-08072-0

The ornament on the title page is from
Domenico Cotugno
PHIL. ET MED. DOCT.
De Aquaeductibus Auris Humanae Internae
ANATOMICA DISSERTATIO.
Naples, 1761

This book was composed in Linotype Baskerville

Printed in the United States of America
by Princeton University Press, Princeton, New Jersey

Preface

THIS IS A HISTORY of a pair of the ideas made manifest in changing patterns of experiment and interpretation during a century of British science. While I believe this particular pair to have been of fundamental significance to the rational dialogue of eighteenth-century natural philosophy, I do not pretend that they comprehend the entirety of the science of that period. This is not, that is, a survey of eighteenth-century science, not even of that science in Britain and her colonies. Hopefully it will provide some of the materials needed for such a study, but my major purpose is something different. The principal barrier to a definitive survey is not so much lack of materials as failure of understanding. For the eighteenth century in British science was not merely a period of empirical experimentation, nor was it simply a fallow season of consolidation, resting between the seventeenth-century geniuses of the mechanical philosophy and the nineteenth-century geniuses of classical physics. As long as these tired generalities continue to provide the frame for discussions of eighteenth-century British science, those discussions will continue to be the catalogues of disconnected fact and episode already available in too ample profusion. Some effort must be made to discover the reasons which relate the apparently unrelated, which identify the sciences of the eighteenth century as characteristic of that Age of Reason. If this work can prompt that kind of effort, I will have achieved my goal.

In the making of a book, there is no end to the help that is needed, and I have been fortunate in all that I have obtained. Case Institute of Technology has constantly encouraged my work. A grant from the National Science Foundation supported the research which went into this book, and a year's leave of absence, made possible, jointly, by the John Simon Guggenheim Foundation and the Institute for Advanced Study, enabled me to put the results in final form.

Necessary as such assistance is, however, in supplying time and money, the major requirement is always in the area of ideas. I hope that the indirect influence on my thinking of Hélène Metzger and Alexandre Koyré will be sufficiently obvious. The direct influence of I. Bernard Cohen must be apparent. I owe to him my initial impulse toward eighteenth-century studies and much subsequent help and advice. The ideas described in this work were first tried on my graduate students in the history of science at Case Institute of Technology. Some few of their names are explicitly mentioned in the text, but all of them should be thanked for their aid in sharpening the concepts. I have also profited from discussions with friends and colleagues, particularly Professors Marshall Claggett, Duane H. D. Roller, Arnold Thackray, and L. Pearce Williams. They did not always agree with me, nor will they now, but they will find this work improved by their criticisms of it.

For permission to abstract, cite, or quote original documents in their possession, I am deeply grateful to: the Trustees of the British Museum; the University Library and the library of Gonville and Caius College, Cambridge; the Linda Hall Library, Kansas City, Missouri; the Library of the University of Texas, Austin; and Wesley C. Williams. I also acknowledge, with gratitude, the use of the unpublished papers and theses cited in my bibliography, without which my work would have been even more incomplete. Naturally, their authors are not responsible for the uses to which I have put those studies. Any historical work is dependent upon the good will of numerous librarians. May they all accept my thanks, expressed by way of direct acknowledgment to those in the library of Case Institute of Technology (now The Sears Library of Case Western Reserve University) and in the library of the School of Historical Studies, Institute for Advanced Study. And finally, for those details which are only routine when one has more than routine help, I must thank my secretary at the Institute, Mrs. Mary Ann Marocco Smith.

To my wife, Mary-Peale, who had her own work to do, and left me alone to do mine, this book is dedicated.

Princeton, New Jersey
1968

Contents

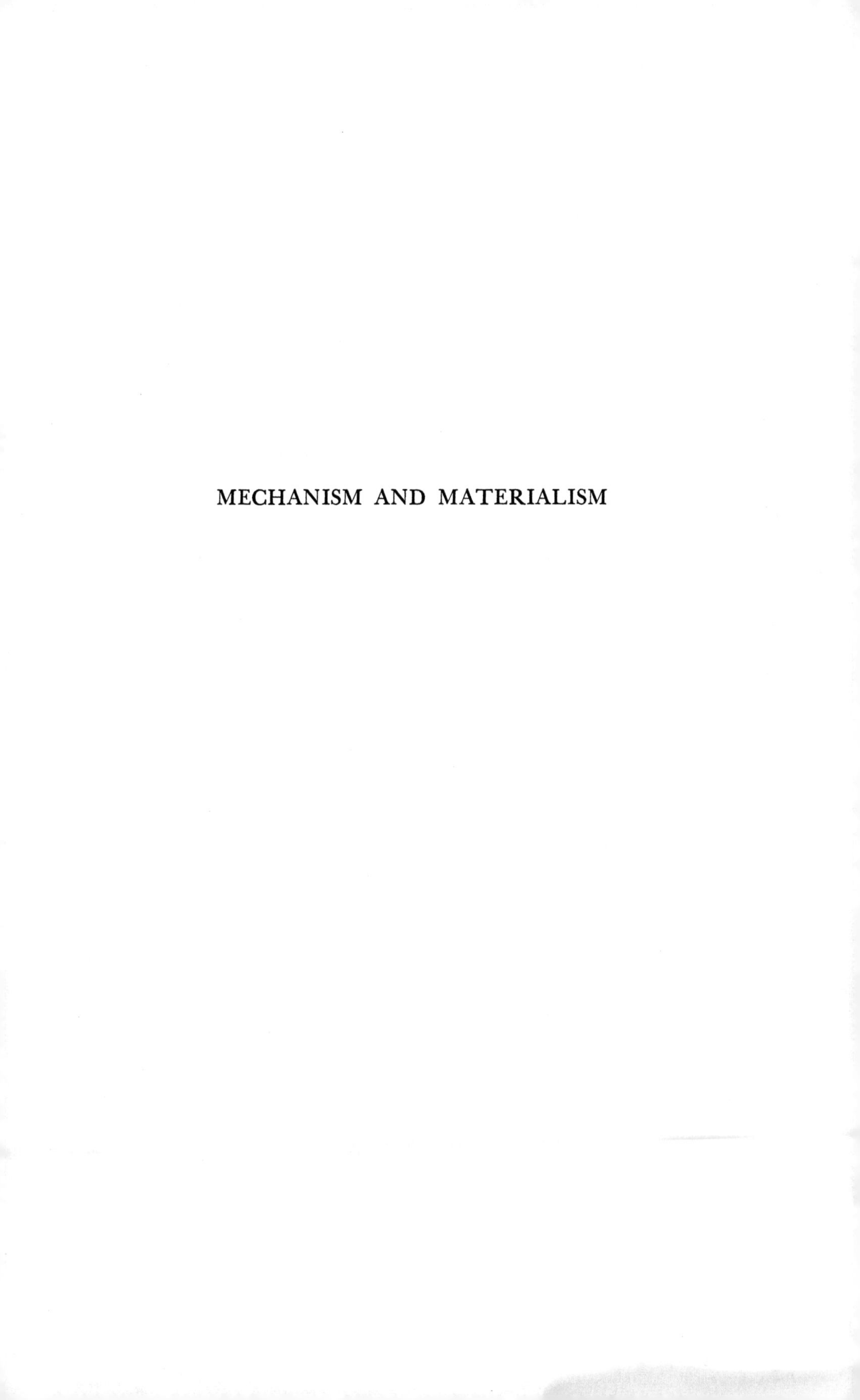

MECHANISM AND MATERIALISM

CHAPTER ONE

Newton's Legacy

THE EIGHTEENTH CENTURY, by common consent, was an Age of Reason, a time when Enlightened men sought the solution to all their problems in the exercise of man's mind. This was true in religion and politics, in the arts and economics; it was also true in science. One might almost say this last was sufficiently obvious, and indeed it is for Continental science, where the mathematicians turned Newton's *Principia* into analytical mechanics and even the experimenters demonstrated, in part through their errors, the uses of theory. But Continental science was inspired by Descartes' equation of thought and existence. In Britain the patron saint of science was Bacon, and there the Age of Reason was moderated by common sense. The standard epistemological declaration in British scientific works, through most of the century, was a ubiquitous denial of feigned hypotheses and a ringing testimonial to cautious and complete induction. Perhaps this is why it has never seemed quite as peculiar as, in fact, it is, that British natural philosophy in an Age of Reason should so generally be regarded as essentially irrational and empirical.[1]

Yet Britain was the country of Newton as well as of Bacon. The eighteenth century was, particularly in science, the Age of Newton as well as of Reason, and if British natural philosophers do not so often proclaim their allegiance to Newton as to Bacon, that allegiance permeates their work and why should they belabor the obvious? It is true that the *Opticks* represented to the British a newer and more vital testament for their exegesis than the *Principia* and that they saw, in the *Opticks*, a continuation of traditional British experimentalism. But the *Opticks* is not an empirical work. In fundamental theory it does not differ from the *Principia*, and, moreover, it was not the experiments of the *Opticks* which inspired their study, but the queries, what Priestley was subsequently to call Newton's "bold and eccentric thoughts."[2]

1 Exhaustive documentation of this view is impossible, as it is nearly universal in the literature relating to eighteenth-century British science. The major surveys—*e.g.*, Abraham Wolf, *History of Science, Technology and Philosophy in the 18th Century* (London: Allen and Unwin, 1952), 2nd edn. by Douglas McKie; or James Partington, *History of Chemistry*, III (London: Macmillan and Co., Ltd., 1962)—are saturated with it, as are the extant biographies of such people as Stephen Hales, Joseph Black, Joseph Priestley, Henry Cavendish, or James Hutton, whose experiments are praised while their possession of guiding hypotheses is denied or decried.

It was under the aegis of Newtonian ideas that British natural philosophers were to design, perform, and interpret their experiments. Any attempt to recover the thread of rationality in eighteenth-century British natural philosophy must, therefore, begin with the legacy Newton left that century in his writings. But the nature of the experiments, and of acceptable interpretations of them, varied continuously throughout the century, while the acknowledgments to Newton's authority remained. Clearly the reasoning which underlay eighteenth-century British science cannot be found by determining what Newton really meant in the *Principia* or the *Opticks*. Indeed, for such a study, Newton's real meaning is essentially irrelevant. What is needed is an investigation of what eighteenth-century Britons thought he meant and how and why their thinking changed through the century.

Some of the ablest and most provocative studies of eighteenth-century thought have employed this kind of iconology, in which the sectarian virtues of the Enlightenment have been traced in the images of Newton it created to represent itself.[3] There are, therefore, worthy precedents for an attempt to illuminate certain aspects of eighteenth-century scientific thought through the changing interpretations of Newtonianism. The existence of such precedents demonstrates, however, the range of subjects on which Newton was regarded as the authority. In science alone these include the nature of space and time, the uses of hypotheses, of mathematics, and of experiments, and even the role of God in the universe. Each of these affords a useful tool for historical analysis of eighteenth-century British science. But there is another, and equally useful, theme to trace, one that frequently comprehends the others and one which can reasonably be regarded as central both to Newton's work and to that of natural philosophy in general. That is, Newton's theory of matter and its action, which pervades both the *Principia* and the *Opticks* and, consciously or unconsciously, most of the scientific writing of the eighteenth century as well.

[2] Joseph Priestley, *Experiments and Observations on different Kinds of Air* (London: J. Johnson, 1775), 2nd edn., p. 259.

[3] See, for example, Hélène Metzger, *Attraction universelle et religion naturelle chez quelques commentateurs anglais de Newton* (Paris: Hermann et Cie, 1938); and I. Bernard Cohen, *Franklin and Newton* (Philadelphia: American Philosophical Society, 1956). Carl Becker's *Heavenly City of the Eighteenth-Century Philosophers* (New Haven: Yale University Press, 1932); and Marjorie Hope Nicolson's *Newton Demands the Muse* (Princeton: Princeton University Press, 1946), discuss the influence of Newton in some nonscientific areas.

The importance of matter theory to the natural philosopher of all centuries has been summarized by J.J. Thomson, himself one of the most distinguished of a generation of matter theorists:

> From the point of view of the physicist, a theory of matter is a policy rather than a creed, its object is to connect or coordinate apparently diverse phenomena and above all to suggest, stimulate, and direct experiment. It ought to furnish a compass which, if followed, will lead the observer further and further into previously unexplored regions.[4]

Except for its experimental bias and the not entirely clear distinction between policy and creed (for surely the latter can unify, stimulate, and inspire as well as the former), this statement can be extended to include all the theoretical sciences, for the importance of establishing a theory of matter and its action has been evident to all natural philosophers since the days of the Greeks. More specifically, amid the epistemological and ontological arguments that distinguished the seventeenth-century flight from Aristotelean Scholasticism, the nature of matter played a major part. Indeed, one of the chief successes of the Scientific Revolution was that it found a new matter theory to replace the old one.

Implicitly or explicitly, Galileo, Bacon, Descartes, Gassendi, Boyle, and the legion of their imitators and disciples rejected the Scholastic view of matter. In this, matter appeared as a common, undifferentiated, substratum of the organized world of nature, a principle of potentiality capable of becoming all substances, exhibiting all qualities, through its union with essential and accidental forms. Change from one substance to another occurred when one essential form replaced another, matter remaining the same. Variation of property: growth, decay, taste, color, temperature, and by the seventeenth century nearly every phenomenological change in nature, was caused by the addition or subtraction, intension or remission, of accidental forms. Whatever their view of the power of reason, the efficacy of experiment, the nature of time and space, or the existence of the void, major seventeenth-century scientists joined in regarding this theory as rhetorical sophistry, for which there must be substituted a new, a mechanical, philosophy.

Individual formulations of this mechanical philosophy of matter varied in structure and detail. The Cartesians equated mat-

[4] J. J. Thomson, *The Corpuscular Theory of Matter* (New York: Charles Scribner's Sons, 1907), p. 1, cited by Cohen, *Franklin and Newton*, p. 285.

ter with extension and, denying the existence of the void, filled space with the variously sized and shaped particles resulting from the infinite divisibility of dimension. To these they added a fixed and finite quantity of motion which was distributed and redistributed among the particles jostling against one another in the universal plenum. The Gassendists, reviving and modifying Epicurean atomism, insisted on a void, within which a variety of uniquely definable particles, atoms with permanently fixed sizes and shapes, moved according to the motions divinely determined for them at their creation. Between the extremes of Cartesianism and atomism is a common element which less dogmatic scientists could use in formulating their own positions. The essential feature to all these views, seen best perhaps in Robert Boyle's eclectic corpuscular philosophy, was the replacing of peripatetic qualities by variations in the sizes, shapes, configurations, and motions of particles.[5]

The various corpuscular philosophies had another, unintentional, element in common. In ridding matter of Aristotelean forms and qualities, the mechanical philosophers rejected volition, sympathy, and appetency from their physical world view. Their unique emphasis on geometrically definable quantities achieved notable successes in Galileo's kinematic mechanics and in geometrical optics, but neither Descartes' rationalism nor Boyle's experimentalism was able to do more than demonstrate the possibility that microscopic physical phenomena might follow from the interaction of specifically designed particles. As a choice between tautologies, there is only slight gain in the suggestion of particle sharpness instead of acid corrosiveness, in coiled shapes instead of innate elasticity, or screw-form effluvia instead of magnetic attractive virtue unless an independent determination of corpuscle shapes and sizes can be made. The neutral parameters of the corpuscular philosophy of matter—size, shape, and motion—proved inadequate to the task of explaining the Aristotelean qualities which soon became materialized in *ad hoc* particulate descriptions. It was at this juncture that Newton added to the kinematics of the corpuscular philosophy the new parameter of interparticulate forces.[6]

[5] The literature on pre-eighteenth century theories of matter is extensive. For some introduction see Ernan McMullin, ed., *The Concept of Matter* (Notre Dame: University of Notre Dame Press, 1963), and, for a recent study particularly applicable to this one, Robert H. Kargon, *Atomism in England from Hariot to Newton* (Oxford: Clarendon Press, 1966).

[6] Examples of corpuscularian attempts to explain microscopic phenomena and their difficulties can be found in Marie Boas Hall, *Robert Boyle on Natural Philoso-*

The appearance of a corpuscular hypothesis in Newton's first published papers, on light and colors, in the *Philosophical Transactions* of 1672 began the development of a Newtonian corpuscularity which continued well past his introduction of aetherial concepts in 1713.[7] Such contemporaries as Robert Hooke and Ignace Pardies objected to the concept of particulate light, and Newton responded with characteristic equivocations which further obscured the nature of a corpuscularity only slightly revealed in its limited application.[8] The revolutionary influence of Newton's writings was, therefore, first manifest through the *Principia*, where dynamic corpuscularity was an integral part of the transformation of kinematic mechanics into dynamics.[9] The fundamentals, even here, are conventional enough. The particulate nature of matter is assumed throughout, and its particles are ultimately defined as extended, hard, impenetrable, movable, and endowed with *vires inertiae*. Their homogeneity is implied in the concept of mass as quantity of matter, although Newton declines to decide whether the mathematically distinguishable parts " . . . may by the powers of nature be actually divided and separated from one another" indefinitely [Book III, Rules of Reasoning III].[10] Although the extremes of Cartesianists and atomists

---

phy (Bloomington: Indiana University Press, 1965), pp. 64, 69-71; and A. Rupert Hall, *From Galileo to Newton, 1630-1720* (New York and Evanston: Harper and Row, 1963), pp. 218-20.

7 The ensuing discussion of Newton's work is not intended to throw new light, nor indeed any light at all, on Newton's real natural philosophy, its development or changes. Rather and more simply, it is an extract of portions of his published "legacy," as read and understood by eighteenth-century Britons, which was the basis for a British Newtonian Theory of matter and its action. Whether this Newtonianism was a "true" expression of Newton's beliefs is a problem for Newton scholars to debate.

8 The most convenient edition of these papers, with a useful introduction by Thomas S. Kuhn, is in I. Bernard Cohen, ed., *Isaac Newton's Papers and Letters on Natural Philosophy* (Cambridge: Harvard University Press, 1958), pp. 27-235. For earlier examples of Newton's corpuscular opinions, see A. Rupert Hall and Marie Boas Hall, *Unpublished Scientific Papers of Isaac Newton* (Cambridge: at the University Press, 1962), *passim*.

9 The terms are used here in their formal sense to distinguish between a system of mechanics which is constructed about the notion of forces and one that is not. Clearly such men as Gilbert, Kepler, Descartes, and Huygens had treated the problem of forces in magnetism, celestial motion, elastic collisions, pendular motion, etc., but their work was limited and frequently unsatisfactory. It was Newton who systematically generalized forces in mechanics and did so in a way involving the corpuscular philosophy.

10 The citations in brackets all refer to the Andrew Motte, tr., Sir Isaac Newton, *The Mathematical Principles of Natural Philosophy* (London: Benjamin Motte, 1729), but quotations and paraphrases have been verified in the earlier editions to assure that the essential sense has not been changed.

might find cause for objection, there is nothing here to which Robert Boyle would object, any more than he would have objected to Newton's corporeality of light as expressed earlier in the optical papers.

The new developments occur in Newton's treatment of particle interaction, for his particles do not act by contact but at a distance. His particles of matter are to be regarded as "endued" with central powers, which vary with the nature, quantity, and quality of their bodies and are diffused or propagated through space to attract or repel other bodies as a function of their distances [Bk. I, Defin. VIII, and Sect. XI, Scholium]. Now Newton repeatedly denies that he means to define the kind, means, or causes of action or that he means that his powers or forces have physical existences [Bk. I, Defin. VIII]. He insists that he is speaking only mathematically; that attraction means only the endeavor of bodies to approach one another, an endeavor which might as well be caused by the action of emitted spirits or of the medium in which they are placed, as by the action of the bodies themselves [Bk. I, Sect. XI, Scholium]. Even in Book III, where he applies his mathematical principles to the physical universe, Newton will not "affirm gravity to be essential to bodies," though "we must . . . universally allow, that all bodies whatsoever are endow'd with a principle of mutual gravitation [Rules of Reasoning]."

Inevitably these denials were not taken seriously by his contemporaries. It is hard to see how they could have been. Newton had disproved the Cartesian vortex theory of celestial motion by a mathematical demonstration that its consequences disagreed with phenomena. Then he had presented a mathematical demonstration that the consequences of a universal attracting force between particles, varying directly with their mass and inversely as the square of their distance, were in agreement with phenomena. What else could he mean but that such a force was true? And if this gravitational force existed truly, by inference so did the other central powers or forces mentioned in the *Principia*: magnetic force, repeatedly used as an example of attraction varying with size and distance [Bk. I, Defins. VI and VIII; Bk. III, Prop. VI, Theor. VI, cor. 5]; the force responsible for reflections, refractions, and inflections of light [Bk. I, Sec. XIV, Prop. XCIV, Theorem XLVIII and Scholium]; or the centrifugal forces of elastic fluids, reciprocally proportional to the distances of particle centers, terminating in those particles next to them or diffused not much farther, by which Newton had mathematically derived Boyle's gas law [Bk. II, Prop. XXIII, Theorem XVIII and Scholium]. Moreover, Newton had declared in the Preface: " . . . all the difficulty of philosophy seems to consist in this, from the phaenomena of motions to investigate the forces of Nature, and then from these forces to demonstrate the other phaenomena." Whatever Newton may have intended or expected, his readers found among the novelties of the *Principia* a new dynamic theory of matter.

Almost at once there were objections that, in his forces, Newton had revived the "occult" qualities of Aristotelean science. Newton denied the charge vehemently and tried to prevent his supporters, members of a growing seminationalistic school of dynamic corpuscularians, from misstating his views. He wrote to Richard Bentley in 1693:

> It is inconceivable, that inanimate brute Matter should, without the Mediation of something else, which is not material, operate upon and affect other Matter without mutual Contact. . . . That Gravity should be innate, inherent and essential to Matter, so that one Body may act upon another at a Distance thro' a *Vacuum,* without the Mediation of any thing else, by and through which their Action and Force may be conveyed from one to another, is to me so great an Absurdity, that I believe no Man who has in philosophical matters a competent Faculty of thinking, can ever fall into it.[11]

But his chief defense, and that of his supporters, was a reiteration of the positivistic assertion that to talk of forces was to speak mathematically without discoursing on causes, and that to define forces and derive laws from them was philosophically sound, if the laws were verified by phenomena, though the origins and causes of the forces were not known.

The publication in 1704 of the first edition of the *Opticks* was no help in maintaining this position.[12] Though Newton writes always of "Rays" of light and not of particles, his readers found his treatment (consistent with that of the *Principia*) to be one of particles and, moreover, of particles acted on by forces at a distance. The issue is confused by the introduction of "easy fits of

[11] Letter III of *Four Letters from Sir Isaac Newton to Doctor Bentley,* in Cohen, *Newton's Papers and Letters,* pp. 302-303. As these letters were not published until 1756, they had little effect on early eighteenth-century Newtonianism, though the tenor of the argument was known and the letters themselves were cited in the latter half of that century and throughout the nineteenth.

[12] [Sir Isaac Newton], *Opticks: or, A Treatise of the Reflexions, Refractions, Inflexions and Colours of Light* (London: Sam. Smith and Benj. Walford, 1704).

reflection or refraction" to explain the production of "Newton's Rings," though he will not say whether these fits are caused by a circulating or vibrating motion of the light ray or of the medium or something else (Bk. II, Part III, Prop. XII]. Nevertheless, the general argument is clear. Reflection, refraction, and inflection are all to be explained by some one power of bodies, very nearly proportional to their density and varying with distance, by which they act on light rays without immediate contact [Bk. I, Part I, Prop. VI, Theor. V; Bk. II, Part III, Props. VIII-X; Bk. III, Obs. I]. Here there are none of those reservations that he does not really mean forces, or that, if he does, they are only mathematical and, in any event, cannot possibly be thought really to act at a distance. Sixteen queries are appended to Book III of the *Opticks*, presumably representing propositions on which Newton was undecided, and two of these (Queries 1 and 4) relate again to the action of bodies on light at a distance. Both queries had, however, been affirmed among those propositions of the text which had been proved "by Reason and Experiments" [Bk. I, Design], and, as all the queries were stated in the affirmative, their readers generally had the impression that Newton really knew they were correct. If Newton believed that the cautionary injunctions of the *Principia* would carry over into the *Opticks*, he was obviously wrong. Here was begun the detailed development of dynamic corpuscularity into a policy statement for eighteenth-century natural philosophers.

The continuing extension of Newton's theory of matter into the coordinating, stimulating, and directing policy for eighteenth-century natural philosophers took place by way of the second edition of the *Opticks*, published in Latin in 1706. There is comparatively little difference in the text of the first 16 queries taken over from the first edition, but seven new queries were added which took in the whole range of eighteenth-century experimental concerns. Newton had introduced the queries with the suggestion that they proposed problems "in order to a further search to be made by others," and the evidence is strong that they did indeed become a starting point for most of the scientific inquiries of the century. I. Bernard Cohen has shown that the *Opticks* was the more widely read and cited of Newton's two major works for eighteenth-century experimentalists and that the queries were the focus of their attention.[13] One might further specify that of these queries, the new ones, numbered 17-23 in the Latin edition and 25 through 31 in the second English edition of 1717, had the

[13] Cohen, *Franklin and Newton, passim.*

major impact, and of these the most frequently referred to were numbers 20-23 (28-31).[14]

Filling nearly three-quarters of the query-space (No. 23 is almost as long as the other 22 combined), Queries 20-23 make explicit the particulate nature of matter within a frame of force. In the beginning God had formed matter in solid, massy, hard, impenetrable, inert, movable particles of determinant sizes and shapes. The smallest of these particles cohere by the strongest attractions, to compose bigger particles of weaker virtue, which, in turn, cohere to compose still bigger particles of correspondingly weaker virtue, and so on, until the progression ends in the bigger particles of sensible magnitude on which the operations of chemistry, colors of natural bodies, etc., depend. All particles are possessed of certain powers, virtues, or forces, by which they act at a distance on one another to produce the greater part of the phenomena of nature. Examples of phenomena that might be produced by attractive forces are given: refraction and inflection of light, gravity, magnetism and electricity, deliquescence, chemical composition and decomposition, ebullition, dissolution, concretion, crystallization, cohesion, congelation, and capillarity. And then, "as in Algebra, where affirmative Quantities vanish and cease, there negative ones begin; so in Mechanicks, where Attraction ceases, there a repulsive Virtue ought to succeed"—and examples follow of phenomena producible by repulsive forces: reflection and emission of light, volatility and evaporation, fermentation and putrefaction, elasticity and disjunction.[15]

True, Newton again reserves judgment on the nature and origin of these forces. "How these Attractions may be perform'd, I do not here consider. What I call Attraction may be perform'd by impulse, or by some other means unknown to me. I use that Word here to signify only in general any Force by which Bodies tend towards one another, whatsoever be the Cause [p. 351]." Equally true, however, he declared:

> To tell us that every Species of Things is endow'd with an occult specifick Quality by which it acts and produces manifest Effects, is to tell us nothing: But to derive two or three general Principles of Motion from Phaenomena, and afterwards to tell us

[14] My quotations, paraphrases, and citations are taken, in the first instance, from the third (second English) edition, second issue, *Opticks: or, A Treatise of the Reflections, Refractions, Inflections, and Colours of Light* (London: W. and J. Innys, 1718), but in all appropriate instances, their import has been verified from the Latin, *Optice, sive de Reflexionibus, Refractionibus, Inflexionibus et Coloribus Lucis libri Tres* (London: Sam. Smith and Benj. Walford, 1706).

[15] *Opticks* (1718), taken from throughout Query 31; quotation from 370, 375-76.

> how the Properties and Actions of all corporeal Things follow from these manifest Principles, would be a very great step in Philosophy, though the Causes of those Principles were not yet discover'd [p. 377].

Little wonder, then, that early eighteenth-century British natural philosophers interpreted these "Principles of Motion" to mean forces or that they found the primary meaning of Newtonianism to be this corpuscular philosophy.[16] And less wonder, considering the authority Newton's opinions had acquired, that they exerted themselves to carry out what might reasonably be thought of as an injunction—from the motions, find the forces; from the forces, derive the laws. Then came the denouement.

After a quarter of a century, during which early eighteenth-century British Newtonians had adopted and extended the notion of dynamic corpuscularity, derived first from the *Principia* and then from two successive editions of the *Opticks,* Newton presented an alternative, aetherial, explanation for phenomena. Although many of his expositors had repeated the reservations on the nature of matter and force contained in the *Principia* (reservations which Alexandre Koyré has described as "pseudo-positivistic and agnostic"), most of them had also elaborated upon the concept that the forces represented the continued activity of God in His universe.[17] Nor was this clearly inconsistent with Newton's opinion. Nothing in his denials that forces are innate in matter precluded their being the product of some immaterial cause. In a discussion with David Gregory in 1705, Newton confessed his belief that "God is omnipresent in a literal sense," filling all space, including that empty of body. The force of gravity (and probably other forces as well) was then the mediate action of God in the universe constantly giving it activity and direction.[18] This belief is consistent with the teachings of the neo-Platonists at Cambridge during Newton's student days there and with the arguments which Newton's spokesman, Samuel Clarke, carried on with Leibniz.[19] Numerous passages in both the

[16] Cohen, *Franklin and Newton,* pp. 181-82.

[17] Alexandre Koyré, *Newtonian Studies* (London: Chapman and Hall, 1965), p. 280.

[18] W. G. Hiscock, ed., *David Gregory, Isaac Newton and Their Circle. Extracts from David Gregory's Memoranda, 1677-1708* (Oxford: for the editor, 1937), p. 30.

[19] See Ernst Cassirer, *The Platonic Renaissance in England,* James I. Pettegrove, tr. (Austin: University of Texas Press, 1953); H. G. Alexander, ed., *Leibniz-Clarke Correspondence* (Manchester: Manchester University Press, 1956); Query 31, *Opticks* (1718), pp. 378-80.

Principia and the *Opticks* make sense in this interpretation. But if this was Newton's belief, for him to say so publicly and explicitly would have involved him personally in an extended discussion of just those theological opinions which this most secretive of persons kept most secret.

There is, moreover, another approach to finding the causes of force, an approach which also clarifies various allusions in the *Principia* and the *Opticks* and which, further, might better provide explanation for phenomena newly described since the first publication of those works.[20] This was the approach Newton now put forward. He had several times previously alluded to the possibility that apparent forces of attraction between bodies might be caused by impulses of the medium in which they are placed, and, though the general tendency of the arguments of the *Principia* is to disprove the idea of motion in a space-filling continuum, there are sections in Book II which treat fluid motion in a continuous rather than a discrete, particulate medium.[21] Newton explains these hints in an added General Scholium to Book III of the second edition of the *Principia* (1713):

> And now we might add something concerning a certain most subtle Spirit, which pervades and lies hid in all gross bodies; by the force and action of which Spirit, the particles of bodies mutually attract . . . and electric bodies operate . . . as well repelling as attracting the neighbouring corpuscles, and light is emitted, reflected, refracted, inflected, and heats bodies. . . .

Thus far there is nothing which necessarily precludes this "subtle Spirit" from being that spirit of God, discussed with Gregory, but later references in the Scholium to its vibrations being propagated along the solid filaments of the nerves, and his designation of it as an electric and elastic spirit are indications of its materiality. And surely it is just this implied materiality which moti-

[20] Professor Henry Guerlac has recently and cogently argued the influence of Francis Hauksbee's electrical experiments for the revival of aetherial concepts in Newton's thinking after 1706. See his "Francis Hauksbee: expérimentateur au profit de Newton," *Archives Internationale d'Histoire des Sciences,* 16 (1963), 113-28; "Sir Isaac and the Ingenious Mr. Hauksbee," in I. Bernard Cohen and René Taton, eds., *Mélanges Alexandre Koyré,* I, "L'Aventure de la Science" (Paris: Herman, 1964), pp. 228-53; and "Newton's Optical Aether," *Notes and Records of the Royal Society of London,* 22 (1967), 45-57.

[21] For relevant sections in the *Principia* see, *e.g.,* Bk. I, Sec. XI, Scholium; and Clifford Truesdell, "Rational Fluid Mechanics, 1687-1765," Introduction to: *Leonhard Euleri Opera Omnia,* Series 2, XII, "Commentationes mechanicae ad theoream Corporum Fluidorum Pertinentes" (Lausanne: Society of Natural Sciences of Helvetia, 1954), pp. xiii-xiv. See also *Opticks* (1718), Bk. II, Part III, Props. XII and XIII.

vates the addition of the General Scholium, enabling one to get away from the appearance of action-at-a-distance.

The way in which such a spirit might act to produce its effects is not described in the Scholium. There is, in fact, the typical *Principia* evasion: "But these are things that cannot be explain'd in a few words, nor are we furnish'd with that sufficiency of experiments which is required to an accurate determination and demonstration of the laws by which this . . . spirit operates." It was in further queries added to the *Opticks,* where Newton regularly felt free to speculate without the constraints of *Principia* formalisms, that the possible nature and action of this subtle, elastic, and electric spirit is discussed. Four years after the publication of the General Scholium, a new edition of the *Opticks* (1717) appeared containing eight new queries inserted as numbers 17-24 between those written for the first edition and those, now renumbered (25-31), for the second.[22] In these, the aether is described as a medium which expands through all space, filling it and pervading the pores of gross bodies by virtue of its great elasticity and because of its extreme subtlety. Its density varies with its distance from bodies, being less within them and increasingly great in free and open spaces. Its rarity and subtlety are such as scarcely to resist the motion of gross bodies, while its elasticity is so great as to produce, by impulse, the effect of gravity as those bodies endeavor to move from a denser to a rarer medium. Refraction occurs in consequence of the different densities of this elastic medium, whose vibrations, produced by the emission of light, are responsible for the conditions of easy fits of reflection or refraction. And this is the medium which, by comparison of subtleties, can probably be identified with the effluvia of electric and magnetic bodies.

Superficially we seem to have come, full circle, to a kind of reconciliation with the Cartesian plenum, but the description of Newton's aether remains incomplete and full of anomalies. It appears to be particulate and of such a subtlety as completely to fill all space not otherwise occupied by gross bodies. Yet it does not act by contiguity and transference of motion, but through its varying density and by means of its force of elasticity—that force of elasticity which Newton previously defined as a power possessed by bodies to repel one another at a distance. By his talk

[22] Perhaps formal, stylistic considerations of length and subject dictated this shuffling of arrangement. In any event, it serves to retain the prominence of the action-at-a-distance queries and leads to endless arguments on which set, 17-24 or 25-31, to take more seriously.

of a space-filling medium with mechanical action of impulse and variation of density, Newton has concealed but not eliminated the prejudicial action-at-a-distance. By the action, at insensible distances, of the single repulsive force between aether particles, Newton has attempted a notable economy of explanation, eliminating all other forces, of attraction or repulsion, between material bodies, which now stand in inert contrast to the one active material, aetherial substance. But Newton does not continue with a mathematical demonstration of how this elastic force between aetherial particles of an incomprehensively full but varying dense medium can produce any of the forces he has enumerated in the other queries of the *Opticks.*[23] He does not (and no other person, of the many who attempted it in the eighteenth and nineteenth centuries, could) mathematically derive from the aether either the force of gravity or that of pneumatic elasticity, whose effects he had computed so effectively in the *Principia.* In his materialistic, mechanistic aetherial hypothesis, Newton was not able to be as precisely mathematical nor as observationally verifiable as he had been in the occultness of his action-at-a-distance. All he could really do was provide conflicting alternatives for eighteenth-century natural philosophers looking for an authoritative, Newtonian, theory of matter.

These alternatives are distinguished, throughout this study, by the terms "mechanism" and "materialism." From the dynamic corpuscularity of the early *Principia* and the first two editions of the *Opticks,* historically evolving from the mechanical philosophy of the seventeenth century, came the conviction of the mechanists—that causation for all the phenomena of nature was ultimately to be sought in the primary particles of an undifferentiable matter, the various sizes and shapes of possible combinations of these particles, their motions, and the forces of attraction and repulsion between them which determine those motions. The materialists believed, instead, that the causes of phenomena inhere in unique substances, each possessing as an essential property

[23] The major logical problem, to modern readers, of a space-filling substance (Query 22. ". . . This Aetherial Medium . . . fills all space adequately without leaving any pores"); possessing variable density (Query 21. "Medium much rarer within dense Bodies . . . And in passing from them . . . grow[s] denser and denser perpetually. . . .") may be resolved in a seventeenth- and eighteenth-century failure to comprehend the convergence of infinities. This possibility is suggested in the statement of the Newtonian mathematician and natural philosopher, John Keill (see next chapter) who starts with the primary definition of matter as extension, adds the infinite divisibility of geometrical dimension, and reaches the position that any quantity of matter, however small, may be diffused through any finite space, however large, so as to fill it without leaving a pore greater than any given size.

the power to convey, in proportion to its quantity, some characteristic quality. Clearly this belief derives ultimately from Aristotelean substantial qualities, but an age which celebrated, in Newton, its final release from Scholastic futilities was not prepared to admit (or even to recognize) such reactionary affinities. The materialists also wanted to claim a Newtonian authority and found, in Newton's aether—that active substance which opposed the passivity of other substances—a substantial model for their fluid matters of heat and electricity, their vital spirit, and their chemical elements. It scarcely mattered to the mechanists that Newtons' dynamic corpuscularity was not the purely physical reductionism which they were ultimately to derive from it; nor did it matter to the materialists that Newton's aether possessed mechanical properties hardly considered in their differentiation of materialized quality. Each group selected from the heritage Newton had left to the eighteenth century those portions most amenable to their exegesis, and thus the scripture of eighteenth-century British natural philosophy was cited, each to his own purpose, by mechanist and materialist alike.[24]

[24] While I recognize that by common practice each of these terms has acquired an accretion of connotative meanings which differ from and even merge the distinctions employed here, it appears that historical understanding may be furthered by a differentiation of these frequently equated approaches. Some terminology is, therefore, required to separate them. My usages are formally consistent with dictionary definitions, there is some logical consistency with the historical evolution of the concepts to recommend those usages, and dynamics is as much a branch of mechanics as kinematics.

PART I

MECHANISM AND DYNAMIC CORPUSCULARITY

1687-1740

CHAPTER TWO

Diffusion of a Newtonian Creed

NEWTON MIGHT HAVE RECONCILED the disparate elements of his theory of matter and trained a school of Newtonians to spread the canonical version of the new science. There is no evidence that he attempted to do so. Possibly his personal views did not contain the published inconsistencies which ultimately were to divide his followers; certainly, though he encouraged and aided defenders of his ideas, he failed to appreciate the advantages of assembling them into a unified and consistent band of disciples.[1] In any event, individual Newtonians, scattered throughout Britain, were left without constraint to present their separate and idiosyncratic versions of Newton's theory of matter and its action.

These early Newtonians were educated and began their careers prior to Newton's introduction of aether concepts. Perhaps it was simply this accident of timing, perhaps also there was some special affinity between mechanism and the deistic, formal, classical tenor of the age; whatever the reasons, the version of "Newtonian" matter theory first assembled was that of dynamic corpuscularity. Once begun, an internal momentum carried this view, uncomprehendingly, through the resistance of the aether. For nearly half a century after the publication of the 1687 *Principia* there was scarcely a "Newtonian" reference, save Newton's own, to any mechanism explaining phenomena other than that of forces acting at a distance. Moreover, the early reputations of works published during this half-century prompted their continued republication throughout the second half-century as well, introducing an additional element of confusion into that age of materialistic, aether explanations.

In the long run, the scientific value of a theory of matter lies in its development as a policy, stimulating and directing specula-

[1] As the creator of the dominant scientific synthesis of the age, and long-time (24 years) president of the Royal Society, there is no doubt that Newton acquired substantial authority over British scientific circles during the first quarter of the eighteenth century. Neither the rhetoric nor the innuendos of Frank E. Manuel's "Newton as Autocrat of Science," *Daedalus, the Proceedings of the American Academy of Arts and Sciences,* 97 (1968), 969-1,001, adequately demonstrates, however, that Newton sought that authority as satisfaction of a neurotic craving for dominance or unduly exercised it for personal aggrandizement. The diverse interpretations of Newtonianism by his followers and their failure to adopt aetherial hypotheses after 1713 argue, at least, for significantly greater freedom of action than Manuel supposes for that "group of scientific adherents . . . joined in absolute loyalty to his person and his doctrine [p. 974]."

tion and experiment. But before this can happen, the theory must first be established as a creed. That is, before it can lead the observer into previously unexplored regions, it must be known to be better than competing theories and its answers to problems already explored accepted without enervating skepticism. Although the people active in elaborating a theory may also be those responsible for its diffusion, this is not necessarily or perhaps even usually the case, for very different talents are involved. The two activities can, at least, be treated as independent aspects in the assimilation of Newtonian matter theory and this chapter will examine the efforts toward establishing the theory of dynamic corpuscularity.

All acceptable theories of matter in 1687 were mechanical and particulate. To establish Newton's theory, it was not, therefore, necessary to prove the basic assumption—that explanation of physical phenomena essentially involved the motion and interaction of the variously sized and shaped particles of matter. It was in the elaborations beyond this common core, the cause of the motions, the mode of the interactions, the nature and limitations of shapes and sizes, that differences occurred. The unique aspect of Newtonian dynamic corpuscularity was the unitary answer suggested for all of these questions. That is, that the varying motions are caused by forces, the forces between particles by means of which they interact and in consequence of which heterogeneous substances are built from homogeneous primitive corpuscles. The establishment of Newtonian matter theory required the demonstration that this answer was superior to competing ones and that it did not require acceptance of philosophically unsound concepts. In practice this meant a polemic against the followers of Descartes, the atomists, and, cautiously, the empirical corpuscularians of the tradition of Boyle; it meant the popularization, by reiteration and simplification, of the *Principia* and the *Opticks*; and, ultimately, it meant the substitution of a modified quantitative, experimental positivism for the mathematical rationalism of Descartes or the qualitative empiricism of the indigenous Baconians. These three elements, in varying proportions, are to be found in all British Newtonian literature of the first 40 years of the eighteenth century.

By an accident of incalculable value in smoothing the way for Newton's ideas, the earliest agency in the establishment of Newtonianism was the church. When Robert Boyle died in 1691, he left an endowment for an annual series of eight sermons, or lectures, "to prove the truth of the Christian religion against infidels, *viz.* Atheists, Theists, Pagans," etc. No stipulation was made that this need be done by way of science, and most of the lecturers have not done so, but Boyle's own example and the presumed threat to religion of a science-supported atheism or deism combined to encourage many early Boyle lecturers to employ science in their defenses of Christianity.[2] Of these early lecturers, the first, Richard Bentley, and one of the last, William Derham, may serve as examples of the spread of Newtonianism as it was thus used.

Bentley was not a scientist, but he was a polemicist of rare power who recognized the value of drawing his arguments from the last word in science.[3] His eight Boyle lectures of 1692, directed against the coffee-house devotees of Epicurus, Hobbes, and Spinoza, were, therefore, drawn in part from the *Principia*, which he read with Newton's guidance. Before the last and most Newtonian two of the sermons were published, Bentley again had the assistance of Newton to insure that the arguments were correct.[4] The success of Bentley's lectures was immense. Published within a year of their delivery, they were adopted as a text in moral philosophy and metaphysics in the colleges; by 1735 six editions had appeared, and another edition was issued as late as 1809.

By the time William Derham delivered his Boyle lectures in 1711 and 1712 the *Opticks* had appeared in the editions of 1704 and 1706, and a few texts in Newtonian science were also available as sources. Derham's lectures could, therefore, be more broadly based in Newtonian philosophy than Bentley's; they were also directed to a different audience, for several intervening Boyle lecturers had familiarized British congregations with theological Newtonianism. Even had he been capable of them, Bentley's rhetorical polemics would have been out of place for Derham, who, moreover, had encyclopedic rather than philosophic ambitions.[5] Derham was a naturalist; what academic training he had

[2] See John F. Fulton, *A Bibliography of the Honourable Robert Boyle* (Oxford: Clarendon Press, 1961), 2nd edn., pp. 197-202, for a list of Boyle lecturers down to the present day.

[3] Richard Bentley (1662-1742), F.R.S., graduate of St. John's College, Cambridge, was chaplain to the Bishop of Worcester when chosen Boyle lecturer. A philologist of massive erudition, he became Master of Trinity College, Cambridge in 1700, and remained there, over the objections of the Fellows, until his death.

[4] Bentley wrote a series of letters to Newton about these sermons, and Newton's answers were preserved. They were published in 1746 and are reprinted, with introductory commentary by Perry Miller, "Newton's Four Letters to Bentley and the Boyle Lectures related to Them," in Cohen, *Newton's Papers and Letters*, pp. 271-394.

[5] See A. D. Atkinson, "William Derham, F.R.S. (1657-1735)," *Annals of Science*, 8 (1952), 368-92. Educated at Trinity College, Oxford; B.A. 1678/9, M.A. 1683,

received in natural philosophy had been acquired prior to the publication of the *Principia,* so that his sources for physical-science arguments were as likely to be Boyle, Hooke, Wallis, or Borelli, or even earlier writers such as Pliny and Seneca, as they were Newton, or the circle of contemporary Newtonians, Bentley, Clarke, John Keill, and others. His lectures are cast in the mode of John Ray's *Wisdom of God Manifested in the Works of the Creation* (1691). His emphasis is on the classic argument from design, which, in published form, he supports in extensive footnotes, drawing from every authority he can find. Few treatises become flat and unprofitable sooner than theologians' apologetics, but Derham's seem even staler now than most. Yet at the time, his lectures were more successful than Bentley's. Published first in 1713 as *Physico-Theology: or, A Demonstration of the Being and Attributes of God, from His Works of Creation,* they were republished more than 15 times, translated into French, German, Italian, Dutch, and Swedish, and were available in a new edition as late as 1798. Derham was encouraged to follow the *Physico-Theology* with a similar work on astronomy in 1715, and *Astro-Theology* was nearly as popular, reaching its ninth edition by 1750 and also being widely translated.

Given their differences in style, intent, and timing, any direct comparison of the texts of Bentley's and Derham's Boyle lectures would seem to be useless, but the matter theory each attempts to establish is surprisingly the same, while the differences reveal as much of the progress of Newtonianism as of the variations in the writers themselves.[6] Bentley is explicitly anti-Cartesian in his attacks on the plenum and on vortex theories [Sermon 7, 272-73]; Derham casually includes Descartes among the non-Moderns whose opinions are not worth enumerating [*Physico,* 26n]. Newton is the "very excellent, and Divine Theorist" to Bentley [Sermon 7, 253], while, to Derham, his philosophy is to

Derham spent the greater part of his life as a country vicar of Upminster, Essex. In addition to the *Physico-Theology* and *Astro-Theology,* his works relating to science include a treatise on clock-making, editions of the letters and writings of John Ray and Francis Willoughby and of the experiments of Robert Hooke, and scattered papers for the Royal Society on natural history, meteorology, astronomical observations, and the speed of sound.

[6] I have used Richard Bentley, *Eight Sermons Preach'd at the Honourable Robert Boyle's Lecture. In the First Year, MDCXCII* (Cambridge: Crownfield, Knapton and Knopstock, 1724), 5th edn., and William Derham, *Physico-Theology,* etc. (London: W. Innys and R. Manby, 1737), 9th edn.; *Astro-Theology: Or, A Demonstration of the Being and Attributes of God, from a Survey of the Heavens* (London: William and John Innys, 1726), 5th edn. Bracketed numbers indicate page references, etc., in these volumes.

be accepted as grounded on phenomena and not upon "chimerical and uncertain Hypotheses [*Physico*, 31]." For both, matter is particulate and passive, its activity ultimately depending upon a non-mechanical principle or force, not inherent or essential to matter but implanted in it by the Creator [Sermon 2, 61-62; Sermon 7, 277-78; *Physico*, 31n]. Bentley's example is gravitation; Derham adds cohesion, capillarity, and the action of bodies on light [*Astro*, 12; *Physico*, 31, 40, 52]. Derham, following Newton, refuses to frame an hypothesis as to the mediate cause of this self-attracting power of matter [*Astro*, 154-56], he declares it congenial and coeval with matter, says the facts of its existence are evident—and leaves it at that. But for Bentley, the existence of such a force provides an opportunity for "a new and invincible argument for the Being of God." Deliberately emphasizing the paradox: attraction cannot be spontaneous in matter, "mere" matter cannot operate upon matter except by contact, but contact is precluded by the necessary existence of the void, Bentley triumphantly offers the escape from impasse—the principle is the immechanical and immaterial living mind of God, informing and actuating dead matter with its constant Divine energy or power [Sermon 7, 277-81].

Clearly, in his spiritualizing of attractive force, Bentley went beyond Derham. As the letters to Bentley from Newton show, Newton had refused to declare whether the agent causing gravity be material or immaterial. Nor, as the letters reveal, was this Bentley's only fall from grace—some of his computations being erroneous and his concept of infinity mathematically unsound—but at least he presented a philosophically consistent if uncanonical view of Newtonianism.[7] Moreover, his work gave a religious sanction for the power of bodies to act on one another at a distance that it never quite lost, through the first half of the eighteenth century.

Derham was not so gifted. He had cast his net wider and with less discrimination; his works reveal a diversity of conflicting theories which must have confused even those students of moral philosophy and metaphysics who used them as texts at Cambridge in the 1730s. Considering his Newtonianism, already described, what is to be made of: the seminal Principle or Plastic Quality which makes minerals grow in the earth [*Physico*, 63], the nitroaerial ferment conveyed to the blood by respiration [*Physico*, 145], the incredible tenuity and sharpness of particles by which poisons effect their actions [*Physico*, 396], the collisions of air

[7] See "Newton's Four Letters to Bentley," p. 303, and *passim*.

particles causing the tremors called sound [*Physico*, 343], or those light and most agile particles of which heat as well as light are composed [*Physics*, 26, 48]? This naïve commingling of kinematic with dynamic corpuscularity, of iatro-chemistry with iatro-mechanics, of tag-ends of the peripatetics with scraps of neo-Platonism is not a deliberate eclecticism, but the same kind of syncretism which later combined Darwin with Lamarck in popular evolutionism or Freud with Adler in popular psychiatry. Derham's lectures demonstrate that popular theological arguments had passed their limit of usefulness in the establishment of Newtonian matter theory. Yet Derham's works, as a contemporary expression of popular Newtonianism, are of interest as much for this indication of attained assimilation as for their acumen, and as much for their sins of omission as of commission. For all his confusion, Derham reveals a confidence in the use of interparticulate force explanations which surely is a reflection of similar attitudes among his fellows; but, in moderating Bentley's claim that these forces were immaterial, he never uses Newton's alternate suggestion—a suggestion that he clearly had come to know. From the disparate elements casually combined in explanation by Derham, there are, in fact, two notable exclusions. In his lectures, "Newtonian" forces do not include repulsion, nor is the aether ever mentioned. Bentley does not refer to them either, but repulsion is not a factor in the phenomena he describes, and the aether was introduced only after his lectures were given, lectures never substantially revised despite their subsequent editions. Derham's works, on the other hand, describe phenomena for which Newton had invoked repulsion, and they were revised for editions published after the 1713 *Principia* and the 1717 *Opticks*. Both the *Physico-Theology* and the *Astro-Theology*, in revised editions, refer explicitly to queries numbered from the second (English) edition of the *Opticks*, and the *Astro-Theology* even refers implicitly, in its refusal to "frame Hypotheses," to the General Scholium of the second edition of the *Principia*. Derham therefore reveals a feature of the establishment of Newtonian matter theory which also reflects the views of his contemporaries. Neither repulsion nor the aether provided any inspiration to early eighteenth-century investigators to extend Newtonian matter theory into previously unexplored regions.

As theology ceased to be an adequate means of spreading Newtonianism, textbooks of Newtonian natural philosophy, already partially responsible for producing that situation, expanded to fill the gap. Like the theological works they replaced, the texts of this period pass in spirit from the heroic to the commonplace, the earliest recognizing their revolutionary character in belligerent anti-Cartesianism, the last complacent in their dynamic corpuscularity, hanging on through a period of aetherial materialism without their users quite understanding that they had outlived their usefulness.

The first three texts: David Gregory's *Astronomiae Physicae et Geometricae Elementa,* John Keill's *Introductio ad veram physicam,* and Samuel Clarke's Latin translation of Jacques Rohault's *Traité de Physique,* with Newtonian annotations, all date effectively from 1702. Gregory's *Astronomiae* was the earliest extensive work in gravitational astronomy, but though described, as late as 1731, as a "noble book" (it was nearly 500 large folio pages), it was "too long and difficult" for beginners.[8] As his published work relates only peripherally to matter theory, Gregory is of interest here primarily for his influence in the spreading of Newtonian natural philosophy exclusive of his formal teaching and writing. As Professor of Mathematics at the University of Edinburgh, Gregory is reputed to be the first person (save Newton) to have lectured publicly on Newtonian science in a British university.[9] In 1691, with Newton's strong recommendation, he was named Savilian Professor of Astronomy at Oxford and there became a focus for Newtonianism in England. The "Memoranda" he kept of his professional life between 1691 and 1708 remain among the best extant records of the early years of the Newton circle.[10] Gregory visited Newton at Cambridge and in London; as early as 1694 he had seen a manuscript of the *Opticks* and some of the proposed revisions for the second edition of the *Principia.* It must be assumed that other members of the group also had access (directly or through Gregory) to much of Newton's thinking before its publication.

The other two texts, Keill's *Introductio* and Clarke's *Rohault,* are more clearly pertinent to a study of the diffusion of Newtonian matter theory, though both have obvious limitations. Keill's *Introductio* was intended as the text of his lectures as deputy Sedleian Professor of Natural Philosophy at Oxford, but the

[8] John Clarke, *An Essay upon Study. Wherein Directions are given for . . . the Collection of a Library, proper for the Purpose, consisting of the Choicest Books in all the several Parts of Learning* (London: Arthur Bettesworth, 1731), p. 158.

[9] David Gregory (1661-1708), born in Scotland and educated at Marischal College, Aberdeen, and the University of Edinburgh. He was a friend and admirer of Newton, the author of several texts in mathematics, astronomy, and optics, and the editor of a scholarly edition of the works of Euclid.

[10] Hiscock, *Gregory, Newton and their Circle.*

published version is notably truncated.[11] Two lectures were added to the initial fourteen for the edition of 1705, and, still later editions added new material from the Latin *Opticks;* but the abrupt ending, without formal conclusion, supports the impression derived from the text that Keill failed to complete his design to cover the entire range of natural philosophy. The first eight lectures comprise a general introduction to natural philosophy, including the "affections" of bodies, and therefore discuss matter theory in a broad sense. But the last eight chapters are confined to elementary mechanics: kinematics, simple machines, Newton's laws of motion, elastic collisions, and finally the motion of bodies under constraint. It is not in Keill's book that one can look for specific applications of Newtonian matter theory to the range of problems which most interested eighteenth-century natural philosophers.

Because Rohault's Cartesian *Traité* does cover these problems, Clarke's translation of it affords him the opportunity of indicating Newton's views on the full range of phenomena. This is particularly true as Clarke develops into an authority on Newtonianism in his own right.[12] Most notorious of the early "Newtonian" texts, Clarke's translation had originally been made at the request of his Cartesian tutor, and the annotations in the first edition of 1697 contain little more from Newton than is to be found in the optical papers. The edition of 1702 is a substantial move toward a direct confrontation with Cartesianism, with notes added from the *Principia* to confute Rohault's text. By the edition of 1710, not only had the 1704 *Opticks* appeared but Clarke had also delivered a set of Boyle lectures involving some exposition of Newtonian philosophy and, under Newton's direc-

[11] John Keill (1671-1721), F.R.S., was educated under Gregory at Edinburgh and followed him to Oxford. There he lectured privately on Newtonian science, was named lecturer in natural philosophy at Hart Hall, and, in 1700, became deputy to Thomas Millington, Sedleian Professor of Natural Philosophy at Oxford. In 1710 he was named Savilian Professor of Astronomy. In addition to words described in the text above, he wrote an *Introductio ad veram astronomiam,* texts in geometry, published papers in the *Philosophical Transactions* and was an editor of the *Commercium Epistolicum,* official support of Newton's claim against Leibniz in the calculus controversy.

[12] Samuel Clarke (1675-1729), graduate of Gonville and Caius College, Cambridge; B.A. 1695, D.D. 1709. A Boyle lecturer, 1704-1705, his *Demonstration of the Being and Attributes of God* went through at least nine editions. He was chaplain to the Bishop of Norwich to 1710, chaplain in ordinary to Queen Anne, Rector of St. James', Westminster, and friend of Caroline, Princess of Wales, later Queen. Reputedly England's foremost metaphysician from Locke's death to his own, he was founder of the Rational (Arian) School of English Theology, editor of works of Caesar and Homer, and author of many works of controversial theology.

tion, had translated the *Opticks*, with the new queries, for the Latin edition of 1706. The 1710 edition of Rohault's *Physica* is essentially the final form, with the polemic against Descartes sharpened by a fairly consistent display of Newtonian footnotes in direct opposition to the text.[13] The limitation here is inherent in the annotation form; for, while Clarke was prepared to use the slightest excuse in Rohault for an elaborate presentation of some Newtonian counterview, his freedom to develop a consistent, independent exposition of matter theory is restricted.

In spite of their limitations, these texts were very influential, setting a pattern for many that followed. Keill's *Introductio* went through six Latin editions by 1741, was translated in 1720 as *An Introduction to Natural Philosophy*, and in this form had appeared in at least six more editions by 1778. Clarke's *Rohault* of 1710 was translated in 1723 as *Rohault's System of Natural Philosophy, Illustrated with Dr. Samuel Clarke's Notes taken mostly out of Sir Isaac Newton's Philosophy*; together, the Latin and English editions were republished more than five times. The works of both Keill and Clarke were adopted for use in colleges of Oxford and Cambridge. As late as the 1730s, Clarke's *Rohault* was the text in natural philosophy read by a majority of Cambridge students, while Keill's book remained standard for college study as late as 1790. An analysis of their expositions of matter theory will, therefore, reveal much not only of that theory as received by British university students during the early years of the eighteenth century but also of some of the constraining influences on them during the later periods.[14]

The form alone of Clarke's annotated *Rohault* is sufficient to indicate its anti-Cartesian character. Keill had to make the same point more specifically and did so, though the decorum of the classroom subdued the pugnacity of his earlier published criticism that Descartes' principles were so ridiculous "that it is a wonder they should be believed by any" and still more "how they came to be so much applauded."[15] Each denies that matter and ex-

[13] See Michael A. Hoskin, "'Mining All Within': Clarke's Notes to Rohault's *Traité de Physique*," *The Thomist*, 24 (1961), 353-63.

[14] The editions used here are: John Keill, *An Introduction to Natural Philosophy: or, Philosophical Lectures read in the University of Oxford Anno Dom. 1700* (London: William and John Innys, and John Osborn, 1720); and Samuel Clarke, tr. and annot., *Rohault's System of Natural Philosophy*, etc., done into English by John Clarke (London: for John Knapton, 1723), 2 vols. Numbers in brackets refer to pages in these editions.

[15] John Keill, *An Examination of Dr. Burnet's Theory of the Earth: with some remarks on Mr. Whiston's New Theory of the Earth. Also An Examination of the Reflections on the Theory of the Earth: and a Defence of the Remarks. . . .* (Oxford:

tension are equivalent, Keill insisting on the large proportion of void in the bulk of any substance [19, 67] and Clarke that impenetrability is at least as essential to matter as dimension [I, 24]. Clarke does not query Rohault's basic corpuscularity, while Keill is detailed in his exposition: all matter is essentially the same, its different forms are but modifications of the same substance, variations in magnitude, figure, texture, position, and other modes of the particles composing bodies [91-92]. Keill is the more explicit in his declaration that these modifications are insufficient to explain phenomena. As he had earlier declared that "loose and general Harangues about Effluviums, Particles, subtle Matter, Modes and Motions signify very little more to explain Nature than the Qualities and Attractions of the old Philosophers . . . ,"[16] so now he insists that most mechanical philosophies are only nominally so, depending on unseen figures, ways, pores, and interstices of corpuscles [iii]. But Clarke implicitly admits the inadequacy of kinematic corpuscularity in his denial that phenomena are caused by impulse of air or subtle matter, by relative rest, or by particle shape. There are, instead, active principles which are not occult properties of matter, but original and general laws of all matter, impressed upon it by God and maintained there by some immaterial efficient power [I, 54-55; II, 96-97]. These principles, or attractive forces, are the cause of gravity, cohesion, capillarity, magnetism and electricity, fermentation, the chemical action of salts, etc. [I, 46, 54-55; II, 137-38]. Keill accepts the forces of attraction, as active principles, and uses gravity as his example (the truncation of his textbook limits his consideration of other phenomena) but refuses to speculate on its cause—whether the action of bodies, agitation of emitted effluvia, or impulsion from a medium—and asks "if the true Causes are hid from us, why may we not call them occult Qualities [3-4]?" The authors agree that light is not motion in a plenum, but real motion of small particles; Clarke adds that these particles are attracted or repelled by bodies at a distance [Keill, 67; Clarke, I, 201, 209-11].

Neither text is completely and consistently Newtonian. Clarke, for example, fails to annotate all sections of *Rohault,* and some of the latter's explanations escape unchallenged. The nature of

H. Clements, and London: S. Harding, 1734) 2nd edn. corr., p. 10. First edition of the *Examination* was 1698 and of the *Examination of the Reflections* (*i.e.,* his defense of the first) was 1699. This second edition is a posthumous reprinting of the first of each.

[16] John Keill, *Examination,* p. 228.

these failures suggests that the omissions are not entirely accidental. Where Newton has authoritatively defined the character of any particular action, Clarke and Keill, insofar as they understand him, dutifully follow, providing long and apposite quotations from the queries to the *Opticks* to support their interpretations. Where there are lacunae in Newton's explanations of natural phenomena, they either do not fill the gaps or they do so in a pre-Newtonian manner. Rohault gives an impulse explanation for the action of the lodestone and explicitly denies attraction [II, 166-67]. Although Clarke had casually included electricity and magnetism among the active principles of matter, he fails to query Rohault's impulse explanation and, significantly, Keill also has a materialistic, effluvial magnetic explanation [67]. Where Newton has explicitly used repulsive forces—as in the elasticity of air, the action of bodies on light, or the chemistry of salts—Clarke admits repulsion, but he ignores the possible extension of this principle to the motion of sound [I, 188], or the elasticity of solid bodies [I, 122], while Keill hardly even mentions repulsion. Another, and more portentous, example relates to heat, which ultimately becomes a critical point in the division between mechanistic and materialist explanations. Keill includes both heat and cold with light, perfumes, and the like, among qualities whose intensions and remissions vary inversely with the square of distance [7]; Clarke, in denying Rohault's Cartesian definition of heat, also denies that heat consists in every motion, it being rather a peculiar motion (of certain particles perhaps) of the small particles of all bodies [I, 164].

Finally, the anomalies resulting from editions revised against the background of changes in the *Principia* and the *Opticks* are even more apparent in the texts than in the theological treatises. In 1714 and 1715, again with the guidance of Newton, Clarke engaged Leibniz in a debate over Newtonian principles, but neither this nor the 1713 *Principia* or 1717 *Opticks* produced any substantial variation in Clarke's *Rohault* as translated into English in 1723. This English edition cites the queries of the 1717 *Opticks*, but none of the "aether" queries, and his sole reference to the aether is an equivocal "if there be any such thing" in connection with subtle matter filling the pores of terrestrial bodies [I, 43]. His flat denial that material impulse in any way causes gravity remains unchanged.[17] The same observations can

[17] For the debate with Leibniz see H. G. Alexander, *Leibniz-Clarke Correspondence*. Clarke later wrote a paper, "Letter occasion'd by the present Controversy among Mathematicians, concerning the Proportion of Velocity and Force in

be made of Keill, and with even greater force. His book was revised after 1706 for the edition of 1715, but without substantive changes in the nature of his explanations by introduction of material from the Latin *Opticks*. This is the more noticeable because Keill was responsible for the first imaginative extension of inter-particulate force discussions in a *Philosophical Transactions* paper (to be discussed later) of 1708. And, although new editions of his *Introductio* appear before Keill's death and after both the 1713 *Principia* and the 1717 *Opticks*, Keill failed to revise them to include references to the aether. Yet Keill was obsessed with the geometrical aspects of material extension; three chapters of his *Introductio* and a paper in the *Philosophical Transactions* for 1714 (substantially included in the English *Introduction* of 1720) are devoted to a discussion of the physical as well as mathematical infinite divisibility of matter, its infinite subtlety and tenuity.[18] His proposition that any quantity of matter, however small, may be diffused through any finite space, however large, so as to fill it without leaving a pore greater than any given size [64] would seem ordained to a subsequent aetherial comprehension—which it never received.

The function of a textbook is the exposition and diffusion of accepted theory, and even the best of them rarely extend to new or controversial interpretations. Three texts: Roger Cotes' *Hydrostatical and Pneumatical Lectures* (1738), Richard Helsham's *Course of Lectures in Natural Philosophy* (1739), and Robert Smith's *Compleat System of Opticks* (1738) have been deliberately selected to represent this norm. They exemplify the use of dynamic corpuscularity in the teaching of science, they are native British texts, and were among the most popular of the century, particularly in the English colleges. So far as matter theory is concerned, the three are very much alike, though they span three subjects and, despite appearances, a quarter-century. Cotes' *Lectures* were first delivered about 1708 in his capacity as Plumian Professor of Astronomy and Natural Philosophy at Cambridge.[19]

Bodies in Motion," *Philosophical Transactions*, 35 (1728), No. 401, 381-88, Attacking Leibniz' views that matter possesses a living soul, thus really giving an occult quality to matter as an essential attribute; there is still no reference to aether.

[18] John Keill, "Theoremata quaedam infinitam Materiae Divisibilitatem spectantia, quae ejusdem raritatem et tenuem compositionem demonstrant, quorum ope plurimae in Physica tolluntur difficultates," *Philosophical Transactions*, 29 (1714), No. 339, 82-86.

[19] Roger Cotes (1682-1716), F.R.S., educated at Trinity College, Cambridge; B.A. 1702, M.A. 1706. A protégé of Richard Bentley, he was named Plumian Professor in 1706 and assisted in the establishment of the Trinity astronomical observa-

The lectures remained unchanged by Cotes' editing of the 1713 *Principia* and were published posthumously by his cousin, Robert Smith, successor to the Plumian professorship, whose optical *System* was published the same year.[20] Helsham's *Course* appears to date from 1729, when he was lecturing at Trinity College, Dublin, as Smith Professor of Natural Philosophy; the lectures were published posthumously by his friend and student, Bryan Robinson.[21] All three texts are listed among those in use at Cambridge in 1774 for studies in mechanics, hydrostatics, and optics (Cotes was used as late as 1809), and all appeared in multiple editions, including translations. None is primarily concerned with the theory of matter and its action, for each is a specialized, though elementary text. And each in its way is a competent treatment. Cotes addresses himself to simple hydrostatics, buoyancy, fluid pressures, etc., at the level of Archimedes, Pascal, and Boyle. The greater part of Smith's *System* is devoted to the design and use of simple optical systems, tracing through many examples the geometrical laws of catoptrics and dioptrics; very little physical theory is discussed. Even Helsham, whose title promises more, wrote essentially a one-subject text, primarily treating elementary (geometrical) mechanics, at the level of Galileo's kinematics and the operation of simple machines, though some hydrostatics and pneumatics and some geometrical optics (rather simpler than Cotes and Smith) are included. All are matter of factly Newtonian, having adopted his mathematical arguments and the mathematical positivism this implies. None could have been written in the spirit it was prior to Newton, but

---

tory. Editor of the second edition of *Principia* and author of a mathematical text, as well as that described above, he was greatly praised by Newton. See also George Huxley, "Roger Cotes and Natural Philosophy," *Scripta Mathematica*, 26 (1961), 231-38. I have used Roger Cotes, *Hydrostatical and Pneumatical Lectures*. Published with Notes, by Robert Smith (London: for the editor; sold by S. Austen, 1738).

20 Robert Smith (1689-1768), F.R.S., educated at Trinity College, Cambridge; B.A. 1711, M.A. 1715, LL.D. 1723, D.D. 1739. Successor to Cotes as Plumian Professor; protégé also of Bentley, whom he succeeded as Master of Trinity in 1742. Author also of a work on Harmonics, he established the Smith Prizes in mathematics at Cambridge. I have used Robert Smith, *A Compleat System of Opticks* (Cambridge: for the Author; sold by Cornelius Crownfield, and by Stephen Austen and Robert Dodsley, 1738), 2 vols.; the second volume is not significant for matter theory considerations.

21 Richard Helsham (1682?-1738), born in Ireland, graduate of Trinity College, Dublin; B.A. 1702, M.D. 1714. He became lecturer in mathematics, 1723-30, Smith Professor of Natural Philosophy, 1724-38; and Regius Professor of Physick, 1733-38. He was a lecturer and physician of high repute in Dublin. I have used Richard Helsham, *A Course of Lectures in Natural Philosophy* (London: J. Nourse, 1777), 5th edn.

few details of their technical treatments would have suffered changes in substance had Newton never written.

In much the way that a modern elementary text might have introductory chapters on the atomic nature of matter, Helsham's *Course of Lectures in Natural Philosophy* commences with three introductory lectures relating to the principles, forces, or powers wherewith all parts of matter are endued, whereby they act upon one another to produce the phenomena of nature [2]. Elements of matter theory appear also in Cotes and Smith but are more generally distributed through the textual material. In all three, the theory of matter and its action is that of Newton. Helsham is, again, the most explicit on this, with his statement that the course was designed to illustrate some of the truths disclosed by Newton in the *Principia* and the *Opticks* [2], but the references to Newton and quotations from his works in the others leave no doubt as to the origin of the concepts presented. If one examines these books in order of preparation rather than of publication, there is a clear evolution in treatment of these concepts. Cotes is the most outspokenly anti-Cartesian and denounces "the famous *Materia Subtilis*" as a "way of juggling" now justly out of credit and altogether to be laid aside [5], this in spite of his having been editor of the edition of the *Principia* in which the General Scholium on the aether first appeared. Helsham does not find it necessary even to mention a subtle matter alternative to action-at-a-distance. But Smith, the last of the three to write, does finally show some influence of the aether revisions of *Principia* and *Opticks.* Discussing the forces acting upon particles of light, he writes: ". . . whether this force be a real attraction, or whether it be an impulse upon light, caused by the spring or elastick power of a subtil fluid which pervades the medium, and being gradually denser without than within it, may impell the light towards the medium . . . be this as you please . . . [90-91]."[22]

For all three, phenomena are caused by particulate matter acting under the influence of attractive and repulsive powers or forces. Cotes, the earliest of them, is the most explicitly positivistic—by attraction is meant no more than some power in nature, from whatever causes it proceeds, by which bodies endeavor to be united to one another [122-23]. Helsham adds a perfunctory "the cause of which is in a great measure unknown, tho' the thing

[22] As editor of his cousin's *Hydrostatical Lectures,* Smith managed also to blur the integrity of Cotes' dynamic corpuscularity, with the inclusion of appendices, one of which relates James Jurin's aetherial interpretation of capillary experiments from a *Philosophical Transactions* paper of 1719.

itself is manifest from experiments" to his discussion of the active principles of matter [2]; while Smith, for all his momentary flirtation with the aether, includes long quotations from Newton's Query 31, on attractive and repulsive forces, including the sections on how repulsive virtues ought to succeed attraction, as negative quantities succeed positive ones [88-89], and how nature, being conformable to herself, would perform small motions as well as great ones, by attractive and repelling powers which intercede the particles [93-94]. Each gives (partially by way of extensive quotations from Query 31) the usual list of attraction phenomena and substantially more attention to repulsion than earlier works, and each shows much more interest in the mathematical form of the force laws. Cotes' interests, naturally, center on the elasticity of air and capillarity phenomena. Experiments on the latter had not progressed far enough, when his lectures were prepared, to discuss force relationships for these, but the elasticity of the air is for him a consequence of repulsive forces between air particles reciprocally proportional to their distances. This hypothesis was demonstrated in the second book of the *Principia* and further confirmed by late additions to the *Opticks*: "Whoever will read those last few pages of that excellent book, may find there in my opinion, more solid foundations for the advancement of natural philosophy, than in all the volumes that have hither to been published upon that subject [123]." Helsham is more catholic in his recitation of attractive and repulsive phenomena. The attractive force of cohesion is exceedingly strong at contact but falls sharply with distance [13]; those of electricity, magnetism, and gravity extend themselves to considerable distances [15]. He has proved by experiment (with lodestone and balance) the inverse-square form of magnetic attraction [20]. Repulsive forces exist in electricity and magnetism, the emission and refraction of light particles, the leaves of sensitive plants, and the elasticity of air [33-34, 242-45]. Smith's major concern is the action, at a distance, of bodies on particles of light. This power is infinitely stronger and decreases much more quickly with distance, or in a greater proportion, than the power of gravity [89-90].

In these works, as in the earlier ones, there is the same lack of originality and, generally, the same ultimate confusion when a departure is taken from the received doctrine of Newton's published writings. Helsham is perhaps the least guilty of confusing, but he pays little attention to the problems of heat, electricity, and chemistry which dominate the experimental natural philosophy of the century. This, and his ignoring of the aether, make it hard to understand the continuing influence of the work, as indicated in repeated editions of 1743, 1755, 1767, and 1777; as late as 1818 an edition of selections from Helsham was published in Dublin. Smith continued to be read for his sound exposition of geometrical optics, particularly in relation to the design of lens systems. His insinuation of aether theory into an otherwise corpuscular frame apparently confused few opticians, whose interests during the eighteenth century did not lie in physical optics. In Cotes, the level of confusion is highest: sound is transmitted by the pressure of contiguous parts of the air, without reference to elastic forces [Lecture 15], and electricity receives an effluvial (and thus basically a materialistic) explanation; but, at least, there are some suggestions of originality. By the time of their publication, his proposals were anachronistic, but Cotes was resident in Trinity College during part of the time that Stephen Hales was at Corpus Christi. If there was any communication between the two, it is hard to imagine that Hales would not have noted Cotes' belief that capillary attraction might be the cause of the ascent of sap in trees and of the secretion of fluids through animal glands [125]. Still more pertinent to Hales' subsequent work is the material recorded by Cotes in his Lecture 16: "Air sometimes Generated, Sometimes Consumed: the Nature of Factitious Airs . . . ," where, for example, he discusses the increase and decrease of the elastic force of air in a receiver, by the addition of oil and sal-armoniac [202-203].

With the exception of Keill's confusing discussion of the infinite divisibility, subtlety, and tenuity of matter, and these suggestions of Cotes, made obsolete by the work of Hales, the textbooks described above give little indication that a theory of matter had been proposed that was capable of suggesting, stimulating, and directing speculation or experiment into previously unexplored regions. There were, however, two texts produced during this period which show distinct originality and awareness of the potential of Newton's matter theories, beyond the mere acceptance of his explicit suggestions. Benjamin Worster's *A Compendious and Methodical Account of the Principles of Natural Philosophy*, first published in 1722, and John Rowning's *A Compendious System of Natural Philosophy*, first published in parts between 1735 and 1743, stand in marked contrast to one another and to the other texts of the period. Neither author had a teaching association with a British university, as had Cotes, Smith, and Helsham, which might reflect prestige on his book. Worster's

Compendious Account contains the text of his lectures in natural philosophy at one of the many private London academies of the period.[23] The school was small and comparatively unimportant, and, though the text was among those in use at Cambridge in the 1730s, it reached only a second edition and was soon superseded by later and more reputable if less sophisticated works. Rowning was, at least once, associated with the instrument maker William Deane in a series of lectures on natural philosophy delivered at Deane's home, the Gardenhouse, New Street, London. It is tempting to suppose that his *Compendious System* also began as the text of a private lecture course, though Rowning had once been a college tutor at Cambridge.[24] Longer and more detailed, though less inclusive, than Worster's, his book was considerably more successful, going through at least eight editions by 1779. It was in use in Cambridge in the 1740s, at Oxford as late as 1798, and was adopted at several dissenting academies, including those at Bristol and Daventry.

Both men have nearly disappeared from the historical record, though each wrote a text at least as good, in general terms, as any written in the period, and their expositions of matter theory are consistently superior to any others. For each, the physical world consists in matter and the void. Matter is particulate and, for Rowning at least, essentially homogeneous in substance [ii-iii; III, 4-5]. The particles have dimension and are, therefore, infinitely divisible as extension is, though Worster insists [3] and Rowning implies [I, 8-9] that this does not mean actual, physical

23 Benjamin Worster (1683?-1730?) is nearly unknown except for the contents of his book. He was educated at Emmanuel College, Cambridge, B.A. 1704/5, M.A. 1708. Sometime later he associated with the principal at Watt's Academy, London, founded as a training school for clerks and accountants and expanded to meet popular demand for education in natural philosophy. I have used Benjamin Worster, *A Compendious and Methodical Account of the Principles of Natural Philosophy* (London: Stephen Austen, 1730), 2nd edn., rev. and corr. with large additions. See E.G.R. Taylor, *Mathematical Practitioners of Hanoverian England 1714-1840* (Cambridge: for the Institute of Navigation, at the University Press, 1966), under "Watt's Academy"; and John Venn and J. A. Venn, *Alumni Cantabrigienses* (Cambridge: at the University Press, 1927), Part I, vol. 4, "Worster," p. 465.

24 John Rowning (1701?-71), educated at Magdalen College, Cambridge, B.A. 1724, M.A. 1728. He was "late fellow of Magdalen" and rector of the college living at Anderby, Lincolnshire, when his book was written. Member of the Spalding Society, he is credited with a genius for mechanical contrivances. An undated syllabus of his "Compleat Course of Experimental Philosophy and Astronomy" is preserved in the Science Museum, Oxford. The bibliography of his book is a complicated one as each part was issued separately in various editions. I have used John Rowning, *A Compendious System of Natural Philosophy* (London: Sam. Harding, 1737-43), prefatory material, 1743, 1st edn.; Part I, 1738, 3rd edn.; Part II, 1737, 3rd edn.; Part III, 1743, 2nd edn.; Part IV, 1743, 1st edn.; note that the parts are separately paginated.

separation of parts. Matter is ultimately hard [Worster, xvi, 3] or impenetrable [Rowning, I, 7], sluggish and inactive. Natural phenomena result from matter acting under forces of attraction and repulsion, immaterial principles or powers which are not occult but manifest in phenomena, established in nature at the beginning; these powers "proceed from the first Cause and Author of all things [Worster, 9-10]," and "are so far from being Mechanical Causes . . . that they are the very Reverse; and . . . can be no other than the continual acting of God upon Matter, either mediately or immediately [Rowning, xxxix]." The attractive forces are those of gravity and of cohesion, which is of very short range, decreasing, according to Rowning, much more than as the squares of the distances increase [I, 14], while Worster says "nearly in reciprocal Proportion of the Cubes . . . and at some determinate small Distance entirely stop [17-18]." All of this is fairly standard, though more precisely and explicitly stated than most. It is in their treatment of the concept of repulsion, interacting with attraction, and, for Worster, in an early, if somewhat limited, use of an aetherial medium that their most notable divergencies from the norm occur.

Worster's use of the "subtile, elastick Medium called Aether" is extraordinarily early. Except for Newton himself, only George Cheyne, in his *Philosophical Principles of Natural Religion* (1715), and James Jurin, in a *Philosophical Transactions* paper, "An Account of some New Experiments, relating to the Action of Glass Tubes upon Water and Quicksilver," of 1719 [30 (1717-19), No. 363, 1,083-96] seem to have used aether notions before Worster, and, for them, the aether essentially acted to introduce perturbing factors into otherwise comprehensible phenomena. Worster does not use the aether as an escape from action-at-a-distance nor to explain the origin of forces. These, as has been seen, are immaterial and unmechanical, while the aether is both material and mechanical. Its particles are probably those of light rays, when their motion is destroyed [28-29], or, equivalently, light is a stream of aether particles, not of the same bigness, emitted from a luminous body and proceeding in a straight line [177-78, 239-40]. These particles are refracted, reflected, and inflected by the attraction and repulsion of other bodies at a distance [183-84, 226-27]. The principal manifestation of the aether is as the medium transmitting the vibrating motion of heat through dense bodies and the vacuum [28]. Heat communicates a repelling force to the particles of all bodies, agitating the parts and the aether included in their pores. The elasticity of the aether is increased until the pores are enough widened to overcome the force of cohesion, whence the solid parts are separated in a fluid or thrown off into a state of mutual repulsion [157-59].

Having introduced the material aether to solve the difficult problem of heat radiation, Worster adapts it as a quasi-materialistic explanation of electrical phenomena. In an electrical body excited by attrition, the particles of aether are driven off with a repelling force strongest near the body. The aether is then denser and its elasticity greater away from the body, and "attracted" substances are, in fact, driven toward the electrical body as a center of lesser elastic aether pressure. Two electrical bodies, each having a sphere of repelling aether, will repel one another, as the aether between them is compressed [29-30].

Rowning was unwilling to accept a material aether, having adopted the concept of immaterial forces of attraction and repulsion, though his refusal meant that he had to acknowledge his inability adequately to explain such phenomena as electricity and magnetism [I, 19] or the vaporization of fluids [II, 137-38]. Even Newton's hypothesis of fits of easy reflection and transmission, involving, as it did, some subtle and elastic substance diffused through the universe, was, Rowning felt, "too much clogged with Suppositions. . . . The Time will come, when the Principles of *Attraction* and *Repulsion* will be found alone sufficient to account for this perplexing Phaenomenon [III, 167]." Light consists of small particles, homogeneous in substance with all other matter, differing in magnitude and density, which are emitted continuously from a luminous body in straight lines [III, 4-5]. Like other small bodies, light particles are subject to the laws of cohesive attraction and repulsion [xviii-xix; III, 8, 160-61]. When they cease to move, they are stifled and lost in other bodies. Heat is the violent motion of bodies rubbing and clashing against one another, augmenting the elasticity or repulsive force of bodies [xvii; II, 40, 133]. But beyond this, Rowning scarcely discusses heat phenomena and does not mention heat radiation at all.

His refusal to speculate on many of the staple problems of eighteenth-century natural philosophy might have made Rowning's work less interesting than Worster's were it not that, on those problems directly involving the action of repulsive and attractive forces, he carried his speculations further than any other person was to do until the appearance of Roger Joseph Boscovich's *Philosophiae Naturalis Theoria,* some 20 years later. Actually, Worster, too, was aware of the possibilities of establishing a state of tension between the repulsive and attractive forces of matter. "When a Particle that is repelled," he writes, "is forced within the

Sphere of Attraction, it shall be attracted; and when a Particle that is held by Attraction, is thrown off by some superior Force beyond the Sphere of Attraction, it shall be repelled [24]." Worster applies this concept toward an understanding of elasticity—both solid and fluid—and even to an ingenious explanation of the shattering of glass lachryma, the so-called Prince Rupert's drops. As the molten glass falls through the air, its outer surface cools and shrinks, bringing the external particles within their attractive spheres and forming a compact skin of hard glass. The inner particles remain fluid, with the distance between particle-centers being but little less than the semidiameter of the spheres of attraction, until the coldness of the water, suddenly decreasing the dimensions of each inner particle and its attracting sphere, leaves all of these particles in a state of repulsion. The equal distribution of elastic pressure is confined by the skin of solid glass until the stem is broken, when the repelled particles are free to move and the drop shatters into small bits. If, however, the drop is slowly cooled, all particles contract within their attractive spheres and the drop will not shatter [162-63].

In almost all cases, however, partly because of the intervening work of Stephen Hales and Jean Théophile Desaguliers, the dynamic corpuscular explanations of Rowning are superior to those of Worster. The latter, for example, again admits a materialistic explanation, *i.e.*, "a certain vivifying Spirit" contained in it, as the reason that air is necessary to the support of combustion and of animal life [152-53], where Rowning inclines to Hales' suggestion that the answer lies in the elasticity of the air [II, 41]. Rowning's explanation of sound is also better than that of Worster, though here the superiority lies in greater clarity and detail. Starting with Newton's exposition [*Principia*, Bk. II, Sec. VIII], Rowning explicitly describes the wave transmission of sound in terms of displaced individual air particles, repelling one another with a force reciprocally as their distances. From this he describes the spread of sound from each particle, as from a center, passing through a tube or hole in an obstacle, and shows how the velocity of sound is independent of its intensity and of the distance from the sounding body [II, 48-50]. His results are not different from Newton's, nor from those of such contemporary writers of general texts as Helsham or 'sGravesande, but the explanations are particularly simple and clear, yet more detailed than the pneumatical lectures of Cotes or of Martin Clare.

Probably it was this elegant simplicity that earned for Rowning's work its continuing popularity as an introductory text in natural philosophy. Certainly the suggestion which, in retrospect, appears the most remarkable has long passed without particular notice. If one reasonably supposes, says Rowning, that the particles of fluids are not in contact (as is suggested by pneumatic compressibility and contraction with cold), then perhaps they are prevented from approaching nearer than a certain distance by a repelling power diffused around each single particle.

> . . . it follows . . . that each Particle of a fluid must be surrounded with three spheres of Attraction and Repulsion one within another: the innermost of which is a Sphere of Repulsion, which keeps them from approaching into Contact; the next, a Sphere of Attraction diffused around this of Repulsion, and beginning where this ends, by which the particles are disposed to run together into Drops; the outermost of all, a Sphere of Repulsion whereby they repel each other, when removed out of that Attraction. [And, of course, there remains the further attractive sphere of gravitation.]
>
> Now, if this *Hypothesis* should be found to be true; and we might . . . suppose the Particles of all Bodies attract and repel each other *alternately* at different Distances, perhaps we might be able to solve a great many Phaenomena relating to small Bodies, which now lie beyond the reach of our Philosophy [II, 5n-6n].

Rowning then indicates that liquefaction and freezing find an easy explanation through this hypothesis.

With Rowning's hypothesis of several concentric spheres of attraction and repulsion, surrounding the particles of bodies, speculation in dynamic corpuscularity reaches a high point soon to be inundated by the rising current of materialistic explanations remotely based on variations of Newton's aether. But Rowning's work continued to be read by a few would-be mechanists; it is not surprising, for example, that Joseph Priestley, who read Rowning while a student at Daventry Academy in 1755 and borrowed his illustrations for the 1772 *History and Present State of Discoveries relating to Vision, Light and Colours,* should have found the work of Boscovich, which also proposed concentric spheres of attraction and repulsion, so immediately appealing.[25]

25 See Robert E. Schofield, ed., *A Scientific Autobiography of Joseph Priestley (1733-1804)* (Cambridge: The M.I.T. Press, 1966), p. 6; and Priestley, *History and Present State of Discoveries relating to Vision, Light and Colours* (London: J. Johnson, 1772), p. x.

CHAPTER THREE

Elaboration of a Theory

THE DIFFUSION of dynamic corpuscularity from Newton's writings into the common sense of early eighteenth-century British natural philosophy was, as we have seen, carried through high and low-level textbooks and even in theological writings. At their best, however, in a Worster or a Rowning, none went much beyond a detailed affirmation of what Newton had left indeterminant in his queries or a simpler exposition of something Newton had himself derived. To establish the theory and to fulfill its function of suggesting, stimulating, or directing inquiry, it was necessary that Newtonian matter theory be translated into terms other than those in which it had originally been conceived. This was first, and most successfully, achieved in speculative elaborations designed for the purposes of chemistry and physiology.

These two subjects were intimately related, in the minds of eighteenth-century natural philosophers, as those of particular concern to physicians, for whose loyalties the theories of iatrochemistry and iatro-mechanics had long competed. It was, therefore, a natural development for all speculative mechanists to attempt to find a Newtonian explanation for chemical phenomena at the same time that they were applying that approach to physiology, though the people earliest and most directly involved came from the school of Newtonians inspired by David Gregory. The writers most significant in proposing a Newtonian force chemistry were John Keill, Gregory's student, and Keill's friend, John Freind. The pre-eminent British iatro-mechanists (and would-be Newtonian physiologists) were Archibald Pitcairne, prompted by Gregory to a study of Newton and mathematics, Pitcairne's students or protégés, Richard Mead and George Cheyne, John Freind, and John Keill's brother James. For the first four decades of the eighteenth century these men and those that they directly inspired were the source of some of the most ingenious speculations and some of the most significant elaborations of dynamic corpuscularity.

The chaotic state of chemical theory in Britain around 1700 is evident from the variety of conflicting systems combed as sources for such compilations as John Harris' *Lexicon Technicum* of 1704 or Ephraim Chambers' *Cyclopaedia, or an Universal Dictionary of Arts and Sciences* of 1728. Robert Boyle was, naturally and nationally, a popular authority and gave to British chemistry an empirical corpuscularian flavor. But rationalist corpuscularians such as Nicholas Lemery and even iatro-chemists such as Thomas Willis and John Mayow, with their theories of conflict between acid and alkali and their ferments of nitro-aerial and other spirits, were also given a hearing. Organized instruction was likely to follow the system of Lemery, whose *Course of Chymistry* was available in as many as five English editions from 1677 through 1720, in two different translations, one of them by James Keill, who had heard Lemery lecture in Paris. Such a course was that taught by John Francis Vigani, first professor of chemistry at Cambridge.[1] A surviving short set of manuscript notes indicates something of the nature of Vigani's teaching.[2] His course was primarily pharmaceutical, with an empirical approach to the compounding of some 48 typical medicines. Vigani refers, once or twice, unfavorably, to van Helmont and Tachenius and several times, favorably, to Lemery, whose Cartesian corpuscularity appears implicitly in occasional mechanistic explanations of chemical operations. The pervasive theory is, however, naïvely materialistic, with the properties of medicines depending simply, in kind and amount, on the substances contained in them.

While Vigani was formally lecturing at Cambridge, a more sophisticated, Newtonian, course of chemistry was being developed at Oxford by John Freind.[3] Freind's *Praelectiones Chymicae*

[1] John (Joan) Francis Vigani (c. 1650-1713), born in Verona, travelled in Italy, France, Spain, and Holland before settling in England, about 1682. He lectured privately at Cambridge for 20 years before receiving an appointment as professor in 1703. Author of a collection of chemical pharmaceutical recipes, *Medulla Chymiae*, first published in 1682 and republished and enlarged four more times, which probably supplied the basis of his course. See E. Saville Peck, "John Francis Vigani, First Professor of Chemistry in the University of Cambridge (1703-1712), and his Materia Medica Cabinet in the Library of Queens College," *Communications of the Cambridge Antiquarian Society*, 34 (1933), 34-39; and L.J.M. Coleby, "John Francis Vigani, First Professor of Chemistry in the University of Cambridge," *Annals of Science*, 8 (1952), 46-60.

[2] Joan Francis Vigani (Veronens), "Cours de Chymi," Cambridge University Library, MS Dd. 12.53 (A), fols. 1-57. Tentatively dated 1699-1700, from the dates 4 December 1699 and 12 April 1700 written on fol. 304 of the same volume, but, as these are separated from the lectures by many blank pages, they may have no direct relationship with the notes.

[3] John Freind (1675-1728), F.R.S., educated at Westminster and Christ Church, Oxford, B.S. 1698, M.A. 1701, B.M. 1705, M.D. (diploma) 1707. Named professor of chemistry at Oxford in 1704, he left the following year for a career in medicine and politics. He became one of London's chief physicians, though, as a Tory M.P. in 1722, he was implicated in the Jacobite schemes of 1722-23 and imprisoned in the Tower for a short time. His most famous work, the first English *History of Physic* (1725-26), was begun in the Tower; he also wrote a monograph, *Emmenologia* (1703), which will be discussed later in connection with mechanistic physiology.

were supposedly read at Oxford in 1704, but as printed in 1709 they contain so much which relates (inferentially) to the queries of the 1706 *Opticks* and to a 1708 *Philosophical Transactions* paper by John Keill, "On the Laws of Attraction and other Physical Properties," that it is hard to disentangle the ideas. Which of these might have been delivered in 1704 and be only derivatively Newtonian and which were taken, before publication, directly from Newton and Keill cannot now be determined. By combining Keill's paper and Freind's lectures it is possible, in any event, to find an elaborated dynamic corpuscular view of chemical operations available to British readers by 1709.

Keill's paper presents a set of 30 theorems, based on three principles which form the "foundation of all physics; viz. 1. A vacuum. 2. The divisibility of quantity *in infinitum.* 3. The attraction of matter."[4] The first two of these principles allow Keill to indulge his penchant for geometrical contradiction with that statement on the diffusion of the smallest part of matter throughout the largest space, already cited.[5] They also lead, however, to a reiteration of the Newtonian discussion, from Query 23 of the 1706 *Opticks,* on the construction of heterogeneous substances from homogeneous particles:

> The most minute and absolutely solid particles of bodies . . . such as have no vacuities . . . may be called particles of the first composition: the moleculae arising from the coalescence of several of these particles may be denominated particles of the second composition: and again the masses made up of several of these moleculae, may be called particles of the third composition; and so on, till at length we come to particles which constitute the ultimate composition of bodies, and into which they may be ultimately resolved [418].

Keill then proceeds to show that the means of this coalescence of particles are the forces of attraction between them, forces Newton has exemplified and demonstrated in gravity. The attractive force which interests Keill is another, which decreases in greater ratio than the square (be it triplicate, quadruplicate, or any other ratio) of the increase of the distances. And from such a force, and the other parameters of particle size and shape, with associated differences in contact and degree of force, Keill indicates how such

[4] John Keill, "In qua Leges Attractionis aliaque Physices Principia traduntur," *Philosophical Transactions,* 26 (1708-1709), No. 315, 97-110; translated in the Hutton, Shaw, Pearson, *et al.* edn. of *Philosophical Transactions of the Royal Society, Abridged* (London, 1809), v, 407-24, from which my observations are taken.

[5] Chap. 2, note 18.

phenomena as cohesion, fluidity, elasticity (involving only attraction), crystallization, dissolution, fermentation, effervescence, precipitation, congelation, etc., are to be explained. The phenomena chosen are essentially those listed by Newton in Query 23, but there is some greater detail in applying the principle.

> The particles of matter, according to their different structure and composition, will be endued with different attractive forces: for instance, the attraction will not be so strong, when a particle of a given magnitude has several pores, as when it is entirely solid [Theorem 14].
>
> There are three things requisite to make a menstruum fit to dissolve a given body: viz. 1. That the parts of the body attract the particles, more than the said particles are attracted by each other. 2. That the body have pores, open and pervious to the particles of the menstruum. 3. That the cohesion of the particles, which constitute the body, be not so great, as that it may not be overcome by the impetus of the rushing particles of the menstruum [Theorem 23].
>
> If the figure of corpuscles, attracting each other, and floating in a fluid, be such as to have a greater attraction, as also a larger contact in some given parts, than in others; these corpuscles will unite into bodies of given figures: and hence crystallizations will arise, and the figures of the component corpuscles may be determined by geometry from the given figure of the crystal [Theorem 27].

There is even a theorem which relates an effluvial with a force theory of electricity, by positing the emission of effluvia with strong attractive forces, which attract bodies to them and, "since the effluvia are much more copious at smaller, than at greater distances . . . the light body will always be attracted toward the denser effluvia, till at length it adhere to the emitting body" [Theorem 30]. And finally, Keill suggests (with obvious implications for the later work of his brother James Keill and Stephen Hales) that these principles may also explain such phenomena as the ascent of sap in plants and trees, the determinant and constant figure of leaves and flowers, the circulation and secretion of animal fluids, and the theory of diseases and effects of medicines [424].

Keill's work is purely qualitative, there is no reference to repulsive forces (and no concern for the cause of forces), but the ground has obviously been cleared for a very different chemistry than that of Lemery or Vigani. It was precisely this different chemistry which John Freind elaborates in his *Praelectiones Chymicae* of 1709.[6] Freind dedicates his book to Isaac Newton and acknowledges his indebtedness to John Keill in his Preface; he declares his dependence on the concept of short-range forces of attraction, "which are not bare speculation, but taken from the very nature of things." Like Keill, Freind never mentions repulsive forces. He rejects the principles "commonly mention'd in Books of Chymistry" as trifling and without foundation, while claiming his own "Mechanical Explication" to be deduced from the experiments themselves. He reduces Keill's 30 theorems to eight and makes larger use of the concept of particle momentum, varying with size and velocity, as a parameter added to the attractive forces, varying with distance and particle size, shape, and texture. His concern is with the mechanisms of chemical operations, as chemistry is the art of conjoining separate parts of natural bodies and of dividing them when conjoined [10]. He is not interested in explaining the formation of particles or the properties of substances, except insofar as these relate to their peculiar fitness for particular operations.

In general, Freind's discussion of the factors involved in dissociation and composition are extensions of Keill's theorems. Hardness arises from mutual cohesion of the parts of body which in turn is proportional to their attraction [17]. Fluidity is achieved when particles of fire insinuate themselves into the body, separating its particles, reducing contact and therefore the force of cohesion [18]. Calcination is the further stage, when those particles of fire are so dispersed and blended throughout the matter as to divide it into minute atoms. The fire particles prevent mutual contact of the atoms and increase the absolute weight of the calx [23-26]. When small corpuscles of a menstruum are carried with considerable velocity, their momentum aids their penetration and, if these corpuscles are elastic, each collision will enhance the motion until the impetus and moment are great enough to break and destroy the hardest bodies [85-88]. Freind's greatest triumph is a semiquantitative explanation of that difficult problem of the dissolution of gold but not silver by aqua regia, of silver but not gold by aqua fortis. Here he invokes the range of parameters open to him in his Newtonian theory: the

[6] Second Latin edition, 1726; English translation, *Chymical Lectures: In which almost all the Operations of Chymistry are Reduced to their True Principles, and the Laws of Nature* (London: by Philip Gwillim, for Jonah Bowyer, 1712), from which my observations are taken.

relative diameters of the pores of gold and silver and of the particles of aqua regia and aqua fortis, the comparative strengths of attractive forces of gold, silver, aqua fortis, and aqua regia, and the moments of each of the acid particles. These finally are generalized in a set of relations:

If the attraction of gold be to that of silver, as *a* to *b*, of silver to aqua fortis as *b* to *d*, and of aqua fortis to aqua regia as *d* to *e*, and if *f* be the magnitude of aqua fortis particles, *r* be that of aqua regia particles, *c* the cohesion of gold, and *g* that of silver; then, should *f* be greater than the pores of gold, aqua fortis will never dissolve gold, however large be *d*. But if *b-d* x *f* exceeds *g*, with *f* less than the pores of silver, then aqua fortis will dissolve silver, while if *b-e* x *r* is less than *g*, silver will not dissolve in aqua regia, whatever its pore size might be. And if *a-e* x *r* be greater than *c*, aqua regia will be able to dissolve gold [96-101].

This is the furthest extent of Freind's reach toward a quantitative chemistry. Only here is there any implication as to how magnitude might ultimately be assigned those parameters he so happily manipulates. Indeed, in his denial that crystallization will reveal the figure of the least particles making up a crystal —as the force of attraction of a particle is not symmetrical and concretion will be greater on the side of the stronger force [147] —Freind retreats from Keill's hope of determining particle shape. Nor can the force of cohesion be estimated by that of gravity, for the "same quantity of matter may be so variously disposed, that in one Body there shall be a much greater Contact than in another [20]." For a time, however, the failure to quantify seems to have made little difference in the enthusiasm with which the work of Freind and of Keill was received. There was a hostile review of the *Praelectiones* in the *Acta Eruditorum* of 1710 [412-16], but the attack is on the very notion of attractive forces.[7] The review of the *Philosophical Transactions* of 1709 [26 (1708-1709), No. 320, 319-23] is a laudatory one. The greater part of Keill's paper was translated and published by John Harris in the 1710 volume of the *Lexicon Technicum,* which also includes a chemical paper, "de Natura acidorum" by Newton, involving the concepts of attraction and of the construction of larger

[7] Freind defended his approach in an answer, first published in the *Philosophical Transactions* of 1710 (Vol. 27, No. 331, pp. 330-42) and reprinted in both of the later editions of his book. Attraction is not a figment; Newton has proved it for gravity and Freind has many experiments to prove the existence of this other kind of attraction. Far more occult is the Leibnizian belief in some subtile, magnetical sphere of fluid surrounding the parts of matter.

particles from smaller ones.[8] In 1723 J. B. Senac published an anonymous French translation (without acknowledgment) of the writings of Freind and Keill on chemistry, as the *Nouveau Cours de Chymie suivant les principes de Newton et de Sthall* (which included also some extracts from the work of Newton and Stahl). Senac's *Nouveau Cours* reached a second edition, also published in Paris, in 1737. Peter Shaw's 1725 methodized abridgement of the philosophical works of Robert Boyle includes a prefatory statement praising Freind's diligent searches into the art of chemistry, which show how a thorough knowledge of Newtonian principles will be useful to chemical theory. Shaw reiterates his praise of Freind and Newton in his 1731 *Scheme for a course of philosophical chemistry*; James Alleyne's *New English Dispensatory* of 1733 extracts from Freind for pharmacological processes; and Edward Strother's 1732 abridgement of Boerhaave's *Elements of Chemistry* (reprinted in 1737) claims that Lemery and Freind together are better than Boerhaave, in presenting a true theory of chemistry, the former accounting physically and the latter mathematically for all the phenomena.[9]

Unfortunately, this flood of praise and reprintings was not paralleled by an equal enthusiasm for extending Freind's work or applying it. The chemical lectures at Oxford fell into desuetude as soon as Freind had left. His appointed successor, Richard Frewin, gave no lectures and the next chemical lecturer there whose views are known was Nathan Alcock, who lectured from the text of Boerhaave, under whom he had studied.[10] There was to be no Freind school of Newtonian chemistry at Oxford. For a time the situation was better at Cambridge. Vigani was succeeded by John Waller, of whose courses from 1713 to 1717 nothing

[8] John Harris, *Lexicon Technicum, or an Universal Dictionary of Arts and Sciences*, II (1710) (New York and London: Johnson Reprint Corporation, Sources of Science, No. 28, 1966); "Introduction," sigs. b3v-b4v, and "Particles," sigs. 5G3v-5H1v. See also Cohen, *Newton's Papers and Letters*, Section III: "Newton on Chemistry, Atomism, the Aether, and Heat," with an introduction by Marie Boas [Hall], pp. 241-58.

[9] For the Strother reference, see Tenney L. Davis, "Vicissitudes of Boerhaave's Textbook of Chemistry," *Isis*, 10 (1928), 44; I was originally led to the Alleyne reference by Wilson L. Scott's "Sources of Boerhaave's Medical Lectures on Physics, with Particular Reference to the Significance of these Lectures to Physical Chemistry," unpub. M.A. diss., The Johns Hopkins University, 1955; and to those of Peter Shaw by Arnold W. Thackray, "The Newtonian Tradition and Eighteenth-Century Chemistry," Ph.D. diss., Cambridge University, 1967, which, Dr. Thackray tells me, is being revised for imminent publication by the Harvard University Press.

[10] Robert T. Gunther, *Early Science in Oxford*, I, "Chemistry, Mathematics, Physics and Surveying" (Oxford: for the subscribers, 1923), pp. 56-57, 200, for a discussion of chemistry there in the eighteenth century.

is known, but Waller was succeeded by John Mickleburgh, a fervent admirer of Freind and of Stephen Hales, the only person effectively to experiment from Freind's principles.[11] Mickleburgh remained professor until his death in 1756, when he was succeeded by John Hadley, a follower of Pierre Joseph Macquer.

During his last 15 years' tenure, Mickleburgh appears not to have lectured, but he gave at least five courses of lectures, in 1726, 1728, 1731, 1733, and 1741, for two of which, manuscript notes of parts are extant.[12] The first set of these lectures (tentatively dated 1731) were clearly in a medical context, with much detail on experiments and medical recipes. Mickleburgh, however, adopts Freind's definition of chemistry and states his axioms, as "the basis on which Philosophy of Chymistry is built [2nd day]." Freind is named as the first man to apply Newton's philosophy to chemistry, and his postulations "are either strict mathematical truths or are sufficiently evinced by clear and undeniable Experiments." The cause of the attractive force between the corpuscles of matter and the particular laws by which it acts are not yet discovered, for want of sufficient data, but experiments prove its existence, and Query 31 of Newton's *Opticks* furnishes many instances of its action [3rd day].

This explicit citation of Newton and his Query 31 may explain the major difference between Freind and Mickleburgh, *viz.*, the introduction of a new postulate on the existence of a *vis repellens,* or repelling force between bodies. Mickleburgh even paraphrases the algebra analogy from the query in support of this force, which causes bodies to fly from one another once cohesion is destroyed [4th day]. But the material of that query had been available, as Query 23, from 1706 without attracting the attention of Keill or Freind, or, indeed, the majority of other Newtonians. Why now should Mickleburgh seize on the concept? It seems clear that the answer is the influence of Stephen Hales,

[11] John Mickleburgh (c. 1682-1756), matriculated 1709 at Gonville and Caius College, migrated to Corpus Christi, Cambridge; B.A. 1714?, M.A. 1716, B.D. 1724. Third incumbent of the Cambridge chair in chemistry 1718-56, he was also a pluralist clergyman and appears to have carried on a business as a dispensing chemist. See L.J.M. Coleby, "John Mickleburgh," *Annals of Science,* 8 (1952), 165-74.

[12] Gonville and Caius College Library, Cambridge, Ms. 619/342, "John Mickleburgh Chemistry Lectures." The first set consists of the lectures of 15 days and contains, on the verso of an "8th day" fol., experiment notes dated 1731 and 1733 indicating, at least, the continued use of the lectures in those years. These notes must be later than 1726 as they refer to the work of Stephen Hales, not published until 1727. The second set contains only the lectures for the 2nd, 3rd, 4th, and 6th days and probably represents variations on earlier lectures. They must date after 1733, from a reference to Hales as "Dr." Hales and, as the notes include a letter of 1739, the date 1741 has been assigned them.

whose *Vegetable Staticks,* making full use of repulsive as well as attractive forces, was first published in 1727. Mickleburgh's most confident use of the concept of repulsion is in connection with just those phenomena involving change of airs from an elastic to unelastic state described by Hales, whose "accurate experiments" are praised by Mickleburgh [7th day].

This new concept introduced into Mickleburgh's lectures some incongruities lacking in Freind. Like Freind, Mickleburgh treats calcination as a last stage of fluidity and here relates increase of weight of the calx to its acquisition of fire particles. But fluidity and calcination are both ascribed to particles, "removed out of the sphere of their attraction," acquiring a repelling force with heat. Fluidity becomes a state of balance between attraction and repulsion; in calcination, repulsion is the stronger [5th day]. The nature of heat now becomes a problem. Is fire particulate as described by Freind, or is it merely the powerful emission of light, fumes, and exhalations from a heated body as Newton suggests [14th day]? Mickleburgh cites Boyle's experiments on the increase of weight by fire, but begins to imply that this increase is due, either to the addition of sulphur particles, or, with Hales, to a fixation of air in an unelastic, compounded state [9th day].

By the second series of lectures (probably 1741), Mickleburgh has resolved this problem. The influence of Boerhaave's chemical lectures is to be seen more clearly in this set of lectures, in frequency of citation (though he has occasionally mentioned Boerhaave in the first set), and in general form, but Mickleburgh denies the claims of "Lemery, Homberg, Hoffman, Borehave and Stahl" that fire is a distinct kind of body. "Our great philosopher Sir Isaac Newton and after him the Learned and Reverend Dr. Hales are of a quite different opinion," and he cites Newton, but particularly Hales, to the effect that heat consists principally in the brisk vibrating action and reaction between elastic repelling air and the strongly attracting acid sulphur, both acting and being acted upon by the aetheric medium of Sir Isaac Newton [2nd day]. Mickleburgh obviously does not know what to make of the aetherial medium and hopes that experiments currently being made on the nature and properties of electricity will throw light upon fire and flame. He has, however, no doubt as to the reality of repulsive forces. The furnaces and utensils made use of by Newton while he lived at Trinity College attest that his concepts were based on experiment and these have since been confirmed by Hales. Moreover, Hales has shown that air is a true, real principle, an essential part of chemistry, without whose repulsive forces to balance the attraction of other bodies the activity of the universe would cease [3rd day].

Mickleburgh's lectures represent an end-point of dynamic corpuscularity in chemistry and the beginning of a new era. Here Freind and Hales, speculation and experiment, are linked in a course which suggests the importance that pneumatic chemistry will have in the coming years. Here the dichotomy between the mechanism of Newton, Freind, and Hales and the materialism of Boerhaave is illustrated in the case of fire, and here implied is the challenge of Newtonian aether and the relation of electricity to heat.

The physiology of late seventeenth-century Europe, like its chemistry, was predominantly mechanistic, though there was not, perhaps, so full an agreement on modes of explanation. Within the camp of iatro-mechanists there were many schools, representing the various influences of the mechanists: Galileo, Descartes, Gassendi, and Boyle; and some vestiges of iatro-chemistry might be found as well. In Britain the most popular variety was that in which the motions and textures of the blood and other body fluids were thought to be of primary importance. This view was so characteristic of British mechanistic physiologists (including the Newtonians) that they might almost be called iatro-hydrodynamicists rather than iatro-mechanists. It seems likely that the origin of the national preference was the work and reputation of William Harvey, but the immediate source of this emphasis on body fluid mechanics was the work of the Italians Giovanni Borelli and Lorenzo Bellini, while the "statical" studies of Santorio Sanctorius struck a particularly responsive chord in those physicians inspired by Newton to quantitative argument. These three were the medical theorists most admired and cited by British dynamic corpuscular physiologists during the early years of the eighteenth century.

The founder of this British school appears to have been Archibald Pitcairne, sometime professor of medicine at the University of Edinburgh and, for one year (1692-93), Professor of the practice of medicine at the University of Leyden, where Herman Boerhaave, Richard Mead, and probably George Cheyne heard his lectures.[13] Pitcairne's writings are very much in the Italian

[13] Archibald Pitcairne (1652-1713), graduated M.A. from the University of Edinburgh in 1671; studied medicine at Paris and at Rheims, where he received the M.D. in 1680. A celebrated and successful practitioner in Scotland, he may also have taught (privately) there; he wrote extensively, including poetry and polemics in medicine, politics, and theology. See G. A. Lindeboom, "Pitcairne's Leyden Interlude described from the Documents," *Annals of Science*, 19 (1963), 273-84.

tradition described above. Though he read the *Principia* (at David Gregory's prompting) and probably saw a version of the *Opticks* in manuscript as early as 1695, Pitcairne demonstrated his Newtonianism primarily in a belligerent anti-Cartesianism, in the use he made of momentum-force relationships, and in an admiration for mathematics which, however, he was able to use only in a qualitative fashion, for the parameters he needed had not been measured.[14]

"All Diseases consist either in a Change of the Quantity of Fluids, or a Change of the Velocity, or a Change in their Quantity and Texture [101]." In demonstrating the conditions and appearances of motion, no help is to be drawn from substantial Forms, subtile Matter, or a Fortuitous Concourse of Atoms [xxii], indeed, an aetherial subtile Matter filling the Pores of all Bodies must be a mere Figment, as Newton has shown from the argument of different densities [4]. The nature of all matter is the same, and the matter of one Body may easily be made that of any other Body [xxiii]. Innate heat is the attrition of the parts of the blood, occasioned by its circulation [20]. The function of respiration is the comminution of the blood, required to facilitate its passage through the pulmonary vessels to the heart, as the business of digestion is the division of the food for its entering into and passing through the lacteal vessels [53]. These heterogeneous particles are then secerned by the glands, certain particles being more easily separated out than others, at any given velocity of circulation, because of their different quantities of motion as measured by bulk and velocity [13, 62-63]. Fevers and most other diseases are produced by untoward changes in texture, or particle size, of the blood, in which the cohesion of the parts of the last composition is changed under the influence of the forces, be they centripetal or centrifugal, between the particles [303-04]. It is the function of medicines, phlebotomy, and physicians to maintain or restore the natural, healthful character of the texture and motion of the blood.

Pitcairne's most famous student was Richard Mead, probably the most successful society physician of his day. Said to have made as much as seven thousand pounds a year, Mead was Newton's personal physician and was self-consciously Newtonian in his

[14] Pitcairne's medical works were collected and published many times. I have used his *The Philosophical and Mathematical Elements of Physick* (London: for W. Innys, T. Longman, and T. Shewell, and Aaron Ward, 1745) of which the first English edition appeared in 1718. This work reprints Pitcairne's Leyden lectures of 1692-93. David Gregory notes, in Hiscock, *Gregory, Newton and their Circle*, p. 3, that Pitcairne offered to translate Newton's *Opticks* into Latin in 1695.

medical writings, which were much admired.[15] Of these writings, his *Mechanical Account of Poisons* and the *Action of the Sun and Moon on Animal Bodies* are notable. The former was first published in 1702, to such applause that it was abstracted the following year in the *Philosophical Transactions* [23 (1702-1703), No. 283, 1,320-28] and reprinted many times. The latter first appeared in 1708. Neither work does much credit to Mead's understanding of Newton or to the critical acumen of his public. Though the *Mechanical Account* praises Newton's force of attraction between particles as the great principle of action in the universe, the explanation of the action of poisons is kinematic, the sharp crystalline particles of poison pricking the globules of the blood, emptying the cases and changing the texture of blood fluid. The *Action of the Sun and Moon* makes a naïve assumption that these bodies have a tidal effect on the sea of air, periodically changing the pressure of the atmosphere and thus producing remitting diseases. Neither barometric data nor period-congruities of physical and physiological phenomena are examined for confirmation of this misplaced ingenuity.

Under the circumstances, Mead's work is best seen as a concession to prestigious Newtonian mechanism. This view is confirmed by his shifting with changing tides of fashion after 1740. In 1747 he published a recanting version of the *Mechanical Account,* apologizing for the youthful enthusiasm with which he had adopted mechanical explanations. Recent experiments in electricity and the opinions of the "divine Newton" in his *Opticks* (by then available for some 30 years) and more especially his letter to the Honorable Mr. Boyle (recently published) have convinced him that poisons act through the nervous fluid, which is probably a quantity of that universal elastic matter, or aether, diffused through the universe.[16]

Of considerably more interest and long-range influence were

[15] Richard Mead (1673-1754) studied at Utrecht, at Leyden (where he heard Pitcairne) and Padua, where he obtained the M.D. in 1695. F.R.S., member of the council in 1705 and from 1707 to his death, he was also vice-president of the Society in 1717. His collected medical writings were published in many editions, of which I have used *The Medical Works of Richard Mead* (Dublin: Thomas Ewing, 1767), as well as his *Mechanical Account of Poisons in several Essays* (London: J. M. for Ralph Smith, 1708), 2nd edn. rev., with additions; and *A Discourse concerning the Action of the Sun and Moon on Animal Bodies; and the Influences which this may have in many Diseases* (London: n.p., 1708).

[16] Mead, *Works,* pp. iv, xiv-xv. Newton's letter to Boyle, dated 1678, but first published in the preface to Thomas Birch's *Works of the Honourable Robert Boyle* of 1744, presents an early application of aetherial notions to chemistry. It was greatly influential in winning acceptance of Newton's aether hypothesis after 1740. See Cohen, *Newton's Papers and Letters,* pp. 249-54.

the physiological speculations of the Oxford school of Newtonians, represented by the work of John Freind and James Keill. Like Mead's *Mechanical Account of Poisons,* Freind's *Emmenologia,* first published in 1703, appeared prior to the *Opticks* and reveals none of that use of inter-particulate attractive forces which characterizes his *Chymical Lectures,* presumably delivered the following year. Though he grasped the essential quantitative requirements of the new mechanical mode of reasoning, Freind's arguments still lack the necessary numerical data. Most remarkable for his time, however, the *Emmenologia* is distinguished by a series of experiments in which Freind attempts to find confirmation of his theories.[17]

Freind's purpose in the *Emmenologia* is to provide an explanation of the menstrual cycle. Chemistry will provide none—such concepts as "ferments" or van Helmont's "Archaeus" are clearly useless; indeed, "chymistry" is rather itself in need of reduction to "Mechanick reasoning" [Preface, 11-13, 49]. The desired explanation is reached as a consequence of Bellini's introduction into Physick of the principles of Mechanicks and Anatomy, and the "statical" studies of Sanctorius. Comparative measures of food intake and excrements have shown that there must be a substantial "insensible perspiration" to maintain normal weight equilibrium. Perspiration is a secretion of fluid from the blood and hence is a function of the velocity of circulation and the orifices of the secretory vessels. Now women have less forceful heart beats and thus less blood velocity and, as their frames are more finely and delicately put together, their capillaries are smaller and the orifices of the secretory vessels finer [17-19]. The quantity of their insensible perspiration must, therefore, be less than in men, and their body fluids must steadily increase to plethora. It is this plethora which occasions the menses, for, if the excess of blood is not used in nourishing a fetus, it periodically breaks through the uterine vessels. These, being curved and twisted, are highly resistant to circulation and, as they are not supported by surrounding flesh, they are less able to withstand a bursting force [23-24, 30]. When the excess fluid is thus removed, the cause of the flux disappears, and the flow ceases until the plethora builds again to the critical point.

This ingenious mechanical hypothesis is supported by an elab-

[17] The *Emmenologia,* first published in Latin in 1703 and republished in that language for editions of 1717 and 1727, was translated into French in 1730 and into English for an edition of 1729, which was republished as John Freind, *Emmenologia,* trans. Thomas Dale (London: for T. Cox, 1752).

orate mathematical discussion of the resistance to fluid-flow of variously shaped and convoluted canals and of the forces by which such resistance may be overcome [25-29]. Quantitative confirmation is sought by statical comparisons of the weight of blood lost during any period, the increase of weight between cycles, and an estimated equivalent change in insensible perspiration [44-46], and from the hypothesis, therapeutic measures are deduced to relieve suppression or inhibit immoderate fluxation. These measures had clinical confirmation in at least nine case studies reported in the book [128-45, 160-67], but laboratory investigations were also performed to test some of them. As changes in blood viscidity will vary its momentum and thus its force of penetration, attenuant and coagulant medicines may be prescribed. In the last thirty pages of his treatise, Freind describes experiments to verify the action of such emenagogues and astringents. Using syrup of violets to show the alkaline and acidic nature of attenuants and coagulants, he mixes varieties of each of these with blood drawn from dogs and then with human blood serum [181-85, 199-204], showing that each does, in fact, liquefy or coagulate. He then injects each into live dogs and, dissecting the animals thus killed, confirms the results *in vita*. Whatever one may now think of the results, Freind's work reveals the increase in sophistication of iatro-mechanics achieved through the use of Newtonian principles, where theory has clearly suggested, stimulated, and directed experiment, and these experiments presage the later and more famous ones of Stephen Hales on the physiological effects of perfusates.

The culmination of speculative Newtonian physiology, based on dynamic corpuscularity, was reached in the work of James Keill.[18] Here the early influences of Harvey, Borelli, Bellini, Sanctorius, and Boyle are integrated with later developments by Pitcairne and other British iatro-mechanists. The result is then informed with insights drawn from Newton's 1706 *Opticks*, as interpreted in John Keill's 1708 paper on the "Laws of Attraction," and the whole developed into a fully articulated anatomy and human physiology. James Keill's theories are embodied in two books, a popular student text in anatomy and a set of essays

[18] James Keill (1673-1719), M.D.F.R.S., educated at Edinburgh, and, in medicine, at Paris and Leyden. Returning to England about 1698, he lectured on anatomy at Oxford and Cambridge at least until his retirement to medical practice in Northampton in 1703, and probably for some time thereafter. He received an honorary M.D. from Cambridge in 1705. See H. M. Sinclair and A.H.T. Robb-Smith, *Short History of Anatomical Teaching in Oxford* (Oxford: for the University Press, 1950), p. 19.

on animal oeconomy. The *Anatomy of the Human Body Abridg'd,* first published in 1698, was initially an epitome of Amatus Bourdon's *Nouvelle Description du Corps Humain* (1679), but as successively revised for later editions it came to have the theoretical flavor of Keill's physiological studies. The *Anatomy* appeared in at least 20 English editions (as well as French, Flemish, and Dutch translations) and provided a persistent source for dynamic corpuscular concepts in physiology as late as 1774. Most of these, however, are more completely described, in quantitative detail, in Keill's essays. These first appeared in 1708 as *An Account of Animal Secretion,* which was favorably reviewed in the same number of the *Philosophical Transactions* which praised Freind's *Praelectiones Chymicae* [26 (1708-09), No. 320, 324-31]. A second edition appeared in 1717 under the title, *Essays on Several Parts of the Animal Oeconomy.* A third (Latin) edition of 1718 was called *Tentamina medico-physica;* it included some statical observations, modeled on those of Sanctorius, which were separately published several times as *Medicina Statica Britannica.* The fourth, posthumous, edition of 1738, again called *Essays on Several Parts of the Animal Oeconomy,* included the *Medicina Statica* as well as translations of *Philosophical Transactions* articles by James Keill and James Jurin disputing a determination of the force of the heart.[19]

Keill's physiological theories start with orthodox iatro-hydrodynamic principles. But when he declares the body to be nothing but tubes or vessels full of blood or the fluids separated from it, he does so in connection with mathematico-experimental arguments by which he obtained a quantitative estimate of the ratio of fluid to solid matter [*Essays,* 18-20]. When he cites Pitcairne's estimate of the force (100 pounds) by which the lungs break the blood globules into particles small enough to enter the heart, or that of the force (250,734 pounds!) by which the stomach breaks the cohesion of the aliment into moleculae small enough to enter the lacteals, he adds accounts of his own computations to verify these estimates [*Essays,* 112-14; *Anat.,* 40, 255-65]. Typically he believes the operations of the body depend upon the secretion of fluids and that this is, in part, a function of the velocity of blood flow. Keill, however, was the first person to present a detailed quantitative estimate of the variation in the flow ve-

[19] I have used James Keill, *Anatomy of the Human Body Abridg'd: or, A short and full View of all the Parts of the Body. Together with their several Uses drawn from their Compositions and Structures* (London: for John Clarke, 1759), 12th edn., corr.; and *Essays on Several Parts of the Animal Oeconomy* (London: for George Strahan, 1738), 4th edn.

locity from aorta to artery to capillary to vein, and the first to compute a rationalized value for the force of the heart. Using an elaborate combination of experimental data—different volumes and cross-sections of tubes, amount of blood ejected with each systole, velocity of blood from a cut artery, amount of branching from an artery, etc.—and calculations based on Newtonian hydrodynamics, Keill concludes that the force of the heart is not more than 16 nor less than 5½ ounces and that blood flows in some capillaries as much as 5,223 times more slowly than through the aorta [*Essays*, 42-49]. Some of his experimental data was wrong, and the rest was inadequate. James Jurin claimed that Keill had misunderstood or misapplied the principles of Newton's hydrodynamics, which, in any event, are today known to be faulty. It is not surprising, therefore, that Keill's values were far from correct, but that for the force of the heart is conspicuously better than Borelli's earlier estimate of 180,000 pounds, based on a supposed relationship between bulk and the strength of muscular fibers. By method and example, Keill had, at least, raised an issue which inspired the subsequent experimental determinations of blood pressure and velocity by Stephen Hales and Thomas Young.[20]

For all the satisfaction engendered by his attempt at quantitative methods, it is in his imaginative application of dynamic corpuscularity that James Keill is most attractive. Admitting that he had adopted the principles of inter-particulate attraction from his brother John, James summarized such of his brother's theorems as he would use. That such a power exists in nature "can be denied by none that duly consider the Experiments and Reasons given for it . . . in the Questions annexed to the Latin Edition of . . . [the] *Opticks* [*Essays*, 104]." John Keill had deduced from this power of cohesion the causes of elasticity, fermentation, dissolution, coagulation, and many other operations of chymistry [*Essays*, 103-104]. James Keill now deduces from it the cause of muscular motion. From his own experiments and those described by Boyle, Keill is convinced that some air must penetrate the substance of the lungs and mix with the blood [*Anatomy*, 148]. This air is strongly compressed by sur-

[20] See John G. McKendrick, "Abstract of Lectures on Physiological Discovery. Lecture I. The Circulation of the Blood, A Problem in Hydrodynamics," *British Medical Journal*, 1 (1883), 654-55. Jurin's "corrections" were justified to the extent that he was a better mathematician than James Keill and better versed in the Newtonianism of the *Principia*, but his substituted value of 15¼ pounds, based partially on his use of fluxions, is also wrong, though nearer Hales' measured value of 51½ pounds than either Keill or Borelli.

rounding, mutually attractive, particles of blood. A muscle contracts when, by effort of the will, animal spirits drop from the nerve ends into the fibers of the muscles. Being of the smallest-sized particles secerned from the blood, animal spirits attract the particles of blood more strongly than these attract themselves; this releases the globules of air to expand, and thus muscle fibers are shortened. As the animal spirits are reassimilated by the blood, mutual attraction of blood particles again compresses the air and muscular contraction ceases unless the supply of animal spirit is renewed [*Anatomy*, 136, *Essays*, 167].

Attractive forces were, however, earlier involved in this process, as they are in all others of the animal oeconomy, in the selection by the various glands of the fluids adapted to their purposes. There are, Keill notes, at least 37 different sorts of glands in the body, each of which separates different humours from the blood [*Anatomy*, 212-13]. But these humours do not initially compose the blood, they are formed there by different modifications and combinations of the particles in it. Supposing that there are but five different particles: *a, b, c, d, e* (based on the number of principles which philosophers agree form all natural bodies [*Anatomy*, 65]), Keill shows that as many as 26 different combinations may be obtained from simple combinations of these five, starting at *ab* and ending with *abcde* [*Essays*, 140]. Other combinations are, of course, possible by increasing the proportions of any one principle, as *aba*, etc. Now in a moving fluid, if the different particles have unequal velocities, their attractions will be variously disturbed and the nature of possible combinations will be determined by the rate of blood flow [*Essays*, 112]. The glands secern particles as a function of size only (smaller particles than those suited to the function of a gland being returned to the blood by a system of branching passages in the gland ducts), but the orifices of the glands are appropriately positioned in the body to receive just those particles most likely to be formed from the blood moving there at the proper velocity [*Anatomy*, 63-64].

For all their several editions published after the 1713 *Principia* and even the 1717 *Opticks*, Keill never revised the *Anatomy* or the *Essays* to discuss the action of a subtle, aetherial matter, nor does he ever avail himself of the concept of repulsion. But he had extended that of attraction to its fullest speculative extent. From the building out of primitive elements of the substances of which the body is made, their mechanical interaction to produce muscular motion, and the explanation from these motions of essential body functions, a view of the animal frame and its

oeconomy had been composed of the same principles used
plaining the motion of the heavens, the operations of cher
and potentially all of physical phenomena. Not again, for
than a century, was there to be so successful an atten
reduce physiology to the dominion of natural law. The l
of speculative dynamic corpuscular physiology after Jame
is, therefore, something of an anticlimax.

Yet the work of one other British writer remains to be exa
George Cheyne was not, at least in science, an original t
nor a profound critic.[21] Though most of his writing ap
in a medical context, substantial sections of it might we
been included among the writings of the Newtonian
theologians. His selection of scientific principles was fre
made for reasons other than logical consistency, and th
syncrasies of his ideas must often be credited to individua
iarities rather than to the nature of contemporary natu
losophy. Nonetheless, he was a member, at least perip
of the school of early Newtonian commentators. In add
having studied under Pitcairne, he talked and corres
with David Gregory, Richard Mead, John Freind, and
Keill, to all of whom he refers in his writings. Further, h
most prolific writer, and his publications span the peri
1701 to 1742. Used with caution, changes in ideas whic
from one of Cheyne's books to another reflect the tides of
in Newtonianisms as well as Cheyne's own odyssey to
ligious enthusiasm, and may serve as a survey of the rise
of dynamic corpuscularity during the first 40 years of th
eenth century.[22]

[21] George Cheyne (1671-1743), M.D.F.R.S., was born in Scotland. He
classical education and originally intended for the ministry, when he w
by Pitcairne to the study of medicine. Receiving an M.D., by diploma, in
King's College, Aberdeen, on Pitcairne's recommendation, he moved
and to Bath, where he developed a large medical practice. One of the mo
writers of popular medical tracts in early eighteenth-century England,
acquainted with most of the famous physicians and writers of his day
confidant of the Countess of Huntington, whose Methodism appealed to
ing religious enthusiasm.

[22] Of the many editions of his many books, I have used: [George Ch
*concerning the Improvements of the Theory of Medicine, with a New
Continual Fevers* (1702), the Harvard University Library copy has a
page; George Cheyne, *Philosophical Principles of Religion: Natural an
. . . Part I. Containing the Elements of Natural Philosophy, and th
Natural Religion arising from them* (London: George Strahan, 1715)
An Essay of Health and Long Life (London: for George Strahan an
1724); and *An Essay on Regimen. Together with Five Discourses Me
and Philosophical . . .* (London: C. Rivington and J. Leake, 1740), 21

Cheyne's first book, on a *New Theory of Fevers* (1701), was a polemic in support of Pitcairne's mechanical theory of medicine, written with all the intemperate zeal of eighteenth-century Scottish medical controversy. Later described by Cheyne himself as a "raw and inexperienced performance," it nonetheless reached a fifth edition by 1740. Having adopted Pitcairne's fluid-mechanical theory of medicine, Cheyne endeavors to prove that fevers result from the constriction or obstruction of the blood vessels, causing stagnation of the blood and disturbing the quality and quantity of secretions [46-47]. His arguments develop the indicators of orthodox British iatro-mechanics: the usual list of authorities, Harvey, Borelli, Bellini, Sanctorious, and Pitcairne; the usual emphasis on a quasi-mathematical discussion of fluid motion; and the usual grossly assumptive and mostly irrelevant calculations. Newton has given the "true causes" of the laws of motion and the true principles of all the effects of nature, to wit, attraction or gravitation [23-24]. What now is needed is a *Principia Medicinae Theoreticae Mathematica,* in which there will be a mechanical account of chymical operations, an explanation of the nature and causes of fluidity, heat and cold, and elasticity, and an understanding of the constituent particles of matter [25-26]. Yet both explanation and deduced therapy are purely kinematic. Stagnated body fluids are corrosive—*i.e.*, their particles are pointed, perhaps consisting of four equilateral triangular planes [119]. Mercurial medicines, by their excessive gravity and smallness, can dissolve the cohesion of the viscous fluids, "break off and plain, the Points and Angles," of the corrosive particles, and carry before them such gross particles as resist them, scouring the passages to leave clear and passable canals [115, 118, 120].

All of this is fairly standard, but two short passages are exceptional. In a discussion of the nature of mercury, Cheyne introduces the concept of particles of the first composition, that is, aggregates of the smallest and least constituent particles of any body [106]. Clearly this is derived from Newton, but Newton's first publication of the concept was in the 1706 *Opticks*. Cheyne may (with Pitcairne) have seen a manuscript of the *Opticks* prior to its publication, but it is unlikely that so early a copy would have contained this concept, and a more likely source for his information was the Newton paper, "De natura acidorum," which, John Harris implies, had been circulated among Newton's acquaintances for years prior to its publication in 1710.[23] Having thus early availed himself of a sophisticated Newtonian corpus-

23 Harris, *Lexicon Technicum,* II, Introduction, sig. b1v.

cular notion, Cheyne then adopts a semi-materialistic explanation of muscular motion. There is, he claims, a *liquidum nervorum* with a special quality whereby its particles are disposed to make a violent inflexion of the membranous fibres of the body [83-84]. This failure to see contradiction in styles of explanation is typical of all of Cheyne's work, but the affinity for material causation is not an accident of ignorance. His ideas move increasingly toward a materialistic, vitalistic universe.

Cheyne's second publication was a mathematical treatise, *Fluxionum methodus inversa* (London, 1703), which, he later admitted, "was brought forth in Ambition and bred up in Vanity [*Long Life*, iv]." Scorned by Gregory and attacked by deMoivre, it provoked Newton into publishing his *Tractatus de Quadratura Curvarum* as an appendix to the 1704 *Opticks*. This was its only virtue and it was quickly forgotten, but ambition and vanity were not, and their next product was a long and famous *Philosophical Principles of Natural Religion.* First published in 1705, the book was denounced by Gregory and his coterie as full of errors and, moreover, stolen, in large measure, from Bentley's Boyle lectures.[24] But Cheyne corrected the errors, revised the first edition to accommodate later work by Newton, and published it again, along with a second volume on the *Philosophical Principles of Reveal'd Religion,* in 1715. Together the two volumes were to be republished in more than five editions, one as late as 1753.

Cheyne's *Natural and Reveal'd Religion* is a curious work. An exercise in Newtonian natural theology, with a title in obvious analogy to Newton's *Mathematical Principles of Natural Philosophy,* it is full of logical inconsistencies and apparent gaps in understanding. The use of argument-from-design is less jejune than in many similar works, but Cheyne is fondest of what he calls argument from analogy. As, for example, there is attraction from the sun to give activity to the planets, so is there a divine attraction, or principle of re-union, to animate the will and actuate the affections to God [50]. Cheyne's use of this kind of analogy, particularly in the employment of mathematics in the volume on revealed religion, is so odd that Hélène Metzger was led to describe him as a neo-Platonist in her *Attraction Universelle et Religion Naturelle chez quelques Commentateurs Anglais de Newton.*

[24] Hiscock, *Gregory, Newton and their Circle,* pp. 24-25. In justice to Cheyne, it should be noted that most of the 429 errors Gregory found in the *Principles* were mistakes of punctuation or misspellings.

Cheyne's treatment of the elements of natural philosophy begins in an orthodox way with denials of the existence of subtle matter and insistence that attraction exists in matter but is not essential to it, being impressed and continued there by virtue of "Omnipotent Activity [41]." According to Gregory, Cheyne claimed among his cronies that Newton's addition to the Latin *Opticks* "were stolen from him by Sr Isaac in private conversation," but this work gives remarkably little evidence that Cheyne understood the fundamental innovation proposed in those queries.[25] He cites the Latin *Opticks* and acknowledges the "judicious corrections and advices" of Dr. Freind; he declares the "whole difficulty of *Philosophy*" to be in investigating the powers and forces of nature from the appearances of the motions given, and then from these powers to account for all the rest [93]. But he personally prefers either kinematic corpuscular or materialistic explanations. Twice he recognizes Newton's mechanical theory of heat [74, 204], but cold and freezing proceed from some *salin* substance, whose particles are thin, double-wedged, and insinuate themselves into the pores of particles making them cohere [61-62]. Bodies cohere because of a force of attraction which is distance-dependent, but a necessary condition of cohesion is a congruity of surfaces. Some of the primary atoms of which bodies are composed are terminated with plane surfaces, still others with surfaces entirely curved, and between these extremes there are infinite combinations of plane and curved surfaces [94-95]. This would appear to be a suggestion that there exists an infinite number of different primary atoms.

Nearly alone among his contemporaries in early adopting the action of a subtle, aetherial fluid, Cheyne neglects to excise his pre-1713 denial of the aether [33-34] and he fails to understand it as a substitute or explanation for attractive and repulsive forces. Instead, its insinuation between the particles of bodies becomes a factor complicating the action of forces of cohesion, perhaps cloaking the inverse-square form which, agreeable to the simplicity of nature, such an attractive force should have [104-06]. Finally, the mind adds a greater force than is natural to animal spirits in muscular motion [137], planets and comets have some secret influences and actions upon one another, different from the bare action of their attractions [156], metals and minerals possess a plastic virtue proper to themselves enabling them to grow like vegetables under the influence of subterranean heat [278], and the nervous fibers are solid and impregnated with a subtle spirit whose law and action are never to be determined, as

[25] Hiscock, *Gregory, Newton and their Circle*, p. 35.

the nature of life, light, and animal motion will be an eternal reproach to mechanism and human invention [315]. Even without the neo-Platonic mysticism of the second volume, it is hard to see Cheyne's *Philosophical Principles of Natural Religion* as much more than an incoherent demonstration that these principles are something other than those of natural philosophy.

It is, therefore, small wonder that his professional Newtonian contemporaries disliked the work, nor that his next publications, some fifteen to twenty years later, on gout (1720) and on the uses of temperance in food and drink to produce *Health and Long Life* should have taken the less philosophical form of practical, popular medical treatises. Cheyne's *Essay of Health and Long Life,* first published in 1724, was the most popular of his writings. Republished on an average of every two years to its tenth edition of 1745, it was translated into Latin, French, and German, extracted in English in 1770, and was still being published in English editions as late as 1827. The nature of the work precluded much explicit theoretical discussion, but implicitly the medical theory underlying the whole is that of Pitcairne. The emphasis is still on kinematic corpuscularity with small advance into greater subtlety of mechanical explanation. There is some discussion of size and cohesion of particles, their momentum and spheres of attraction, but equivalent attention is given to the essential difference between spiritual beings: active, self-motive, and self-determining; and material bodies: inert, impenetrable, and active only according to laws and limitations established by the "Author of Nature [147-49]." Only in this renewed emphasis on the dichotomy of mind and body is a change from mechanistic physiology overtly signaled, but covertly there is a hint of the new, or renewed, materialism in the references to nitrous effluvia of the air [11] and especially in a tendency to classify substances by their functional properties rather than their microscopic forces and structures [25-27 ,64].

In 1740 Cheyne returned to the concocting of philosophy and medicine in his *Essay on Regimen.* This was the least successful of his writings, and he had to indemnify his publishers for losses suffered on the first edition, but it subsequently reached at least a third edition by 1753. The first third of this work contains Cheyne's final view on the nature of matter and the human body. In many ways it represents his most consistent natural philosophy. The structure of matter is here derived from its least and last particles, and the concepts of attraction and repulsion, the distance dependencies of these forces, even the algebraic analogy where attraction becomes repulsion, are displayed at length from Newton's Query 31. But the shapes of the primitive particles of specific bodies are probably diversities of the three most simple figures, spheres, cubes, and equilateral prisms [92-96], and aggregates of these different particles into elements (water, air, light, salt, and earth) have specific operational or functional qualities by which they may be distinguished and which they never lose, however much the quality may be concealed by combination with other particles. Chemistry is thus transformed from a mechanical problem to one, potentially at least, of classification.

Heat is a brisk vibrating action and reaction of the elastic repelling nitre of the air with the strongly attracting acid sulphur; and heat is conveyed and propagated by the subtle, elastic fluid which pervades all bodies. But the particles of fire, which concoct the active virtues of plants, metals, and minerals into chemical medicines, do so only as they are also impacted and transubstantiated in them so as to divide their particles and impart to them so strong a force of attraction as to rend, tear, and destroy the fibers of the body [xxi]. Somehow, mechanism has been subordinated to a special, material substance, and this materiality is repeated, though somewhat spiritualized, in the assertions, first that the body is an hydraulic machine, and second that it operates only as a consequence of a spiritual animal matter, more subtle and elastic than the finest aether [5].

This picture of a mechanized world view transmuted by practical or religious considerations into a material, vitalistic one is confused by the last two-thirds of the book which repeats the earlier neo-Platonism, in a Christian frame so foreign to his age that some readers accused him of heresy. Christian theology apart, it is difficult to assert Cheyne's materialism in the face of his claim that substances are merely pictures, emblems, and miniatures of the deity [198-99], as space, time, motion, and its velocity are abstracted ideas only and have no external reality. Yet these are comparatively minor instances of extreme idealism in a section whose major purpose is the demonstration of the matter-spirit dichotomy, with the spirit, directly reflecting the active energy of the divine will and goodness, necessary for even the simplest of human physiological phenomena [126, 148-49]. So far as physiology is concerned, the introduction of a substance whose activity precludes mechanical explanation is a change toward materialism, however spiritual such a substance may be. The net influence of Cheyne's last philosophical work is, therefore, to bring to final focus the materialistic tendencies implicit in Cheyne's earlier writings.

CHAPTER FOUR

Experimental Newtonianism

GEORGE CHEYNE's progress from kinematic mechanism toward vitalistic materialism, by way of Newtonian dynamic corpuscularity, was, in part, occasioned by religious considerations. At the end of the period, however, other natural philosophers had moved in a parallel direction for better, or at least more scientifically defensible, reasons. For, in the end, it must have seemed that dynamic corpuscularity had failed in an essential regard. The most convincing aspect of Newton's work had been its quantitative agreement between theory and observation. The Cartesians had been mathematical, the Boyleians experimental, but the two methods had not joined in a rigorous natural philosophy before Newton. The agreement had, however, been complete only in celestial mechanics, for which centuries of observation had provided both clues and confirmation to the form of gravitational force. During the first four decades of the eighteenth century, there was scarcely another such triumph for experimental Newtonians.

This is not to say that there had been no experimentation, but merely that the Newtonian injunction—from the motions find the forces, from the forces derive the phenomena—had not been served. Newton himself had not succeeded. He had derived another force law, for the elastic repulsion of the air, from Boyle's law; but the speed of sound deduced from this elastic model was not in agreement with observation. The experimental mode of the *Opticks,* so compatible to British natural philosophy, had indicated the existence of other forces but left their determination to future generations. The theoretical inspiration of the *Opticks* is clear in speculative elaborations of dynamic corpuscularity in the work, for example, of the Keills, John Freind, and John Rowning. Its experimental inspiration is scarcely less clear. For 40 years and more, the problems, the experimental design, and the interpretations of British experimenters were likely to derive from suggestions of the *Opticks* or of those speculations elaborated from it.[1] And the result of these years of experimentation was a demonstration that the demands of dynamic corpuscularity were too sophisticated for the conceptual grasp or experimental abilities of most eighteenth-century scientists. Yet in the

[1] See Cohen, *Franklin and Newton,* which completely discusses this aspect of the *Opticks.*

process, there had also been introduced a set of new and exciting problems and a hint as to how these might be solved.

This pattern of Newtonian-inspired experiments was set from very early in the period. Between 1687 and 1704, few of the volumes of the *Philosophical Transactions* even mention Newton, save for his mathematics. Papers relating to matter and its action continue in the tradition of Boyleian kinematic corpuscularity or even of Aristotleian qualities. This is not particularly surprising, as the early *Philosophical Transactions* was primarily a journal of experiments and observations, and Newton's work did not become conspicuously experimental until the publication of the *Opticks*. But for years after 1704, very little change occurred in the nature of *Philosophical Transactions* papers. The pervasive theory underlying experimental interpretation was still that of Boyle. The first important sign of change is to be seen in the work of Francis Hauksbee, who, from the death of Robert Hooke till his own in 1713, dominated the meetings of the Royal Society and the pages of its *Transactions* with his experiments.[2]

Hauksbee first appeared on the public scene in 1703, performing an experiment before the Royal Society on the day Newton first presided as its president. Until his death, Hauksbee acted, though without formal title, as the demonstrator of experiments for the Society, and his work and Newton's have been thought to be intimately related. Most of Hauksbee's early papers, however, are either unrevealing of their theoretical presuppositions or suggest an invincible affinity for the kinematic corpuscularity that he learned, Henry Guerlac has suggested, as a laboratory assistant for Robert Boyle.[3] This does not mean that he had failed to react to Newtonian concepts, but rather that his earliest reactions, at least, were negative. This is shown in his three papers of 1707 and 1708, on compressing air so violently, or over so extended a period, that its "Springs or Constituent Parts" were incapable of restoring themselves [*Phil. Trans.*, 25 (1706-07), No. 311, 2409-11, 2412-13; 26 (1708-09), No. 318, 217-18]. From the dates of their appearances, there can be little doubt that these papers are intended in deliberate contradiction to the just published Query 23 of the Latin *Opticks*, where Newton declares that the elasticity

[2] Francis Hauksbee (b.?-1713), F.R.S., electrical experimenter and constructor of first frictional electric machine. For many years experimenter for the Royal Society, his experiments and papers were described collectively in his *Physico-Mechanical Experiments*, 1st edn., 1709; augmented 2nd edn., 1719.

[3] Henry Guerlac, "Sir Isaac and the Ingenious Mr. Hauksbee," in I. Bernard Cohen and René Taton, eds., *Mélanges Alexandre Koyré*, I, "L'Aventure de la Science" (Paris: Herman, 1964), pp. 228-53.

cannot be explained "by feigning the Particles of Air to be springy and ramous, or rolled up like Hoops, or by any other means than a repulsive power." Other papers, as well, support the pre-Newtonian tradition: water density, for example, changing with temperature, because there is a substance between its particles possessing an elastic quality capable of expanding with heat [26 (1708-09), No. 318, 221-22; No. 319, 267-68] and hot iron emitting an effluvia divesting air of its power to support life or flame [27 (1710-11), No. 328, 199-203]; but a more positive response to Newton's suggestions can be seen in Hauksbee's paper on the refractive indices of fluids, where bodies refract light from some quality peculiar to themselves, whether their inflammability, different textures, or figures of component parts [27 (1710-11), No. 328, 204-07].

After a time, and somewhat inconsistently, Hauksbee adopted some part of Newtonian force language. Corrosive menstruums, for example, separate the parts of bodies by the greater vigor of their attraction, or affection, for the bodies' particles than of either substance's attraction for its own particles [26 (1708-09), No. 320, 306-08]. From 1709, his experiments tended to concentrate on the phenomenon of capillarity, and here, increasingly, his interpretations are in terms of short-range attractive forces. Nowhere are his manipulative and observational skills more handsomely displayed than in these experiments. He notes, among other things, that the attractive force must act only between corpuscles in contact, or at infinitely short distances and that the direction of attraction is perpendicular to the sides of the cylindric glass.[4] In the last several of his experiments, on the capillary action between two plates, he even provides the data from which Newton derived the statement inserted into Query 31 of the 1717 *Opticks*, that "the Attraction is almost reciprocally in a duplicate Proportion of the distance of the middle of the Drop from the Concourse of the Glasses, . . . [and] by reason of the spreading of the Drop, and its touching the Glass in a larger Surface . . . the Attraction . . . within the same quantity of attracting Surface, is reciprocally as the distance between the Glasses [368-69]."

These experiments were clearly influential, and capillarity became a major example of a Newtonian attractive force. Brook

[4] See W. B. Hardy, "Historical Notes upon Surface Energy and Forces of Short Range," *Nature*, 109 (1922), 375-78; and Francis Hauksbee, *Philosophical Transactions*, 25 (1706-1707), No. 305, 2,223-24; 26 (1708-1709), No. 319, 258-66; 27 (1710-11), No. 332, 395-96, No. 334, 473-74, No. 336, 539-40; 28 (1712-13), No. 337, 151-56.

Taylor was inspired to follow Hauksbee's lead and observed, in a paper of 1712, that the curve formed by liquids rising between glass plates was an hyperbola. In another part of the same letter, not read until 1721, he reported an experimental demonstration that the cohesive attraction between a plane of wood and the surface of water was about 50 grains.[5] Hauksbee was on the verge of having led to the experimental determination of the magnitude and form of a Newtonian attractive force. Then James Jurin introduced a perturbing factor. In the first (1718) of two sets of experiments on capillarity, Jurin criticizes Hauksbee's conclusion that attraction was a function of the contiguous surface, and in the second (1719) he suggests that the cohesion may, in part, depend upon the pressure of a medium subtle enough to penetrate the receiver and "as to the Existence of such a Medium, I shall content my self to refer to what has been said by our *Illustrious President* in the Queries at the latter end of the last Edition of his *Opticks*."[6] Not until Thomas Young treated capillarity in a *Philosophical Transactions* paper of 1805 was there again to be so near an approach, by a British scientist, to a quantitative discussion of capillary attraction. Jurin's early introduction of the Newtonian aether acted to stop the attempt, from capillary motion, to find a force from which other phenomena might be derived. The only set of Hauksbee's experiments from which a positive influence on dynamic corpuscularity might have been obtained was forestalled, the influence vanishing into a fog of Newton's own devising.

There may be a certain wry retribution in this, for Henry

[5] Brook Taylor, "Part of a Letter . . . Concerning the Ascent of Water between two Glass Planes," and "Extract of a Letter . . . Giving an Account of some Experiments relating to Magnetism," *Philosophical Transactions*, 27 (1710-12), No. 336, 538; 31 (1720-21), No. 368, 204-08. Brook Taylor (1685-1731), educated at St. John's College, Cambridge, LL.B. 1709, LL.D. 1714. Reputedly the only British mathematician of his day, after Newton and Cotes, capable of holding his own with the Bernoullis. Correspondent of John Keill, his best work was in the application of fluxions to difficult physical problems. He is known for the discovery of Taylor's Theorem," on the expansion of functions of a single variable in infinite series.

[6] James Jurin, "An Account of some Experiments . . . with an enquiry into the Cause of the Ascent and Suspension of Water in Capillary Tubes," and "An Account of some new Experiments, relating to the Action of Glass Tubes upon Water and Quicksilver," *Philosophical Transactions*, 30 (1717-19), No. 355, 739-47; No. 363, 1,083-96. James Jurin (1684-1750), M.D.F.R.S., educated at Trinity College, Cambridge, B.A. 1705, M.A. 1709, M.D. 1716, and, in medicine, at Christ's Hospital, London, and at Leyden. He was Secretary of the Royal Society for the last six years of Newton's presidency and edited four volumes of the *Philosophical Transactions*. His forte in science was as a commentator and critic, not in innovation; but he was one of the earliest persons to adopt aetherial concepts from Newton, who is said to have instructed Jurin personally in mathematics.

Guerlac has recently argued that it was the influence of Hauksbee's electrical experiments that prompted Newton's return to aetherial explanation in the *Opticks* of 1717. These are the experiments on which Hauksbee's lasting reputation was built, and they uniformly reveal no dynamic corpuscularity. From the first, of 1706, to the last, of 1711, Hauksbee always talks in terms of the effluvium emitted by the glass when rubbed. These effluvia lay hold of threads, giving the appearance of attraction [25 (1706-1707), No. 309, 2,372-77]; they are capable of acting as solid bodies, forming a stiff continuum, even through glass, such that "when any part of them are pusht, all that are in the same Line suffer the same Disorder," and the largeness of their effect ("of Light or Attraction") proceeds from their number and strength [26 (1708-1709), No. 315, 83-88]. Here are to be found the seeds of later materialistic explanations of electricity. A substance, obtained from bodies by a particular operation, produces a result which is dependent upon the special nature of the substance and on the amount of it present.

The significance of these experiments, with their semi-materialistic explanations, was much larger than their number warrants. Even ignoring their possible influence on Newton—and the ease with which Newton's inconvenient conceptual revisions could be ignored has already been demonstrated—they combined with Hauksbee's other experimental studies to give a collective impression which was not really Newtonian.[7] More important, they were on a subject which became increasingly popular as the century wore on. Electricity was one of the most exciting problems for experimental investigation over the greater part of the century. That the man who introduced it to the serious attention of British natural philosophers should consistently have adopted a quasi-materialistic explanation for it could not but have an influence on the experimenters who followed him. And this is the more significant as the other major electrical experimenter of the first four decades of the century had also a similar interpretation of his results.

The biography of Stephen Gray is still insufficiently known to determine the origin of his philosophical preconceptions.[8] There

[7] In the collected F[rancis] Hauksbee, *Physico-Mechanical Experiments on Various Subjects. . . .* (London: by R. Brugis, for the Author, 1709), and in the 1719 posthumous edition as well, for all the prefatory praise of Newton's "general Laws of Attraction and Repulse, common to all Matter."

[8] Stephen Gray (1666/7-1736), F.R.S., onetime assistant to Roger Cotes at Trinity College observatory, later resident with J. T. Desaguliers in London, and finally a pensioner in Charterhouse. Author of papers in the *Philosophical Transactions*

is reason to believe that he sided with John Flamsteed in the latter's quarrel with Newton, but that need have no necessary connection with the non-Newtonian character of his work. Though the greater part of his electrical papers was not published until after 1731, an early (1707/8) letter to Hans Sloane suggests that his theoretical interpretations were independent of the similar views of Hauksbee. Gray was exceedingly careful to avoid, in print, any obvious commitment to theoretical explanations. He uses the terms "attraction" and "repulsion," but the reference is to the virtues of electrified bodies and not to forces. His intent seems to be descriptive, though it implies a theory, and the notion of electrical effluvia moving or being transferred from or through bodies, which pervades his papers, reveals the material-mechanistic bias of his work. The letter to Sloane is still stronger evidence of his combination of material substance with mechanical action. Glass emits a current of luminous and electric effluvia, which either in its emission and reflection or, more probably, "as all Bodies Emitt soe they Receive part of the Effluvia of all other Bodies that Inviron them" occasions the appearance of attraction, which is made "according to the current of these Effluvia." The mechanism suggested is the kinematic corpuscular one of contact; there is no reference to forces or to action-at-a-distance.[9] Not yet has there been achieved the full materiality of an electrical substance, effective by virtue of its presence and to the extent of its quantity, but the impression of a material substance is there to be assimilated, as later experimenters turn from the quantitative inadequacies of Newtonian forces which have not been measured to the greater satisfactions of an aetherial substance whose quantity can be conserved.

As the greatest speculative influence of Newtonian matter theory was felt in chemistry and physiology, it is appropriate, if not entirely to be expected, that some of the most significant experimental advances should occur there also. This was the work of Stephen Hales, in whom was vested a substantial part of the Newtonian legacy. Here is to be seen, in its fullest detail, the

---

on microscopic and astronomical observations as well as the electrical studies for which he is now best remembered. See Robert Chipman, "An Unpublished Letter of Stephen Gray on Electrical Experiments, 1707-1708," and "The Manuscript Letters of Stephen Gray, F.R.S. (1667/7-1736)," *Isis*, 45 (1954), 33-40, and 49 (1958), 414-33, and I. Bernard Cohen, "Neglected Sources for the Life of Stephen Gray (1666 or 1667-1736)," *Isis*, 45 (1954), 41-50.

[9] Gray's most important electrical papers appeared in the *Philosophical Transactions*, 37 (1731-32), No. 417, 18-44; No. 422, 227-30; No. 423, 285-91; No. 426, 397-407.

power of dynamic corpuscularity to suggest, stimulate, and direct experiment. Hales was the author of two major works: the *Vegetable Staticks* (1727) and the *Haemastaticks* (1733), as well as numerous smaller treatises on ventilators, dram-drinking, the purification of sea-water, etc. Through these works he may well have had an influence second only to Newton's on the development of eighteenth-century natural philosophy.[10]

Although the *Vegetable Staticks* was published first, Hales notes in its preface that his "haemastatical" experiments had preceded those on plants by some twenty years. As these experiments on animals mark the beginning of Hales' scientific career, the study of the nature of his scientific contribution is best begun with an analysis of the *Haemastaticks*.[11] First published in 1733 as volume two of the *Statical Essays* (volume one being the *Vegetable Staticks*), the *Haemastaticks* was republished, with its companion volume, at least four times in English by 1769 and was translated into French, Italian, Dutch, and German. During the eighteenth century, it was to influence the researches of such physiologists as Robert Whytt, Albrecht von Haller, and Thomas Young, and it has since been accepted as one of the great experimental classics of physiology, fit to place beside *de Motu Cordis* of William Harvey. Yet Hales seems to have little, if any, formal training in anatomy or physiology. In the *Memoirs* of William Stukeley, from which have been derived what little is known of Hales' scientific interests at Cambridge, there is some indication that Hales had joined in anatomical dissections and, with "Mr. Rolf, dissecting there" and soon to be "declar'd professor of anatomy in the

[10] Stephen Hales (1677-1761), educated at Bene't College (Corpus Christi), Cambridge, B.A. 1699/1700, M.A. and Fellow 1703, B.D. 1711; D.D. (hon.) Oxford 1733. About 1710 he became resident perpetual curate of Teddington, Middlesex, and ultimately also rector of Farringdon, Hampshire, clerk of the closet to the princess-dowager and chaplain to her son, George, Prince of Wales, later George III. F.R.S. 1718, Copley medalist 1739; one of eight foreign associates of the French Academy of Sciences, 1753. A founding member of the Society of Arts (now Royal Society of Arts), he was one of its vice-presidents, 1755. The standard biography by A. E. Clark-Kennedy, *Stephen Hales D.D.F.R.S.* (Cambridge: at the University Press, 1929), is useful but now sadly out of date.

[11] I have used the facsimile of the first edition, Stephen Hales, *Statical Essays: Containing Haemastaticks; or, An Account of some Hydraulick and Hydrostatical Experiments made on the Blood and Blood-Vessels of Animals. Also An Account of some Experiments on Stones in the Kidney and Bladder; with an Enquiry into the Nature of those anomalous Concretions. To which is added An Appendix, containing Observations and Experiments relating to several Subjects in the first Volume* (New York and London: Hafner Publishing Company, New York Academy of Medicine, History of Medicine Series No. 22, 1964). Discussion of the last two sections will be postponed until consideration of the *Vegetable Staticks.*

University," gone botanizing with Stukeley into the country.[12] There is no reference to reading or lecturing in physiology, but there can be little doubt that then, or shortly thereafter, Hales came to know the work of the Oxford physiologists. Perhaps he heard James Keill lecture at Cambridge (Stukeley notes [42] a visit to Keill at Northampton in 1708); the 1710 volume of the *Lexicon Technicum*, to which he had subscribed, contains several lengthy discussions of the work of both the Keills; certainly by the time the *Haemastaticks* was published, Hales was familiar with Keill's essays in the *Tentamina Medico-Physica* of 1718, for he constantly refers to that work. Indeed, the work of the *Haemastaticks* can only properly be understood as an experimental solo against the speculative obligato of James Keill, and, to a lesser degree, of Pitcairne, Cheyne, Mead, and John Freind.

The *Haemastaticks* has been described as a treatise "without form or order," in which "Hales simply rambles on and on, each experiment suggesting another, so that he was led almost imperceptibly to study a great many physiological phenomena."[13] There is less injustice to Hales in suggesting, instead, that the form and order of his work had been pre-established by that of James Keill. In this work Hales is to be seen as the experimenter who provided the "Number, Weight, and Measure" previously lacking to assumptions on the hydraulic system of the blood. With that experimental genius which is characterized by the design and execution of investigations anyone could have done (and did not), Hales measured, one by one, just those parameters so essential to the iatro-hydrodynamic theories of the British school of mechanistic physiologists—and, primarily by implication, substantially reduced their credibility.

First with a crude end-pressure manometer and later with a slightly more sophisticated lateral-pressure one, Hales measures the pressure of the blood in the arteries and veins of horses, dogs,

[12] W. C. Lukis, ed., *The Family Memoirs of the Rev. William Stukeley, M.D. and the Antiquarian and other Correspondence of William Stukeley, Roger and Samuel Gale, etc.* (Durham: for the Surtees Society, 1882-83), I, 33, 39. William Stukeley (1687-1765), educated at Bene't College, Cambridge; M.B. 1708, M.D. 1719. He studied medicine under Richard Mead at St. Thomas's Hospital, London, was ordained and held several livings, including ultimately one in London. But he was better known as an antiquary who early made valuable drawings of Stonehenge. F.R.S. 1718. Stukeley also wrote articles and pamphlets on medicine, anatomical dissections of animals, and the causes of earthquakes (electricity!). See Stuart Piggott, *William Stukeley, An Eighteenth-Century Antiquary* (Oxford: Clarendon Press, 1950).

[13] Percy M. Dawson, "Stephen Hales, the Physiologist," *Bulletin of The Johns Hopkins Hospital*, 15 (1904), 237; a survey of later secondary work on Hales reveals the continuation of this assessment.

sheep, oxen, and deer, and in different conditions of rest, excitement, age, and health of the animals studied. Clearly this pressure is inadequate as an explanation of muscular motion [58-59]. Progressively he drains the blood, measuring the pressure at various stages during the process, and determines, roughly, the percentage by weight of blood in animals. His results, indirectly, cast doubts on Keill's estimate of the fluid content of the body and directly challenge the conclusion (of Freind, for example) that perspiration is a function of increasing blood pressure [7]. Pouring molten beeswax into the heart, Hales measures the volume and surface area of the cast produced and the diameter of the aorta and its orifice. Though this is not a completely valid measure of heart capacity, it is closer than anyone previously had come, and from it and his measures of blood pressure and calculations of blood velocity at the aorta [e.g. 21-22, 37-43], he determines the force of the heart—substantially (and correctly) increasing the estimate of Keill. By measurement of the diameters of artery, vein, and capillary, by forcing "blood warm" water through various organs of the body, and by direct microscopic examination, Hales determines the differences of blood velocity and pressure in various parts of the body. He notes that this difference must vary with the state of the blood and with the motion, rest, food, evacuations, heat, and cold of the animal involved, so that "nature has wisely provided, that a considerable Variation in these, shall not greatly disturb the healthy State of the Animal [56]." The notion that health is directly to be determined by variations in the hydraulic system of the blood is thus weakened. Using this technique of forcing water through various parts of the animal anatomy, Hales measures the resistance to flow and bursting pressure of arteries, veins, capillaries, intestines, and stomachs of dogs. Borelli and Pitcairne are wrong in the assumption that digestion is produced by mechanical action of the stomach on food [177-80]. Freind, by implication, is wrong in assuming that pressure of the blood can periodically burst the uterine vessels, and Keill is manifestly wrong in assuming the dilatation of the lungs and force of air is sufficient to promote the passage of blood through the lungs and into the heart [77-78, 106].

Although his results ultimately challenged the conclusions of the mechanists, Hales' experiments were guided by their preconceptions, which he generally accepted. And as these preconceptions frequently led him to valuable results, so, occasionally, they led to erroneous ones. The injection into the body of various fluids: warm and cold water, brandy, pyrmont water, decoctions of Peruvian bark, "Chamomel flowers," or cinnamon, constricts the body fibers and increases their resistance to blood flow. This, presumably, invigorates the animal by increasing the force of arterial blood and (as Freind had earlier suggested) serves to explain the effects of such substances when taken as medicines [126-39]. Bleeding may be beneficial, by increasing the velocity of flow in the part near the opened vein, though it will also abate the force of the blood [165-66]. The body acquires its heat by the friction of blood particles in the capillary vessels, particularly in the lungs. It is (perhaps) the chief use of the red blood globules, that their firmness and elasticity make them more susceptible to acquiring heat by friction while their high sulfur content (as their color intimates) makes them more retentative of the heat acquired [90-93]. The chief purpose of the lungs must, therefore, be the blending of inspired air with the blood to promote cooling. One of Hales' simple and most ingenious experiments was made, in support of this hypothesis, to determine the degree of refrigeration that blood acquires in the lungs. From the amount of blood passing through the lungs per minute, the quantity of air inspired during the same time, and the difference in temperature of inspired and expired air, Hales obtained the data from which his friend J. T. Desaguliers computed that the effect of the lungs was to reduce the temperature of all the blood in the body by 0.101928 of a degree Fahrenheit in two minutes [97-106]. Number, weight, and measure can conspire to support as well as confute misconceived hypotheses.

In two regards only are there clear indications that Hales' work in the *Haemastaticks* had been influenced by ideas derived later than his days in college, and both of these may be misleading. Boerhaave's *Elementa Chemiae* (first published in 1732) is cited for evidence to support the refrigatory function of the lungs. Hales obviously respects Boerhaave (and had earlier read an unofficial version of Boerhaave's chemistry), but the blood's heating by friction and cooling by respiration are standard physiological assumptions by the late seventeenth century, and there is little sign in Hales of Boerhaave's material theory of heat. The other instance is more serious, for here Hales seems to depart from mechanistic concepts in the direction of materialism. After rejecting blood pressure as the cause of muscular motion, he suggests that some "more vigorous and active Energy, whose Force is regulated by the Nerves," is required. Is this energy the same as that exhibited by "electrical Powers"? Hales hints that it may be and cites the *Philosophical Transactions* papers of Stephen Gray to show the transmission along the surface of animal fibers, and therefore the nerves, of "a vibrating electrical Virtue" which acts there freely and with considerable energy [58-59]. Hales later performs experiments to see if the body might acquire electricity by agitation of the blood [93-96]. The Gray citation may be misleading chronologically, for Stukeley reports having witnessed, in Hales' company, electrical experiments performed by Gray during his years at Cambridge. In any event, even accepting the possible influence of Boerhaave and Gray, the *Haemastaticks* is a retrospective work, and its late appearance is deceiving in a study of Hales' ideas. Important though it was, in exercising a quantitative control on mechanistic physiology, the mechanism it examines is comparatively unsophisticated. Later workers, reading only the hydraulic and hydrostatical experiments would, at best, see Hales as an unstructured experimentalist; they might, worse, conceive incorrectly that he had abandoned the notion, hardly mentioned in that work of inter-particulate forces which had played so central a role in the earlier published investigations.

The creative design and execution of beautifully conceived experiments is no mean accomplishment, but it must be admitted that Hales can no longer be regarded, on the evidence of the *Haemastaticks,* as quite so prescient or original as he once was thought to be. The evidence of the *Vegetable Staticks,* on the other hand, enhances the reputation of Hales, both as an experimenter and as a theorist of the highest quality.[14] The theoretical insight is not immediately evident from the experiments on plants. Roughly half of the *Vegetable Staticks* relates to studies of the transport of liquids from the earth through plants and leaves into the air. The conception of such studies is clearly an extension of the "statical" investigation on humans by Sanctorius and by James Keill, to whose *Medicina Statica Britannica* Hales himself draws a parallel, and the technique by which he measures the sap pressure in vines is an extrapolation of that developed for measurement of blood-pressure. The ingenuity of his experiments is characteristic of Hales, as is the variety of parameters, heat, humidity, season, age, and state of the plant, which he examines. Only Nehemiah Grew, Marcello Malphigi, and John Ray appear earlier to have studied plants in anything like the systematic spirit exhibited by Hales. His results were to be used by generations of

[14] I have used the reprinted edition, edited by M. A. Hoskin, *Stephen Hales, Vegetable Staticks: Or, An Account of some Statical Experiments on the Sap in Vegetables: Being an Essay towards a Natural History of Vegetation. Also, a Specimen of An Attempt to Analyse the Air, By a great Variety of Chymio-Statical Experiments. . . .* (London: Oldbourne Book Co. Ltd., from the 1727 edition, 1961).

plant physiologists. The French translation of the *Vegetable Staticks* was made by Buffon; Christian Wolf wrote a preface for the German translation. Bonnet, Duhamel du Monceau, Dutrochet, Humboldt, and Leibig, among others, were influenced by Hales' experiments.

Experiments must, however, be interpreted, and the theoretical abilities shown by Hales in interpreting those of the *Vegetable Staticks* have been neglected. As Julius von Sachs, himself a distinguished plant physiologist, wrote, Hales' successors "quoted and repeated Hales' experiments and observations again and again, but forgot that which in his mind bound all the separate facts together."[15] The uniting concept was, for Hales, that of dynamic corpuscularity, derived, ultimately, from Newton, whose corpuscular queries from the *Opticks* he quotes more than any others, as he cited Newton more than any other authority. More immediately, his conceptual sources were John Freind, who from the "properties of the principles of matter . . . has . . . given a very ingenious *Rationale* of the chief operations in Chymistry [xxvii]," and James Keill, whose *Tentamina Medico-Physica* is explicitly cited and is obviously the source for Hales' belief that ". . . the great work of nutrition, in vegetables as well as animals . . . is chiefly carried on in the fine capillary vessels, where nature selects and combines. as shall best suit her different purposes, the several mutually attracting nutritious particles which were hitherto kept disjointed by the motion of their fluid vehicle . . . [83]." Perhaps he also acquired, from Roger Cotes, who was lecturing at Cambridge while Hales was in residence, the explanation of sap movement by capillary attraction and the idea of some of the experiments on diminishing the elastic force of airs, as he may have got from John Keill the same explanation of sap rise and an additional notion of explaining, by attraction, the "determinate and constant figures of leaves and flowers."[16] Whatever the immediate sources of his ideas, Hales wove them together with Newton's emphasis on quantitative data and added the experimental skills of a Boyle or a Hooke. But this is not all that he added. Hales was the first dynamic corpuscularian to see the potential of Newton's concept of repulsive forces.[17] The earlier Newtonians, Bentley, Clarke, Pitcairne, Mead, Cheyne, Cotes, even John and James Keill and John Freind, had referred to repulsion not at all, or with reluctance and only as Newton had ex-

[15] Julius von Sachs, *History of Botany* (Oxford: Clarendon Press, 1890), p. 481.

[16] For Cotes, see Chap. 2; and for Keill, Chap. 3.

[17] Benjamin Worster's text of 1722 (which I have not seen) may also discuss repulsive forces as the edition of 1730 does (see Chap. 2). In any event, Worster's book had nothing like the influence that Hales' had.

plicitly described its uses. John Rowning and J. T. Desaguliers, who follow Hales, use the concept with greater freedom and, perhaps, greater ingenuity. But it was Hales who saw that the tension between attracting and repelling forces was essential in maintaining the balance of nature:

> . . . if all the parts of matter were only endued with a strongly attracting power, whole nature would then immediately become one unactive cohering lump; wherefore it was absolutely necessary, in order to the actuating and enlivening this vast mass of attracting matter, that there should be every where intermixed with it a due proportion of strongly repelling elastick particles, which might enliven the whole mass, by the incessant action between them and the attracting particles . . . [178].

The use of this concept of interacting forces is not particularly noticeable in the hydrostatical and hydrodynamic sections of the *Vegetable Staticks.* Forces appear in these sections as the attractive principle, operative in the nutrition of vegetables "all whose minutest parts are curiously ranged in such order, as is best adapted by their united force, to attract proper nourishment [54]," or to explain capillary action in wood, in ashes and sponges, and as the cause of the ascent of sap in plants. The use of repulsive forces appears significantly only in Chapter VII, "Of Vegetation," in the last two sections of the *Haemastaticks* (appendices, really, to the *Vegetable Staticks*), and particularly in Chapter VI. "A Specimen of an attempt to analyze the Air by a great variety of chymio-statical Experiments, which shew in how great a proportion Air is wrought into the composition of animal, vegetable, and mineral Substances, and withal how readily it resumes its former elastick state, when in the dissolution of those Substances it is disengaged from them." Had Hales done no more than the work in this chapter on the analysis of air, his place in the history of science would still be secure. It was the starting point for the chemical studies of Joseph Black; Henry Cavendish read and profited from it; Joseph Priestley began his pneumatic investigations as a continuation of "Dr. Hales' inquiries concerning air"; and, on the Continent, Hales was read by Rouelle and, through him, by Lavoisier. The whole of pneumatic chemistry, next to electricity the more exciting topic of eighteenth-century scientific investigation, had its rise in this chapter, of 92 pages, in Stephen Hales' *Vegetable Staticks.*[18]

Hales' formal training in chemistry was but little better than

[18] See, for example, Henry Guerlac, "The Continental Reputation of Stephen Hales," *Archives Internationales d'Histoire des Sciences,* 4 (1951), 393-404.

that he appears to have acquired in anatomy and physiology. Stukeley reports attending the chemical lectures of Vigani and performing experiments with Hales, and Hales seems to have heard Vigani also [112]. At least by the time he had completed the chemical experiments described in his Chapter VI (begun about 1725, according to the preface), he was also familiar with the work of Boyle, Boerhaave, Lemery, George Wilson, John Mayow, and John Freind and with the chemistry described in the *Lexicon Technicum*, particularly "de Natura Acidorum" of Newton. In none of these, however, would he have found the chemical role of the air described as he was to do.[19] For to Hales the elastic repulsive powers of nature inhered in the air, whose particles strongly repel one another in their normal, fluid state. The particles of other matter are strongly attractive and can seize and bind to them the elastic air, changing those particles from an elastic repulsive to a strongly attracting state. In their fixed state, the particles of air form part of the cohering substance of bodies, but are ready, under suitable circumstances, to regain their elasticity, disjoining the bodies in resuming their earlier fluid form and giving to those bodies their active energy [e.g. 165-68, 178-80].

Hales' experiments commenced when he noted the quantity of air ascending through the sap in his manometers [58]. From this he was led to examine the amounts of air "imbibed" by plants, and, in turn, the possible role that air might play in the oeconomy of nature. He devised a technique (water displacement) to measure the quantities of air released or absorbed ("i.e. changed from a repelling elastick to a fix'd state by the strong attraction of other particles [91]") under various conditions of heat, pressure, combustion, fermentation, respiration, etc. He determined that most of the great quantities of air thus obtained was permanently air by observing that it retained its elasticity [116], though he had, sometimes, to prevent its resorption and fixation by fumes raised with the distillations [105]. He "demonstrated" that the air was "true air, and not a mere flatulent vapour" by two crude experiments comparing the specific gravity and elasticity of "air of tartar" with those of common air [105-06]. From the nature of his experiments, Hales must have prepared carbon dioxide and monoxide, ammonia, hydrochloric acid, hydrogen, oxygen, and

[19] Mayow, perhaps, comes closest, as, in keeping with his iatro-chemical beliefs, he held that the air contained an aero-nitrial ferment of spirit which fed fire and animal life. Boerhaave's recognition of the chemical role of air, as reported in the *Elementa Chemiae*, was chiefly derived from Hales and published after the *Vegetable Staticks*. Hales read the earlier, unofficial version, probably in the 1727 English translation.

nitric oxide, but the conceptualization of his work prevented his differentiating between them. By the dichotomy he established between attracting solid matter and elastic repelling air, Hales moved toward the reifying of repulsion in a particular substance, but it would not have occurred to him that airs might differ. With his usual attention to quantitative detail, he reports the weight of substances heated and the volume of the air obtained. He sometimes also gives the weight of the generated air, but that was apparently determined from the decrease in weight of the residue; his data are too inadequate for specific gravities. What interested Hales was the variation in mechanical action revealed by the release of a great elastic volume of air from the small volume of its unelastic condition. Apparent differences in properties were easily to be explained: ". . . our atmosphere is a *Chaos*, consisting not only of elastick, but also of unelastick air particles, which in great plenty float in it, as well as the sulphureous, saline, watry and earthy particles, which are no ways capable of being thrown off into a permanently elastick state, like those particles which constitute true permanent air [179]."

From his experiments on air, certain conclusions seem to follow. Sal tartar explodes with greater violence than nitre, partly because it contains one-fifth part more air, but especially because it is more fixed, requiring more heat to separate the air from its strongly adhering particles, giving the air a greater elastic force [102-103]. It is probable that some air, together with the acid spirits with which it abounds, is conveyed to the blood through the thin partitions of the vesicles of the lungs, where the attracting sulfurous particles of the blood may change air particles from their elastic repulsive to a strongly attracting state [136-37]. It seems unlikely that Boyle, Boerhaave, Homberg, Lemery, or Nieuwentyt are correct in imagining fire to be a particular, distinct sort of body. "The heat of a fire consists principally in the brisk vibrating action and re-action, between the elastick repelling air, and the strongly attracting acid sulphur" of combustible bodies [161-64].[20] And, with manifest and almost explicit relevance to the subsequent work of Joseph Black he states:

> From the great quantities of air that are found in . . . Tartars, we see that unelastick Air particles . . . [are] by the same attractive power apt sometimes to form anomalous concretions, as the Stone, etc. in Animals, especially in those places where

[20] In this place Hales also introduces a rare reference to the Newtonian aether, as a medium contributing to the intenseness and duration of heat by its vibrations.

> any animal fluids are in a stagnant state, as in the Urine and Gall Bladders. . . . This great quantity of strongly attracting, unelastic air particles, which we find in the Calculus, should rather encourage than discourage us, in searching after some proper dissolvent of the Stone in the Bladder, which, upon the analysis of it, is found to be well stored with active principles, such as are the principle agents in fermentation [110].
>
> From this manifest attraction, action and re-action, that there is between acid, sulphureous and elastick aereal particles, we may not unreasonably conclude, that what we call the fire particles in Lime, and several other bodies, which have undergone the fire, are the sulphureous and elastick particles of the fire [?air?] fix't in the Lime; which particles, while the Lime was hot, were in a very active, attracting and repelling state; and being, as the Lime cooled, detained in the solid body of the Lime, at the several attracting and repelling distances, they then happened to be at, they must necessarily continue in that fix't state. . . . But when the solid substance of the Lime is dissolved, by the affusion of some liquid, being thereby emancipated, they are again at liberty to be influenced and agitated by each other's attraction and repulsion, upon which a violent ebullition ensues . . . till one part of the elastick particles are subdued and fix't by the strong attraction of the sulphur, and the other part is got beyond the sphere of its attraction, and is thereby thrown off into true permanent air . . . [162-63].[21]

In Chapter VII, Hales epitomizes his dynamic corpuscular convictions:

> We find by the chymical analysis of vegetables, that their substance is composed of sulphur, volatile salt, water and earth; which principles are all endued with mutually attracting powers, and also of a large portion of air, which has a wonderful property of strongly attracting in a fixt state, or of repelling in an elastick state . . . and it is by the infinite combinations, action and re-action of these principles, that all the operations in animal and vegetable bodies are effected [182].

The major experiments of the chapter relate to the influence of the leaves in "imbibing elastick air"; the possible role of light, convertible as Newton suggests into gross bodies, in vegetation;

[21] The substitution of particles of "air" for those of "fire" is suggested as a less ambiguous approach to what Hales really means, for he denies that fire is a substance *sui generis* and intends by "sulphureous and elastick particles of fire" those particles of air and fuel most vigorously agitated by heat.

and the nature of vegetable and animal growth. Here nature "selects and combines particles of very different degrees of mutual attraction, curiously proportioning the mixture according to the many different purposes she designs for it [194]," while the force of dilating sap and air, included in the vesicles, is sufficient for extending of shoots and expanding of leaves [196]. This combination of mechanisms is applied to explain the shape of trees, alteration in size and shape with shade and crowding, and the nature of the fruit. Hales even conjectures that plants may be impregnated through the attraction of the sulfurous "farina foecundans" for elastic air particles, which, being united, may perhaps be inspired when all the parts of the plant are in a strongly imbibing state [203]. And, as sulfur also attracts light strongly, the "result of these three by far . . . more active principles in Nature, will be a *Punctum Saliens* to invigorate the seminal plant: and thus we are at last conducted, by the regular Analysis of vegetable nature, to the first enlivening principle of their minutest origin [203]."

The appendices, added by way of the *Haemastaticks,* contribute little new to this picture. Direct examination of human calculi confirms their aerial content and suggests the use of fermenting mixtures to release the fixed air and dissolve the stones. The experiment is unsuccessful, but introduces the proposition that there may need be "certain harmonic proportions between the Vibrations of the fermenting Liquor and the Tone or Degree of Tenseness of the parts of the Calculus [205]." Electrical experiments provide a model example of particles being sometimes in an elastic and at other times in a fixed state, while water also exhibits the same property when heated and cold [279-80]. Hales does not here clearly generalize this observation to all matter, using it only as demonstration that air particles might be similarly endowed, but he later hints at such a generalization, and, at the same time, pronounces the ultimate frustration of dynamic corpuscularity. In criticizing an experiment of Musschenbroek which seemed to prove that heat is a "real inherant elemental fire," Hales suggests that the acquired heat of effervescent mixtures may be due to the intestine motions produced by the vastly great attractive and repulsive force of many of the particles of matter, near the point of contact. "But as we cannot pry into the various Position of these Particles, in their several Combinations, on which their different Effects depend, so it will be difficult to account from any Principle, even though a true one, [for] . . . the very different Effects of effervescent Mixtures [318-19]."

Jean Théophile Desaguliers, in his own way and one very unlike that of George Cheyne, also served as a connecting link between the heroic age of dynamic corpuscularity and the winter of its discontent.[22] A student of John Keill at Oxford, Desaguliers succeeded Keill as lecturer in natural philosophy until his departure for London, where he followed Francis Hauksbee as curator of experiments for the Royal Society during the last thirteen years of Newton's presidency and some thirty years beyond it. Assisted by Hauksbee's nephew, Francis Hauksbee the younger, and later, perhaps, by Stephen Gray, Desaguliers also developed a popular private course of lectures in natural philosophy, illustrated by experiments, which, it is said, he gave more than one hundred and fifty times before his death.

He appears to have earned his living primarily from his Royal Society fees, his lectures, and from translating and writing on various subjects in natural philosophy. Among the many works with which he was associated were translations of Pitcairne's medical *Works* (1727), Mariotte's *Motion of Water and other Fluids* (1718), Gregory's *Elements of Catoptrics and Dioptrics* (1735), and 'sGravesande's *Mathematical Elements of Natural Philosophy* (1720), which had its origin in Desaguliers' lectures heard by 'sGravesande on a visit to London in 1715. Desaguliers wrote a prefatory letter to the translation of Bernardus Nieuwentyt's *Religious Philosopher* (1718), appears to have been involved in revising Harris' *Lexicon Technicum* for its later editions, the preface to Pierre Coste's French translation of Newton's *Opticks* (1722) acknowledges Desaguliers' assistance. His own text, *A Course of Experimental Philosophy*, first published in two volumes between 1734 and 1744, is based on John Keill's *Introduction to Natural Philosophy* and on parts of Stephen Hales' *Vegetable Staticks*, for which he had written the lengthy and laudatory essay-review that appeared in two parts in the

[22] Jean Théophile Desaguliers (1683-1744), born in La Rochelle and smuggled to England by his Huguenot-minister father at the Revocation of the Edict of Nantes. Educated at his father's school near London, where he later assisted, and, at his father's death, at Christ Church, Oxford; B.A. 1710, M.A. 1712, LL.D. 1718. Lecturer in natural philosophy at Hart Hall, Oxford, 1710-13, curator of experiments for the Royal Society and F.R.S. 1714; Copley medalist 1742. Ordained priest in 1717, and holder of two livings *in absentia*, he devoted his life to private lecturing, translating and writing, experimentation and engineering consultation. The most successful private lecturer in London for nearly 36 years, by 1719 he was sufficiently prominent to induce the publication of a spurious text of his lectures as *A System of Natural Philosophy*. There is unfortunately no English biography, but my investigations have been aided by those of Mrs. Carolyn Vetter Lindstrom, graduate student in the history of science, Case Western Reserve University.

Philosophical Transactions for 1727 [34 (1726-27), No. 398, 264-91; 35 (1727-28), No. 399, 323-31]. It would be hard to find a man more involved in all aspects of British natural philosophy, theoretical, experimental, and practical, for the years from 1714 through 1744, or one whose work would serve more conveniently as a survey of experimental physical science over that period.

Desaguliers' experimental work can be divided into two phases. Before 1728 his papers in the *Philosophical Transactions* are nearly always explicit defenses of some controverted Newtonian doctrine. The *Acta Eruditorum* denies the validity of some experiments of the *Opticks*; twice (1716, 1722) Desaguliers carefully repeated the experiments in confirmation of Newton. Cassini and Mairan deny the figure of the earth is as Newton describes it; during 1727, Desaguliers published four papers in support of Newton's conclusions. Two papers prove the existence of vacuum interspersed in matter, one (1717) by a repetition of the "guinea and feather" experiment in a glass tube nearly eight feet long and the other by showing that warm mercury communicates more heat than an equal bulk of equally warm water, "but how much more matter there is in the *Mercury* is not determin'd by this Experiment alone [31 (1720-21), No. 365, 81-82]." Desaguliers even designed and performed experiments to prove that *vis viva* equals MV rather than MV^2 [see, *e.g.*, 32 (1722-23), No. 375, 269-79], though he later concedes in his book that this argument is primarily a dispute over words. The work is carefully performed and well reported (showing, perhaps, the influence of his lecturing), but it is scarcely original. Only in his experiment attempting to measure the force of cohesion between freshly cut surfaces of two lead balls [33 (1724-25), No. 389, 345-47] is there a significant attempt to extend and define quantitatively part of the accepted Newtonian doctrine.

Possibly it was release from the awesome presence of Newton presiding which freed Desaguliers to more creative thinking and research, but another event, besides Newton's death, intervened between the commonplace repetitions and the signs of originality. Midway through 1727, Desaguliers read and reviewed Hales' *Vegetable Staticks,* finding that the author has "illustrated, and put past all Doubt, several Truths mention'd in Sir *Isaac Newton*'s Queries; which tho' believ'd by some of our Eminent Philosophers, were call'd in question by others of an inferiour Class, who were not acquainted with those Facts and Experiments upon which Sir Isaac had built those Queries." It is surely significant that Desaguliers' first emphatic use of the concepts of both attractive and repulsive forces, acting at a distance ("without considering whether it be any real Virtue in the said Surface, or the Action of a Medium. . . ."), appears in 1728 in his experiment-review of a book on optics for the *Philosophical Transactions* [35 (1727-28), No. 406, 596-629], and that his major, non-electrical papers published after this all use both attraction and repulsion and refer admiringly to Hales.

All of Desaguliers' original work appears in two forms, first as papers for the Royal Society (or some other society) and then in his *Course of Experimental Philosophy* either integrated into the text or, for the more significant items, as revised and extended reprintings of the complete papers. The book is, as I. Bernard Cohen notes in his discussion of its great influence on Benjamin Franklin, a charming work, well written and nicely illustrated.[23] Except for optics (omitted, Desaguliers says, because the section on popular optics in Smith's text fulfills his purpose), the *Course* was intended as a complete elementary text in natural philosophy. Its bias is toward the practical, with a long section on the mechanical advantages of lever, pulley, wedge, screw, etc., and another on practical machines (water wheel, windmill, even steam engines), and it deliberately avoids "difficult geometrical demonstrations and algebraic calculations." Its two volumes bracket the conclusion of this first period of eighteenth-century British natural philosophy and the beginning of the next, and, as they include a substantial discussion of Desaguliers' views of the nature of matter and its action, it seems appropriate to abstract this as a background to the treatment of his last and original studies.

Matter is defined by extension and resistance. Quantity, and consequently matter, is conceivably divisible *in Infinitum*, but actual division is not possible beyond those *Atoms* or extremely small Parts, solid, without pores, firm, impenetrable, perfectly passive, and moveable, created by the "Wise and Almighty Author of Nature" as the original particles of matter. Matter is the same in all bodies, the variety in bodies and changes in them depend upon situation, distance, figure, structure, powers, and cohesion of the parts that compound them. The compounding is achieved by the union of the original particles into parts of the first composition, these into parts of the second, and so on, to bodies of sensible magnitude [245]. There are other properties

[23] Cohen, *Franklin and Newton*, pp. 243-61. I have used the third edition of the text, Jean T. Desaguliers, *A Course of Experimental Philosophy* (London: A. Millar, 1763), 2 vols.

of matter, not essential yet universal and, in one sense, inseparable from it, not occult properties or supposed virtues, but manifest in experiment and observation, acting always in the same manner under the same circumstances [6-21]. These properties are the forces of attraction and repulsion between particles. They include: gravitation, which acts proportionable to the quantities of matter in bodies and in reciprocal duplicate proportion of their distances; and cohesion, stronger than gravity at contact and decreasing in a proportion not fully known, but believed with some reason to be a biquadratic ratio of increased distance [10, 16-17]. James Jurin's experiments on capillary attraction are cited as an example of cohesive attraction [38-39], and Desaguliers points out, in a later discussion of friction in mechanical engines, that it is possible to polish bearing surfaces so smoothly as to increase friction by the increased force of cohesion [191].

Another attractive force is that of magnetism, not so strong as cohesion but stronger than gravity. Brook Taylor and Musschenbroek have performed experiments on the distance relationship of magnetic force, Taylor finding an inverse cube of distance and Musschenbroek no regular proportion at all [41], but Desaguliers prefers "nearly as the Cube and a Quarter of increased distance." Magnetism is also an example of a repulsive force and other kinds of bodies, *e.g.*, electrical, attract under some circumstances, or distances, and repel under others. Repelling forces are best seen in the production of air and vapors, where particles forced from bodies by heat and fermentation, beyond their sphere of attraction, recede with great force [16-17]. Finally, it is not to be understood that the effects of gravity, light, and heat are derived from the same causes nor that all sorts of attractions in bodies have the same kind of laws, as the attraction of cohesion and that of the loadstone do not act in the same manner [31].

This is all fairly standard dynamic corpuscularity, though perhaps more clearly set out than most and maybe a bit overconfident in a text of 1734. Desaguliers departs from the usual in two major applications of these concepts, to the elasticity of solids and to vaporization. His small treatise on elasticity, taken over and expanded from a *Philosophical Transactions* paper of 1739 [41 (1739-40), No. 454, 175-85], is probably his most ingenious work. Accepting as fact Newton's conjecture that the elasticity of air results from repulsion of its particles, Desaguliers criticizes Hauksbee's experiments and approvingly describes, at length, those of Stephen Hales. He then addresses himself to the elasticity of bodies other than air, denies that this can result from attraction only, and proposes a solution considering both attraction and repulsion. May not, he suggests, springy solids be composed of lamina, or strata, parallel to one another, each consisting of a line of spherical particles in cohesive contact? If the particles are bipolar, like magnets, with the poles aligned perpendicular to the points of linear contact, mutual attraction of unlike poles will keep the strata together, while flexure of the strata will bring like poles along each layer closer together, increasing their repulsion, which then will tend to restore the original shape. Spherical particles will not make a tough spring, but, in annealing, the particles might be flattened, increasing their surfaces of cohesive contact and therefore the resistance to breaking, though decreasing the springiness. By variation in the heat treatment of metals, one could, presumably, produce the requisite combinations of toughness and springiness. Desaguliers elaborates this theory to an explanation of the vibration of strings, wire, and glass, but he does not, and cannot, mathematize it even to the explication of Hooke's law and, though it is based on dynamic corpuscularity, it had finally to assume the existence of a special kind of particle which Desaguliers never relates to his earlier declaration of the ultimate homogeneity of matter. His contemporaries had no better explanation to offer—the major alternative was that of Père Maziere, who won a prize of the Academie Royale des Sciences in 1726 with a Cartesian vortex theory of springs—but few of them took seriously to Desaguliers' suggestion. A *Philosophical Transactions* article of 1744 [43 (1744-45), No. 472, 46-71] by James Jurin, "Concerning the Action of Springs," declares that the elasticity of air is a power of a different nature and governed by different laws than that of a spring, and then develops a mathematical theory of springs, based on the work of Hooke and never even mentioning Desaguliers. Thomas Young was to be the next British scientist seriously to return to the idea of elasticity of solids being the result, jointly, of cohesive attraction and repulsion.[24]

Desaguliers' discussion of vaporization follows a treatment of heat, in which he denies the opinion of "those philosophers [including Nieuwentyt] who assert the Being of an elementary Fire." Hales has shown this to be wrong, and Desaguliers and Brook Taylor have performed experiments "whereby it appear'd to us that *actual Heat* was to *sensible Heat,* as Motion is to Velocity [II, 296]." The precise meaning of this curious phrase is obscure,

[24] Thomas Young, *Course of Lectures on Natural Philosophy* (London: Joseph Johnson, 1807), Vol. I, p. 141.

though it rather obviously is an attempt to inject quantity into a motion theory of heat, by way of mass (density), and may relate back to his experiments with water, mercury, and heat transfer. In any event, Desaguliers does not extend its quantitative overtones to cover his discussion of the action of heat, its production of motion, and the rise of vapors:[25]

> When the particles of water are separated by any Cause whatever that puts them into motion, the attraction of cohesion yields by little and little, and acts no longer at a Distance something sensible; and then a second repellant Force [the first being one within the cohesive sphere, preventing contact of water particles] may succeed to the Attraction of Cohesion; and the particles acquire a Force . . . by which they repel each other, and fly off. . . . This will happen by the action of that degree of Heat which makes water boil . . . [II, 336ff].

This discussion is followed by another in which Desaguliers firmly declares his belief in Newtonian forces of attraction and repulsion, as confirmed by Hales:

> . . . when Sir *Isaac Newton* publish'd a second Edition of his Opticks, in year 1717, he added Queries to the 3rd Book. Those, together with the rest of the Queries, contain an excellent Body of Philosophy, and upon Examination appear to be true; though our incomparable Philosopher's Modesty made him propose those Things by Way of Queries, which he had Observations enow to satisfy himself were true; he was unwilling to assert any Thing that he could not prove by Mathematical Demonstration or Experiments. This made a great many People consider what he says in them as mere Conjectures; and I know very few, besides the Reverend Learned Dr. Stephen Hales and myself, that look upon them as we do on the rest of his works [II, 403].

Desaguliers then outlines some of the consequences of this doctrine as revealed in Hales' and his own experiments and, following the design of the *Opticks* itself, concludes with a set of queries, the most potentially fruitful being Query 14: "When the Air changes from an elastick into a fix'd State, what becomes of that prodigious repulsive Power in the Air that no mechanical Force

[25] First developed in a *Philosophical Transactions* paper, 36 (1729-30), No. 407, 6-22, and here reprinted in *Course*, II, pp. 306ff, it is more elaborately presented in a second dissertation on the rise of vapors, *Course*, II, pp. 336ff, and complicated by the presentation of a conflicting, electrical theory sandwiched between them, II, pp. 332-36.

could surmount? Is it not overcome by a stronger Power of Attraction in Contact; if so as to remain still in the Body, tho' inactive? And may it not be call'd a latent Power of Repulsion [409]?"

The possibilities for fruitful extension of dynamic corpuscularity, even to quantitative results, in some of these speculations seem enormous: Actual heat to sensible heat as motion to velocity; three alternate, concentric spheres of repulsive and attracting forces; latent power of repulsion—even Rowning achieved little if anything more. But Desaguliers was unable to exploit his achievement. At the end of his first discussion of vaporization, he confessed his inability to "shew by any experiment how big the Moleculae of Vapour must be which exclude Air from their Interstices and whether those Moleculae do vary in proportion to a Degree of Heat by increase of repellant Force in each watery Particle, or by a farther Division of the Particles into other Particles still less . . . [II, 313-14]." Though he affirms that the rarity of vapor is proportionable to the degree of heat, he is frustrated by an inability to find a firm quantitative support to his hypotheses. Indeed, by the time his first essay on vaporization was republished in the second volume of the *Course*, he had already gone on into the study of electricity and weakened his argument by a paper in the *Philosophical Transactions* of 1742 (also reprinted in the *Course*) relating the rise of vapors to electricity. Desaguliers had delayed his studies of electricity until after the death of Stephen Gray, though "I can excite as strong an electricity in glass by rubbing it with my hand, as any body can" because Gray would have given up his work "if he imagined that any thing was done in opposition to him."[26] Commencing in 1739, by 1742 he had written a "Dissertation concerning Electricity" which won a prize of the Academy of Bordeaux. This, his other papers, and extracts from those of Gray, Hauksbee, DuFay, and others, were scattered in various forms and places in volume one or two of his *Course*. The ideas expressed seem as incoherent as the places in which they appear, but though he declares he will not endeavor to guess at the cause of electrical action [332], the general impression is one not easily reconciled with a strict dynamic corpuscular view of homogeneous matter, differing only by situation, distance, figure, structure, powers, and cohesion. There is electricity inherent in some bodies, not perceptible until they are given a vibratory motion which throws off the electrical effluvia or emanations, which fly off and return in a circle carry-

[26] Quoted by Cohen, *Franklin and Newton*, pp. 376n-377n.

ing with them the little nonelectrick bodies they meet in their way, on their return [317, 330]. But DuFay has shown there are two electricities. May not Hales' mutual attraction between air and sulfurous vapors be that between the permanently vitreous-electric air and the resinous-electric sulfur, the combination producing a nonelectric and therefore unelastic particle? Perhaps vaporization is produced by the attraction of nonelectric particles of water by the vitreous-electric particles of air flowing over the surface, producing electric air-water particles which repel one another and the surrounding particles of air, driving them off and away from the water [333-36]. The consequence of his electrical ideas is a chaos in which, nonetheless, the coalescing tendency is clear. The singularity of Hales' repulsive air particles has now been reified into their electricity and this has mechanical, even (occasionally) forceful qualities, but they remain unrelated to those he has elsewhere discussed relating to the same phenomena.

Desaguliers does not associate quantity of electric action to quantity of effluvia, nor does he associate electrical effluvia with Newton's aether. Indeed, though he admits the aether into his discussion of Hales' notions on the transmission of heat [II, 368-69], he was not happy with aetherial generalizations. In a letter of 1731 (four years after the Hales discussion appeared in the *Vegetable Staticks* and fourteen after the appearance of the aether queries of the 1717 *Opticks*), he discusses the attempt of a Mr. Brown to "deduce from the phenomena and prove the observations and experiments" that the aether acts upon bodies according to their quantity of matter, that its properties are such as to cause elasticity and the cohesion of attraction, and that the same cause extends to electricity and magnetism. "As for giving encouragement to the gentleman by declaring whether such an attempt seems likely to be successful, I can speak only for myself and say that I shall never think myself equal to it and therefore never attempt it. . . ."[27] Desaguliers' successors, however, were not to find the aether so foreign to their sensibilities and, in adopting that, found it easy to bring to quantification the materialism implicit in his electrical speculations.

As electricity had resisted the inductions of Desaguliers' (and Newton's) dynamic corpuscularity, so it became one of those subjects in which the next generation found a capacity for materialism.

[27] J. T. Desaguliers to Hans Sloane, 4 March 1730/1, Sloane MSS 4051, Vol. XVI, Fol. 200-01.

PART II

AETHER AND MATERIALISM

1740-1789

CHAPTER FIVE

Second Thoughts and the New Revelation

BETWEEN 1735 AND 1745 there was a shift in the social and intellectual temper of eighteenth-century Britain. About the causes of this change, "I will not feign an hypothesis" nor conjecture on their relationships, but the phenomenon is manifest in a wide variety of indicators. Arthur O. Lovejoy refers to a "romantic" movement which "began pretty definitely in England in the seventeen-forties," and quotes the opinion of Edmund Gosse that "Joseph Warton's youthful poem, *The Enthusiast,* written in 1740 . . .[was] the first clear manifestation of the great romantic movement. . . ."[1] Other literary historians are not, perhaps, so precise in date or name; they may have other candidates: Thomas Gray, whose first *Odes* and *Sonnets* were published in 1742, Samuel Richardson and his *Pamela* of 1740, Edward Young and *Night Thoughts* of 1742, or William Collins and his *Persian Eclogues,* again of 1742. Nevertheless, there is as reasonable an agreement as one is likely to get, between such determined controversialists, that English literature moved, during this period, toward an increase of sensibility and concern with greater naturalism and away from Augustan conventions of rationality.

Lovejoy also sees the beginning of an "incipient Romanticism" in the revival of Gothic architecture, represented, for example, in Batty Langley's *Ancient Architecture Restored and Improved . . . in the Gothick Mode* (1742) and the designs of Sanderson Miller, beginning about 1744 [101, 150-51]. David C. Douglas finds a notable change (decline, he feels) in English historical studies after about 1730, and Stuart Piggott parallels this change by another in antiquarian pursuits, influenced "by the contemporary growth of romanticism" to find Druidic mysteries in ancient megaliths.[2]

It was in 1742, with the first performance of "The Messiah," that Handel signaled his final turning from composition in the artificial Italian-opera mode to the oratorios for which he is now more famous. In a chapel in Aldersgate Street in 1738, John Wesley felt his heart "strangely warmed," and issued forth to begin the Methodist movement, with its emotional and revivalist overtones. Finally, David Hume's *Treatise of Human Nature,*

[1] Arthur O. Lovejoy, *Essays in the History of Ideas* (Baltimore: The Johns Hopkins Press, 1948), pp. 235, 237.

[2] Piggott, *Stukeley,* pp. 183, 16-17.

denying the rationalist foundation of contemporary aesthetics and learning, was first published in 1739-1740.

One must not overemphasize particulars, especially individual dates. James Thomson's *The Seasons* (1726-30) might be cited as proto-Romantic; the Gothic mode of architecture was preceded by one for the "natural style" in gardening, represented by William Kent (c. 1730). "The Messiah" was not Handel's first oratorio, and the loss of popularity of his operatic works can be dated at least as early as 1735; William Law's *Serious Call to a Devout and Holy Life* (1728) is a foreshadowing of evangelical Christianity, while Griffith Jones and the religious revival in Wales, starting about 1735, provided a pattern for Wesley and Methodism. Hume's work, as he tells us, had its origins in the study of Bishop Berkeley's works, especially the *Principles of Human Knowledge* (1710). But all this is merely to say that intellectual and cultural revolutions have their preparations, and the appeal in finding precursors should not blind us from recognizing that winds of change, blowing perhaps from many directions, were disturbing the Augustan compromise.

The forces of reason and classical restraint were still strong enough to ignore, assimilate, or sometimes even reverse many of the dislocations implicit in these fitful gusts. Samuel Johnson was still to enforce the dictates of rationalism and humanistic traditions on late Augustan literature. Already by 1753 Gothicism had lost much of its vigor, though enough remained (in Horace Walpole, for example) to spark a re-revival toward the close of the century. Hume's *Treatise* fell "still-born" from the presses and found its major recognition in the responses of Emmanuel Kant, seriously regarded in Britain only in the early nineteenth century. Nonetheless, in spite of their containment, these variant modes had demonstrated possibilities of different attitudes, requiring some response, in almost every phase of the intellectual and cultural life of mid-century Britain, and natural philosophy was naturally not exempt. The chapters in this section will examine in some detail the nature of the changes in natural philosophy and the responses to them in the period from 1740 to 1789.

The most obvious change in British natural philosophy about 1740 was a social and institutional one. In 1741 Martin Folkes succeeded to the presidency of the Royal Society and held that position until 1752. From Newton's election in 1703 until Sir Hans Sloane's retirement in 1741, the president had been a man of scientific or scholarly distinction. Folkes was neither scientist nor scholar. Under his presidency, the Society came increasingly to be dominated by dilettantes, whose major interests lay in antiquarian pursuits; meetings of the Society are described as literary rather than scientific. Weld claims that a "greater proportion of trifling and puerile papers" appeared in the *Philosophical Transactions* during Folkes' administration than at any time before or after. The Society became a laughing stock to the critics.[3]

There is, however, a more significant, because more general, change in the social composition of working British natural philosophy, which becomes apparent about this time. The great majority of significant contributions to textbooks, theory, or experiment had, during the period just studied, been achieved by persons connected in some way, as teacher, student, or both, with one of the two English universities. By 1744, and the death of Desaguliers, almost every one of those who had argued in the Newtonian mechanical mode were gone from the scene: Newton, Gregory, Bentley, Derham, Clarke, John and James Keill, Freind, Cotes, Taylor, and Worster were dead; Hales had turned his attention after 1733 almost entirely to the practical application of his earlier studies; Rowning is unheard; and Robert Smith gave up his scientific studies in 1742 to succeed Bentley as Master of Trinity. Only Richard Mead, who recanted his mechanism in 1745, and James Jurin, who disturbed capillary mechanism with aetherial hypotheses, remained of those trained at English universities. Helsham, of Trinity, Dublin, was dead, as were Hauksbee and Gray, neither of them university-trained or connected and neither a convincing exponent of dynamic corpuscularity. It is surely significant that of the group that took their places during the rest of the century, only John Michell and Henry Cavendish were educated at English universities and both (as will be shown in Section III) were mechanists. The others who were to make significant original contributions to British natural philosophy during this period—Cullen, Black, Watson, Franklin, Brownrigg, Priestley, Hutton, Herschel, etc.—were educated at Leyden, one of the Scottish universities, at a dissenting academy, or were self-taught.

Considering the value of the work done by these men, it is impossible to take their lack of English university training as criticism. Indeed, in what appears to have been a singularly low period in the state of these universities, with the melancholy reports of

[3] C. R. Weld, *History of the Royal Society* (London: John W. Parker, 1848), I, 424.

one professor of chemistry, anatomy, or natural philosophy after another failing to lecture, it is perhaps as well that these new scientists had not been students at Oxford or Cambridge. Certainly science had much to gain in the relatively untrammeled approach and even in the practical concerns of this new generation of natural philosophers, but something was lost as well. Given the studies of a Gregory or Maclaurin, it cannot be said that mathematics could not be cultivated at a Scottish university while, as has been seen, repeated republication of the early texts indicates that the mechanists could be studied well into the second half of the century. But when mathematics was cultivated, and if these works were studied, it appears generally to have been done in a spirit markedly different from the earlier one. That peculiar combination of classics, logic, and mathematics which had tempered the mind to abstract studies was missing in this second period. It is significant that the philosophical works of Francis Bacon, presumably the inspiration of British science, had gone without publication in Britain from 1687 to 1730, while four editions were to be published between 1730 and 1750 and at least four more by 1803.[4]

Few British scientists since the time of Bacon have dismissed the claims of experiment finally to validate a scientific theory, and, as we have seen, experimental confirmation was declared for nearly every mechanistic theory from 1704 on. It appears, however, that lack of sympathy with mathematical abstraction and the pragmatic bent of these new natural philosophers, in interaction with the revival of Baconianism, combined to produce a "Newtonianism" in which the "exact" but "experimental" character of the *Opticks* became a dominant feature.[5] This new experimental Newtonianism found its theoretical base in tangible interpretations of Newton's aether hypotheses, and it found, in the increasing influence of Continental scientists (particularly the Dutch), substantial support for its materialism.

A detailed study of this materialistic influence will follow in a later chapter, but one, somewhat anomalous, mode can better be indicated here. During the three years between 1735 and 1738 that Carl Linnaeus was in Holland, he published some twelve works describing his new system of classification. The impact of these works, and their revisions, on botanical taxonomy is well known, but the *Systema Naturae*, first published in 1735 and in a revised sixth edition by 1748, had also discussed classifica-

[4] Reginald Gibson, *Francis Bacon: a Bibliography of his Works and of Baconiana to the Year 1750* (Oxford: Scrivener Press, 1950), *passim.*

[5] Cohen, *Franklin and Newton*, esp. Chapter 6.

tion in zoology and entomology.[6] Before long a rage for classification was to infect Europe. Linnaean classification of plants became one of the most exciting pursuits for the scientific amateur and proto-professional alike. The latter, however, indulged a penchant for the binomial classification not only of plants or animals but of minerals, diseases, and chemicals as well. Essential to the system of Linnaean classification, and a major ingredient in its popularity, was the simple character of its approach. One placed the objects examined in their proper niches on the basis of a few, external characteristics and came to feel that essential knowledge of those objects had been acquired by simple enumeration of its obvious properties. The influence of Linnaeus could not but encourage a turning from the mathematical analysis of motions, to find forces, and toward a neo-Aristotlean contentment with the creation and categorizing of different qualities from experimentally observed characteristics.

There was one essential difference in this neo-Aristotleanism, however, for these new "substantial qualities" were, in fact, made substance; light, heat, electricity, magnetism, and the different chemical properties—never quite rationalized into motion and force—were all to become material, and in becoming material they all, for the British at least, were assimilated into Newtonianism. It is the purpose of this section to trace the way in which substantial forms become material representations of Newton's work. For the major change in British natural philosophy, concurrent with those in the other intellectual and cultural activities of the 1740's, was this change in the mode of scientific interpretation.

The contrasting extremes, neither, perhaps, fully represented by any individual, between which the change occurred are these: The complete mechanist would argue that all phenomena of the physical universe (including physiological) were ultimately to be explained in terms of undifferentiable, homogeneous matter, existing in small indivisible particles, solid, hard, and inert. To these particles were added, by the Creator, certain immaterial, central forces of attraction and repulsion under whose influences the primary particles acted upon bodies, at a distance, and were variously arrayed into particles of a first composition. These particles, possessing in virtue of their sizes and modes of arrangement, their own, lesser forces of attraction and repulsion, also acted upon bodies at a distance and were also combined to form

[6] See Norah Gourli, *The Prince of Botanists: Carl Linnaeus* (London: H. F. and G. Witherby Ltd., 1953).

still larger, more porous, and less forceful particles, which acted upon other particles and combined in yet larger particles, and so on, until the substances of the sensible world were achieved. Insofar as these substances might exhibit regular properties and react to phenomena in predictable ways, they provided clues through which the natural philosopher could analyze to the various and, finally, ultimate particles and forces.

For the materialists, although such ultimates might exist, it was useless to attempt (at least in their time) to find them. Regularities in property behavior and reaction, instead, provided the distinguishing categories by which one substance might be differentiated from another. All physical phenomena were to be explained by the possession, or absence, of a substance carrying the necessary distinguishing characters, and (and here is a major strength in this view) the amount of substance present gave a measure of the phenomenon it causes.

In actual practice, of course, these different modes are usually less clearly distinguished. Most of the changes in argument are more subtly expressed and are revealed primarily in changes of attitude. The materialized, substantial causes are almost all imponderable, highly tenuous fluids, and most of these are partially characterized by their possession of varying forces of attraction and repulsion. But there is a major difference between these apparently-mechanistic properties of the materialists' fluids and those of the mechanists' fluids. To the materialists, these are descriptive and observational terms, not the essential operational properties. The fluid is imponderable only because it cannot be weighed, it is tenuous only because it insinuates itself through and into bodies; its forces may be the mechanism by which its primary function is distributed, but all of these properties are superposed on the essential character of the substance, which is its power to produce a particular phenomenon. For the materialist, the possession of certain mechanistic properties does not initiate analysis into how these properties cause a phenomenon; the cause of that phenomenon is simply the presence of the substance. The fluid of heat, for example, does not heat because it is imponderable, tenuous, and possesses certain attractive and repulsive powers; it is imponderable, tenuous, and possesses these particular powers because it is the fluid of heat. The heating is a function only of its addition or subtraction from a body.

Just as there were certain indicators which, without analysis, exhibited the change occurring in literature, architecture, music, religion, and philosophy, so are there such indicators for natural philosophy. In a series of *Essays Medical and Philosophical,* first published in 1740 and reissued three times by 1792, George Martine denounced mechanistic physiology:

> It is now above a Century since the illustrous and immortal *Harvey* taught the world the Circulation of the Blood. . . . And about the same time men began to get a taste for Experimental and Mechanical Philosophy. Great things were expected from such a concourse of circumstances. . . . But what a disappointment did the world meet with? . . . Since these lights there have perhaps been more absurdities given out and defended concerning the natural and morbid state of the Human Body, than had been from the beginning of the world to that time.[7]

Under misapplied mechanical causes, Martine includes "vortexes of subtile matter," "universally pervading elastic aethers," "harmonies of conspiring vibrations," attraction, magnetism, and electricity [138-39]. The ancient physicians did not study philosophy, languages, criticism, poetry, history, mathematics, natural history or natural philosophy, chemistry, botany, or anything else save diseases, symptoms, changes, and periods. On this clinical, empirical base they erected their practice [19-20], and Martine recommends that "physicians of the present age" proceed in the same way.

A rather different indicator is to be found imbedded in an otherwise undistinguished text of 1748, *A System of Natural Philosophy* by Thomas Rutherforth.[8] Rutherforth had lectured in mechanics, optics, hydrostatics, and astronomy at Cambridge, and his book, containing the substance of his lectures, is what one might expect of a Cambridge text—but of at least a decade earlier—except in one respect. Matter is extended, solid, mov-

[7] George Martine, *Essays Medical and Philosophical* (London: A. Millar, 1740), p. [sig.] A4r, Martine (1702-41) was educated at St. Andrews, Edinburgh (1720), and Leyden, M.D. 1725. Author of medical and anatomical works, and essays (included in the above and later published separately) on the construction and gradation of thermometers and of experiments on degrees of heat in bodies, he practiced medicine at St. Andrews until 1740, when he became an army physician, dying of a fever contracted at Carthagena.

[8] Thomas Rutherforth (1712-71), educated at St. John's College, Cambridge, B.A. 1729, M.A. 1733, B.D. 1740, D.D. 1745. Taught physical sciences privately at Cambridge; F.R.S. 1743. Appointed Regius Professor of Divinity at Cambridge, 1745, was chaplain to Frederick, Prince of Wales, held two livings, and was archdeacon of Essex, 1752. Author of several controversial tracts, including some against Samuel Clarke, Hume, and Wesley, and others in support of clerical subscriptions. His *System* was his one foray into natural philosophy and reached only the single edition: Thomas Rutherforth, *A System of Natural Philosophy* (Cambridge: by J. Bentham, for W. Thurlbourn and J. Beecroft, London, 1748), 2 vols.

able, infinitely divisible, has a *vis inertia.* To it are added forces of attraction and repulsion, not due to pressures of a fluid, not essential properties (for they are relative, requiring the presence of other bodies to act, and contradict the essential property of inactivity), and not occult, for they are manifest to our senses. By these forces one explains gravity, cohesion, and the emission, inflection, reflection, and refraction of light. "But these attractive and repulsive forces, though they may be proved to exist in nature, are frequently applyed in philosophy to cases, which they do not sufficiently explane [8]." And giving as an example the differential-attraction, force explanation of chemical dissolutions, he continues, "These answers have something in them very like the occult qualities of the Peripatetics. For as they used to assign as many different occult qualities as there are different appearances to explane; so the philosophers who give these answers, introduce as many different sorts of attraction as there are bodies to be dissolved and fluids to dissolve them [9]."

Admittedly Martine is crudely empirical—and unhistoric as well—while Rutherforth has no conception of the composition of particle forces into new force distributions. But that two relatively unimportant figures (one of them from the citadel of Newtonianism) should so clearly find dynamic corpuscular explanations inadequate manifests the diffusion of that change in attitude and interpretation which was described above. As these changes are the substance of this book, hypotheses as to causes cannot here be avoided, though single-cause hypotheses are still not to be feigned. Some of the causes can be developed only in the course of detailed examination into the writings of the new natural philosophers as they grapple with new problems. Some, however, have already been described, or alluded to, in the social or scientific development of early eighteenth-century British natural philosophy. Social causative factors in the changing of scientific conceptualizations are less precisely definable and, consequently, more historiographically tentative than one would like, but the probabilities remain to be faced. Does Martine, for example, represent the new pragmatic Baconianism, shifting the emphasis from the abstract, as well as a simple reaction to failure? Henry Guerlac, in describing Newton's changing reputation in France, notes a midcentury anti-mathematics tendency, accompanied by pleas for more attention to experiment and natural science. Taking place roughly a decade after the similar change postulated here for Britain, its champions were primarily Diderot and the pragmatically oriented Encyclopedists, who replaced Newton by Franklin as their hero:

> The drift away from Newton is clearly evident in the writings of Diderot and d'Holbach. The "scientific materialism" of d'Holbach's *Système de la Nature* owes little to Newton, toward whom he is by no means friendly. Instead, his materialism is an odd amalgam of chemical speculation and veiled Cartesianism; and it shows a revived interest in the materialist philosophers of classical antiquity. These same influences . . . are found in the scientific speculations of Diderot. Newton's thought seemed, to these men, too mathematical, too abstruse, and too clearly tied to the deism of the older generation.[9]

This reference to deism has some pertinence also for Britain. Guerlac has suggested that the taste for atomistic speculations in late seventeenth-century England accompanied the revival of aristocratic influence and the power of the church establishment after the restoration of 1660.[10] We must, as he insists, regard this proposition with some caution, but surely it finds support in the decline of Newtonian atomism in Britain as deism comes under the attack of evangelical Christianity. George Cheyne's progress toward materialism as he progressively adopts a methodistic religious enthusiasm has already been noted. Among the most persistent anti-Newtonian English philosophers of the first four decades was Robert Greene, whose weird semi-materialism seems to have found its only serious contemporary recognition in the discussions of the mystic John Byrom and its only serious imitators in the religiously motivated, anti-Newtonian, materialist Hutchinsonians, whose work was much admired by John Wesley.[11] No doubt there is a paradox, not quite resolved by the clear relationship between deism and scientific abstraction, in the association

[9] Henry Guerlac, "Newton's Changing Reputation in the Eighteenth Century," in Raymond O. Rockwood, ed., *Carl Becker's Heavenly City Revisited* (Ithaca: Cornell University Press, 1958), pp. 21-24.

[10] Henry Guerlac, "Newton et Epicure," Conférence donnée au Palais de la Découverte le 2 Mars 1963, D 91, Histoire des Sciences (Paris: Université de Paris, Palais de la Découverte [1963]), pp. 18-19.

[11] Both Greene and the Hutchinsonians will be discussed more fully in Chap. 6. For Byrom and Greene see Richard Parkinson, ed., *The Private Journals and Literary Remains of John Byrom*, I, part 2, *Remains Historical and Literary connected with the Palatine Counties of Lancaster and Chester*, XXXIII ([Manchester]: for the Chetham Society, 1855), 397-98, 405, 433. For Wesley and the Hutchinsonians see Robert E. Schofield, "John Wesley and Science in Eighteenth-Century England," *Isis*, 44 (1953), 331-40.

of religious mysticism and materialism. It is difficult to go beyond the assertion, with examples, that the phenomenon exists, but a parallel can, at least, be suggested between the favorite evangelical use of science, in the argument from design, and the materialistic mode of scientific interpretation. Both identify proximate cause with result. In what amount almost to caricatures of Molière's caricature of scholasticism, the cause of heating is a substance whose essential property is its heating, the cause of the moon's giving light at night is its giving of light at night.[12]

Happily, more clearly definable scientific justifications for the changes in scientific temper exist. A major one of these, repeatedly noted in the previous chapter, was the inability of the dynamic corpuscularian to assign magnitude or a determinant form to any of the various forces of attraction and repulsion which they used with such ingenuity in their speculations. In 1740 the situation was as Newton had left it in 1687; form and magnitude of gravitational attraction were known, and no others. Of the other forces suggested by Newton, only capillarity, with Hauksbee and Taylor, and cohesion of lead, with Desaguliers, had even reached toward the requisite state of mathematical precision, and both were still but crude. Magnetism seemed to defy exact measurement, and electricity had yet to be tested. As for the elastic repulsion of the air, this near-triumph for Newton's forces remained a problem. In the form given the force law by Newton—inversely proportional to the distance between centers and terminating at adjacent particles or near them—Boyle's law was deducible, but not the velocity of sound. Its curious, almost un-Newtonian form, not dependent simply on distance and quantity of particular matter but, in a sense, saturable, made further analysis difficult with the mathematical tools available. Moreover, the work of Hales, though firmly in the tradition of dynamic corpuscularity, had only served to confuse the issue, for his treatment of elastic air particles did not so much extend the form of Newton's force law as ignore it, in establishing the tension between attraction and repulsion. Particles capable, in some circumstances, of binding together instead of repelling, while retaining, in Desaguliers' phrase, "a latent power of repulsion," were not obvious candidates for simple mathematical reduction. Hales' measurements in animal and vegetable physiology

[12] Incidentally, John Wesley was the author of one of the most widely distributed scientifico-theological argument-from-design books of the eighteenth century, in his *A Survey of the Wisdom of God in the Creation*, first published in 1763 and enlarged and reprinted by his Methodist press four more times during the century.

also complicated matters by their wide variation from the values expected (and needed) by the Newtonian mechanistic physiologists if their speculations were to seem correct.

Such failures lend a cutting edge to comments such as Rutherforth's. Without measurement, it does appear that forces multiplied like peripatetic qualities, and that appearance was not then to be masked in the proliferation of parameters implicit in the multiple, concentric spheres of attraction and repulsion of a Rowning or Desaguliers. Given the failure finally to relate mathematics, abstract ingenuity, and empirical evidence, there is small wonder that the frustrated natural philosopher should, as if by instinct, return to the simpler mode of Aristotlean qualities, objectified, perhaps, in the sizes and shapes of the kinematic corpuscularians. And such a return was made the more respectable as the two foremost Newtonian mechanists, Newton himself and Stephen Hales, had provided models for such endeavors. When Hales gave to the air, uniquely, the property of repulsion, without an analysis as to how that property might ultimately be resolved into the particles and forces common to all matter, he created a material distinction which others were to adapt to heat and electricity as well as to exploit in chemistry. The view of air as a component, as well as an instrument, in chemical operations led to a pneumatic chemistry whose purpose it was to distinguish, by their properties, a variety of different airs.

The model Newton provided is a subtler one. He was always to insist on the materiality of light—actual particles, emitted by shining bodies, differing in size and unsymmetric in shape. Newton's particles of light interacted with and were convertible into other bodies; their forces of attraction and repulsion were different only in consequence of their very small sizes.[13] Part III of Book Two of the *Opticks* even demonstrates how the colors reflected by bodies might enable a determination of the sizes of the corpuscles of which the different bodies were compounded. But here the matter had rested. Neither the sizes, shapes, nor forces between light particles or between them and other matter had been determined, and it was comparatively easy, given a convincing impulse, to see in light, possessing "original and unchangeable properties [Query 27]," a uniquely differentiable substance.

That impulse, and what provided the primary materialist thrust

[13] See Queries 5 and 30 of the 1717 *Opticks;* the section on attractive and repulsive force varying with size of light particle is omitted from the translation of the Latin Query 22 (English 30) as this section is cut and transposed to the English Query 21.

of Newton's work, came through his introduction of the aether. Here was the active substance to set against the inherent passivity of all other matter. Through the action of the aether, the conveyance of heat *in vacuo*, the explanation of those confusing "fits of easy reflexion and refraction," reflection and refraction themselves, even gravitational attraction were potentially to be explained. In his discussion of the aether, in the new Queries of the 1717 *Opticks*, Newton associated its subtlety with that of the effluvia of electric and magnetic bodies [Query 22], and his "Hypothesis explaining the Properties of Light," delivered to the Royal Society in 1675 and first published in Thomas Birch's *History of the Royal Society* of 1747, goes so far as to suppose there might be several aetherial spirits compounding the aether and representing such different principles as gravity and electric and magnetic effluvia.[14] To Newton, the aether was, however, a dynamic conception. In the same paper in which he proposed the several aetherial spirits, he also suggested that "the whole frame of nature may be nothing but various contextures of some certain aetherial spirits, or vapours, condensed as it were by precipitation . . . and after condensation wrought into various forms. . . . Thus perhaps may all things be originated from aether." Though the aether was material, it was the matter of all substance; perhaps its particles were those ultimate particles of nature described later in the queries of the *Opticks*. Though the aether was active, its activity was that of all matter. Clearly it was not imponderable, nor did it effect its results by, and in proportion to, its presence. Although its action was not to be established in detail, the mechanistic properties, size of particle, repulsive force, and variable density, are the parameters proposed for ultimate analysis. Such an analysis did not occur. Newton's aether became, instead, the prototype for every imponderable fluid. I. Bernard Cohen has provided a convenient summary of the uses to which fluid theorists were to apply Newton's aetherial hypothesis. The porosity of solids provided the lodging for elastic fluids, existing, like the aetherial medium, between the corpuscles and trapped there, as light was trapped, as a function of the structure and composition of bodies.[15] Once the materialist frame was adopted, acceptance of the aether gave direction to the elaboration of the structure.

The adoption of Newton's aether hypothesis provides, therefore, one of the scientific justifications for the change to materialist

[14] Reprinted in Cohen, *Newton's Papers and Letters*, pp. 178-90; see especially page 180.

[15] Cohen, *Franklin and Newton*, pp. 161-62.

modes of interpretation. But the adoption is a result of changing tempers as well as a cause, for Newton's aether had remained an unexploited concept for nearly a quarter of a century before its rapid expansion to dominate the meaning of theoretical Newtonianism. This delay in its acceptance was not wholly expected by Newton's contemporaries. With ill-concealed malice, the writer of the *Newsletter* of 19 December 1717 had noted: "Sir Isaac Newton had advanced something new in the last edition of his Optics, which has surprised his physical and theological disciples."[16] The surprise, however, was well concealed. Previous chapters have shown how little aether suggestions changed the nature either of explanation or of experimental interpretation after the 1713 *Principia* and the 1717 *Opticks* down to 1740. Of the fourteen writers discussed who might have been affected by the proposal of the aether, only four, Cheyne, Desaguliers, Jurin, and Worster, allowed it to influence their work and, of the four, only on Worster can it be said to have had a positive influence. The record does not suggest that Newton had made, in the aether, a proposal of significance to natural philosophy.

A measure of the change in attitude toward the aether is to be found in the differences between two popularizations of Newtonian philosophy, published 20 years apart and separated by the years of transformation. Henry Pemberton's *View of Sir Isaac Newton's Philosophy* (1728) and Colin Maclaurin's *Account of Sir Isaac Newton's Philosophical Discoveries* (1748) are among the best popular eighteenth-century accounts of Newton's thought. Neither contributes any new ideas to the science of the period, but each attempts, clearly and exactly, to report to the general, non-scientific reader what Newton had achieved. Pemberton, though largely self-taught in mathematics, had been intimately associated with Newton while editing the third (1726) edition of the *Principia*.[17] Newton had, Pemberton declares, read and approved a greater part of the treatise just before his death—

[16] Quoted in Portland Manuscripts (Norwich: Historical Manuscripts Commission Reports, 1899), v, 550.

[17] Henry Pemberton (1694-1771), F.R.S., educated in London, and, in medicine, at Leyden, under Boerhaave, in anatomy in Paris, in clinical practice at St. Thomas' Hospital, London, and then again, as Boerhaave's guest, at Leyden; M.D. 1719. His ill health restricted the development of a London practice, but he was an industrious writer on medical and other subjects. One of the earliest of his several articles in the *Philosophical Transactions* recommended him to the friendship of Newton, who selected him to superintend the third edition of the *Principia* (1726). Gresham Professor of Physic (1728), his lectures in chemistry and in physiology were published posthumously by James Wilson, in 1771 and 1779, respectively. See Cohen, *Franklin and Newton*, pp. 209-14, for a full description of Pemberton's *View*. I have used the only English edition, Henry Pemberton, *A View of Sir Isaac Newton's Philosophy* (London: by S. Palmer, 1728).

which may, perhaps, account for its pedestrian quality. Maclaurin, a professional mathematician, brought special geometrical talents to the appreciation of Newton's mathematical achievements, and his work is also graced with an historical introduction which contributes greatly in setting Newton's work in proper perspective.[18]

The primary object of each book is the interpretation of the *Principia*, although Maclaurin occasionally refers to the *Opticks* and Pemberton devotes the last quarter of his *View* to it. Both men emphasize the exact, final quality of the *Principia*, and both see the greater difficulty involved in discovering both the phenomena of light and the laws that govern them. Personal differences between the authors are, perhaps, revealed in Pemberton the physician's, emphasis on the Baconian nature of Newton's work, where Maclaurin, the mathematician, is more inclined to stress the formal alternation between synthetic and analytic modes of reasoning. But, with one major exception, the general nature of the two works is the same: each is strongly anti-Cartesian and denies the existence of a plenum; each repeatedly emphasizes Newton's care, or modesty, in proposing anything for which he had not adequate proof; each adopts the Newtonian language of attractive and repulsive forces and gives the usual examples of their operations; each denies these forces to be occult. Pemberton adds that Newton often complained to him that he had been misunderstood, that "attraction" was "not intended by him as a philosophical explanation of any appearance, but only to point out a power in nature not hitherto distinctly observed, the cause of which, and the manner of its acting, he thought was worthy of a diligent enquiry [407]."

The fundamental difference between the books is in their attitude toward the aether. Pemberton's is permeated with aether phrases, e.g., he speaks of "aetherial space" far beyond the moon, but he is cautious in explicit references to it. Gravity is demon-

[18] Colin Maclaurin (1698-1746), educated at the University of Glasgow; M.A. 1713. Student of divinity for a year previous to his appointment as professor of mathematics at Marischal College, Aberdeen, 1717. Traveled in Europe as a private tutor, 1722-25; named deputy professor of mathematics, University of Edinburgh, in 1726, on the recommendation of Newton, who contributed toward his salary. Author of several mathematical papers in the *Philosophical Transactions*; two memoirs: on the percussion of bodies (1722) and on the gravitational theory of tides (1740), which won prizes of the Academie Royale des Sciences; a *Treatise on Fluxions* (1742) and a *Treatise of Algebra* (1748), as well as his *Account*, which was begun in 1728, but completed just before his death and published posthumously. I have used Colin Maclaurin, *An Account of Sir Isaac Newton's Philosophical Discoveries* (London: J. Nourse, W. Strahan, J. and F. Rivington, *et al.*, 1775), 3rd edn.

strated to be a property universally belonging to every part of all matter, its power penetrates the bodies themselves, acting alike on every particle, but Newton makes no "pompous pretense" of explaining its cause, giving only a hint, in Query 21, of his supposition [22, 62]. "Whether some of the qualities and powers of particular bodies, be derived from different kinds of matter entring their composition" cannot yet be decided, though, in fact, there seems no reason to doubt that all bodies are framed of the same kind of matter and their different qualities occasioned only by individual variations in the general powers of all matter caused by the varieties of their structures [21-22]. The deduction of Boyle's law from a law of repulsive forces "does not determine" that the air is endued with such a power to act out of contact. Newton leaves that for future examination and philosophical discussion, but Pemberton sees little reason to deny to bodies such powers to act at a distance, and cites gravity and magnetism as examples [150-51]. Newton has not discovered the cause of the power whereby light and bodies interact, but in general has hinted his opinion that "probably it is owing to some very subtle and elastic substance diffused through the universe, in which . . . vibrations may be excited by the rays of light. . . . He is of the opinion, that such a substance may produce this, and other effects also in nature [377]." At the end of his treatise of optics, Newton expressly declared his conjecture, "that this power is lodged in a very subtle spirit of great elastic force diffused thro' the universe producing not only this, but many other natural operations. He thinks it not impossible, that the power of gravity itself should be owing to it [406]."

To Pemberton's confusion and hesitancy contrast Maclaurin's confident assertions. All attempts to explain the powers of nature mechanically, as by effluvia, environing atmospheres, or vortices have hitherto proved unsatisfactory [114-15]. For these Newton has proposed the term "attraction," but he never affirms or insinuates "that a body can act upon another at a distance, but by the intervention of other bodies [115]." He has "plainly signified that he thought that those powers arose from the impulses of a subtile aetherial medium that is diffused over the universe, and penetrates the pores of grosser bodies. It appears from his letters to Mr. Boyle, that this was his opinion early; and if he did not publish it sooner, it proceeded from hence only, that he found he was not able, from experiments and observation, to give a satisfactory account of this medium, and the manner of its operation, in producing the chief phenomena of nature [116-17]." Maclaurin does not himself use the aether, nor does he show how this might be done, but he has little doubt that it could be. "If . . . the most noble phaenomena in nature be produced by a rare elastic *aetherial medium*, as Sir *Isaac Newton* conjectured, the whole efficacy of this medium must be resolved into his [God's] power and will. . . . This, however, does not hinder, but that the same medium may be subject to the like laws as other elastic fluids, in its action and vibrations: and that, if its nature were better known to us, we might make curious and useful discoveries concerning its effects, from those laws [408]."

Maclaurin was a moderate man, and restrained by the text he was expounding, yet that text was the same as that previously discussed by Pemberton. The most reasonable explanation of the differences in readings is the publication, between 1740 and 1745, of four discussions explicitly enhancing the view of the aether and Newton's use of it. One of these, Newton's letter to Robert Boyle, is mentioned by Maclaurin. First published in Thomas Birch's *History of the Royal Society* in 1744, this letter dates from 1678/9 and is here joined with another, of 1675/6, to Henry Oldenburg which refers to Newton's "Hypothesis" paper of the previous year and its notion of the precipitation of various aetherial spirits.[19] The letter to Boyle expounds, at some length, the way in which a subtle aetherial substance, capable of contraction and dilation, strongly elastic, and diffused through all space, might act to cause cohesion of metals, refraction and inflection of light, capillarity, chemical activity (though here he introduces also "a certain secret principle in nature," by which substances are sociable to some things and unsociable to others), and even gravity. The mechanism proposed is chiefly that of aether pressure and variable density which is discussed in nearly the same detail in the General Scholium of the 1713 *Principia* and the aether queries of the 1717 *Opticks*; the major influence of the published letter was to suggest that Newton's aether hypotheses were of considerably earlier development than had previously been supposed. The year after Birch's publication of the letter to Boyle, it was republished by Bryan Robinson in *Sir Isaac Newton's Account of the Æther* (1745). By that time Robinson was making a career of aetherial discussion, having already published a lengthy *Dissertation on the Æther of Sir Isaac Newton* (1743); he was soon to become the focus of attention for fluid theorists in heat and electricity, several of whom traveled from London to Dublin to visit him.

[19] These two letters are republished in Cohen, *Newton's Papers and Letters*, pp. 250-54.

There is no clear evidence that Maclaurin had even read Robinson's works, let alone joined the pilgrimage to Dublin, but there is an indication that he read a still earlier, anonymous work on the aether, *An Examination of the Newtonian Argument for the Emptiness of Space and of the Resistance of subtile Fluids* published in 1740.[20] The first part of this tract is an argument that Newton has failed adequately to demonstrate, from unresisted celestial motion, the existence of a vacuum [7-8]. The author agrees that seemingly solid bodies are porous, citing the passage through them of "magnetic virtue," "streams of *Bolonian* stone," electric forces, light and fire [9-10]. Then he supposes a kind of matter might exist, so subtle, figured, and attenuated as freely to permeate all other matter, and therefore be unresistant to bodies' motion through it [11-12]. He will not insist that such an "aetherial medium" exists nor maintain an absolute plenum, but alleges only that the possibility of such a substance invalidates Newton's arguments [13]. Later in the treatise, the author declares, "As I am no mathematician, I may possibly be mistaken in all this, and I should be fond to be instructed by better Judges. In the mean time, an invisible, imperceptible and penetrating *medium* seems to be, at least its existence is supposed by all; the common phaenomena seem to require it [19-20]." It appears that Maclaurin responded as the mathematical "better Judge" to instruct the author. His *Account* includes a long section replying to the objection, "In a small piece published on this subject, a few years ago by an ingenious gentleman," that a fluid freely penetrating the pores of bodies would not resist the motion of bodies through it [72-76].

If, as seems clear, Maclaurin did read the *Examination of the Newtonian Argument*, and if, as then seems likely, he was encouraged by it to take the aether seriously, he was remarkably restrained in his response. In the continuation of the *Examination*, the anonymous author develops most of the major tenets of the aetherial, imponderable fluid, materialists. One would not, from the first 16 pages of the total of 22, suppose that their author had ever read or heard of Newton's proposal of an aether, but when he commences a discussion of why there may be matter without weight, Newton's aether is cited, along with the "aetherial

[20] The British Museum has ascribed this work to George Martine, author of the *Essays Medical and Philosophical* previously discussed in this chapter. The grounds of the designation are unclear and the abstract nature of the argument is most unlike Martine's other works. I have used the only edition: *An Examination of the Newtonian Argument . . .* (London: T. Cooper, 1740).

elements of fire" proposed by Homberg, Gulielmini, and Boerhaave, as being just such a substance: "Newton's aether . . . is described by himself as not subject to gravity [16-17]." The author even suggests that some bodies have rather "a contrary tendency . . . endeavouring to recede from one another by a repelling, instead of . . . [a] mutually gravitating force, as in the phenomena of magnetism, electricity, the mixture of heterogeneous liquors . . . inflexions and reflexions of light . . . etc. [15n-16n]." Here we first see the amalgamation of "the *ignis* of some folks, the *aether* of others, the *materia subtilis* of others" into "one and the same thing . . . producing the wonderful phaenomena of attraction, gravitation, cohesion, magnetism, electricity, etc." And here also, for the first time, we find the dichotomy clearly marked between the Newton of attractive and repulsive forces and that of the aether. For, according to the author, Newton had failed to consider the aether, in his first edition, because he was then "bent on demolishing the *Cartesian* system of *Vortices* and subtile matter" and showed a great keenness for establishing atoms; but "in his riper days," he became less sanguine that way "and filled the heavens with rays of light . . . is fond to introduce an omnipresent aether, as the great agent in nature, causing attraction, cohesion, electricity, the motions and modifications of light and heat, animal sensations and motions, etc. [20-22]."

The connection between Newton's aether, imponderable (even levitating) fluids, and light, heat, electricity, and magnetism had now been made, but as part of a panacean claim, without proofs, asserted by an unknown author—admittedly no mathematician —and in an anti-Newtonian frame. Before Newton's aether could be taken seriously it would have to be re-introduced under more honorable auspices. Most of this requisite respectability was to be achieved within three years by the publication of *A Dissertation on the Æther of Sir Isaac Newton* by Bryan Robinson.[21] The *Dissertation* was neither Robinson's first publication nor his earliest published reference to the aether. As early as 1704 he had demonstrated his mathematical abilities with a translation of de la Hire's *New Elements of Conick Sections,* dedicated to his friend John Harris, while his much respected *Treatise of the*

[21] Bryan Robinson (1680-1754), educated at Trinity College, Dublin, with Richard Helsham; M.B. 1709, M.D. 1711. Anatomical lecturer at Trinity, 1716-17; 1745 appointed professor of physic. Three times president of the College of Physicians in Ireland, also a member of the Royal College of Surgeons. Developed a thriving practice in Dublin, where he probably attended Swift's "Vanessa." Author of several admired medical works and an *Essay upon Money and Coins,* as well as the treatises discussed above.

Animal Oeconomy, first published in 1732, leavens an otherwise commonplace mechanistic physiology with a section on the aetherial cause of muscular motion.[22]

Robinson begins his *Animal Oeconomy* with the standard claim that he has avoided hypotheses and explained the body by reason and experiment only [iii]. He demonstrates his mathematical facility with nearly eighty pages of hydrodynamical computations, propositions, corollaries, "proof by experiment," and scholia on the motion of liquids in variously shaped, sized, and convoluted pipes. Much of the remainder discusses the physiological effects of the motion of the blood as James Keill would have done, though without as much detail. The force of the heart is proportional to the capacity of the system of blood vessels, and the force generated by the motion in the vessels is as the whole force of the heart [116-17]. The constituent solid parts of animals (especially the glands), according to their several natures, are endued with peculiar attractive powers by which they draw out of the fluids moving through them like parts in appropriate quantities, as capillary tubes attract different fluids with different degrees of forces [220-42]. There is a minor anomaly in the proposition that the life of animals is preserved by the same acid parts of the air which preserve fire and flame, these parts mixing with the blood in the lungs to attenuate it and preserve its heat, keeping up the motion of the heart [187-92]. But the chief difference of the text appears as part of Section II, "Of Muscular Motion . . . ," where, in 15 of a total of 250 pages, the aether is mentioned. With ample supporting citations from Newton's aether queries, Robinson maintains that the nerves are solid, not pipes, nor is any fluid "Animal Spirits" drawn off from the blood to the brain. Muscular motion is, instead, excited when the power of the will causes vibrations in a very "Elastick Æther" lodged in the nerves. These vibrations travel through the nerves, exciting a like motion in the aether lodged in the muscle membranes, increasing their expansive force, and producing a corresponding contraction in the fleshy fibers. This view is verified by the rapidity with which motion responds to the action of the will—as is the behavior of the vibrating motion of a very elastic fluid. ". . . and since the other Phaenomena of Nature absolutely require such an elastick Fluid, as the Æther described by Sir

[22] Bryan Robinson, *A Treatise of the Animal Oeconomy* (Dublin: George Grierson, 1732). This work involved Robinson in a controversy with George Cheyne, from which evolved a larger, revised edition of the *Animal Oeconomy* in 1734, and still another edition appeared in 1738; neither of these later editions have I seen.

Isaac Newton; and since Causes are not to be multiply'd without necessity: Therefore it must be granted, that this Motion begun in the Nerves at their Origin, is the vibrating Motion of that Æther . . . [92-93, 96-98]."

Inserted as this section is, in the midst of an otherwise dynamic corpuscular text, one wonders if Robinson was then aware of any distinction between the two types of explanation. That question has added meaning when it is noted that seven years later he edited, for its posthumous first publication in 1739, the *Course of Lectures in Natural Philosophy* of his friend Richard Helsham, which contains no sign of aetherial conjecture. There is a further complication in the implication of the Appendix to Robinson's *Sir Isaac Newton's Account of the Æther.*[23] Here Robinson suggests that his interest in the aetherial cause of muscular motion was the result of an answer received from Newton to a question posed by "a gentleman [presumably Helsham] who asked him how muscular motion is caused by vibrations of the aether [33]." Newton's answer does not vary significantly from the explanation given in the *Animal Oeconomy,* though it gives Robinson an opportunity to describe some anatomical observations and to note that fat or serous moisture in the body would weaken the vibrations of the aether, as may be observed in the weakening of sound vibrations in a moist atmosphere, the flatter note of moistened vibrating strings, and the weakening, if not destruction of "the electrical virtue of bodies, caused by the vibrating motion of the aether, . . . by water and watery moisture [37-38]." The *Account* is primarily devoted to reprinting Newton's letter to Boyle and the aether queries. The first affords Robinson the evidence for his declaration: "Sir Isaac Newton discovered the *Æther* soon after he became acquainted with the properties, action, and motions of corporeal things by experiment and observations [Preface]." The remainder of the work adds little of significance to what Robinson had already written in the *Dissertation on the Æther of Sir Isaac Newton* two years before.[24]

Robinson's *Dissertation* was the pivotal work on the aether hypothesis, doing for it what John Keill and John Freind had earlier done for inter-particulate forces. With a reputation that the anonymous author of the *Examination* did not, and a mathematical skill he could not, command, Robinson was able to present

[23] B[ryan] R[obinson], *Sir Isaac Newton's Account of the Æther, with some Additions by way of Appendix* (Dublin: G. and A. Ewing, and W. Smith, 1745).

[24] Bryan Robinson, *A Dissertation on the Æther of Sir Isaac Newton* (Dublin: Geo. Ewing and Wil. Smith, 1743).

an argument for the aether sufficiently convincing to get it adopted into the common thinking of mid-century British experimental natural philosophers. The work is, besides, a masterly piece of special pleading. With consummate artistry (one might almost say sophistry), Robinson weaves together exact, accepted, mathematical calculation, experimental observations, and Newtonian quotations and assumptions, inserting, at just the right moments, references to the aether such as to cloak the new concept in the authority of what had preceded. There is no necessity of proving the existence of the aether, it "being a very general material Cause, without any Objection appearing against it from the Phaenomena, no Doubt can be made of its Existence: For by how much the more general any Cause is, by so much the stronger is the Reason for Allowing its Existence [Preface]." Robinson's problem is, therefore, to show in detail where Newton had only hinted just how the aether causes universal attraction and gravity, repulsion and elasticity, the various phenomena of light, heat and rarefaction of bodies, muscular motion, coherence, and fermentation.

The text is too long (124 pages) and intricate to do more than illustrate its quality with a few examples. Robinson begins with a section leading to a determination of aether particle size and distribution. In a static model of elastic fluids, with exceedingly small particles of equal diameters and densities, endued with a centrifugal force, terminating in adjacent particles, proportional to particle density and inversely to particle diameter and distance between centers, the elastic force of the fluid is directly proportional to the density of the fluid and inversely as the fourth power of the diameter [2-5]. The diameter of the particles is then the fourth root of the density divided by elasticity, and the distance of centers is as the diameter of the particle and the cube root of particle density divided by fluid density. From Boyle's law one knows that air is really composed of such particles, with such a force, for its properties are not to be explained in any other way [6]. Newton has estimated that the density of air is 1/870th that of water, and its elasticity 870 times that of water, while its particle size must be less than $1/4 \times 10^{-5}$th of an inch (say $1/6 \times 10^{-5}$) from Newton's chart of corpuscle sizes. Now the particles of aether must be assumed to be perfectly dense and void of pores, and their density, therefore, is 40 times that of water, as Newton has declared water to have 40 times more pores than solid parts. Supposing the density of an air particle to be twice that of water, and the rarity and elasticity of the aether at the surface of the earth to range from 7×10^5 to 10^6 times that of the air, then: the diameter of aether particles must be between $1/6\text{th} \times 10^{-8}$ to $1/5\text{th} \times 10^{-8}$ of an inch and the distance between particle centers between $1/1.84 \times 10^{-5}$ and $1/1.73 \times 10^{-5}$ of an inch [7-13].[25]

Proceeding to an aether explanation of gravity, Robinson first demonstrates that, were bodies surrounded by a subtle aether, everywhere of the same density and endeavoring to recede from those bodies with a force equally the product of the quantity of matter in the body and the density of the surrounding aether, then the density of the aether would spontaneously change until its increment at any distance from the center of the body would be as the quantity of matter and inversely as the square of the distance [26-27]. From this it follows that the effect of the interaction of aether and body will be an attraction, or mutual tendency of bodies toward one another, to positions of lesser aether density, with a force proportional to the quantity of matter in each body and the increment of density of the aether at its center of gravity, caused by the other body. That is, as the product of the masses and inversely as the square of the distance [34-35]. Robinson follows this demonstration with some twenty pages of computations showing, from weights, laws of falling bodies, periodic times of planetary revolutions, etc.—without reference to the aether—that this is the law of gravitational attraction. He concludes: "Having shewn how the *Æther* causes an universal Attraction and Gravity, I shall . . . shew how it causes some other Phaenomena, not explicable by any other cause [54]."

He proceeds now to the assertion that capillarity is caused by the aether's being denser within a tube, or between glass plates, than without—relating Hauksbee's measurements to "the Increment of the density of the aether, which is the Measure of the Force that moves the Drop," that is, "the duplicate Ratio of the Distance of the Middle of the Drop from the Concourse of the Glasses [61]." One might suppose there would be a problem in explaining repulsion from this hypothesis, but Robinson avoids that through simple confidence. If attraction is caused by the rarity of the aether between bodies, then repulsion must be caused by its greater density, and he asserts: ". . . as the Force of Gravity or Attraction at the Surfaces of Globes, is measured by the Increment of the Density of the *Æther* at the Surfaces of the Globes, so the Force of Repulsion in the Particles of Air is meas-

25 I have obviously modernized the notation, without, however, doing violence to the results except by reducing the number of figures to which the calculations were taken.

ured by the Increment of the Density of the *Æther* at the Surfaces of the Particles [64]." The heat in bodies consists of the vibrating motion of their parts, excited by the vibrating motion of the aether within the bodies—which is proved because bodies grow hot from light falling on them, which is known to excite aether vibrations, or whenever the aether is put into a vibrating motion by agitation of the parts of the body by friction, percussion, fermentation, or any other cause [93-94].

The treatment of the interaction of light and aether follows five pages of discussion of the motion of bodies, as expressed in the kinematic equations of Galileo and Newton's laws of motion. Given this authority, we can now proceed to: "If the moving Force of the Body be as the Increment of the Density of the *Æther* at its Surface, and that Increment be proportional to the Density of the Body, *or to the Quantity of Light contained in the Body*; the Quantity of Light . . . in proportion to its whole Quantity of Matter, will be as the Velocity generated by that Increment in a Given Particle of Time [72-73, italics added]." From the known velocity of light, it is demonstrated that the force of interaction between light and aether is immeasurably larger than could be caused by the gravitational action of the aether [73-76]. Newton's theory of refraction indicates that the velocity of light is increased on refraction. Thus, the "Force with which the *Æther* refracts, reflects, or inflects a Ray of light incident on a Body . . . is something greater than the Force of the *Æther* which emits it from the Sun," and is (as Newton has demonstrated) nearly proportional to the density of the body, except that it is greater in unctuous and sulfurous bodies, which are known to contain more light, in proportion to their densities, than other bodies [76]. Therefore, the power of bodies to act on rays of light is nearly proportional to the quantities of light contained in the bodies [81], which is to say that the increment of aether density at the surface of the body is a measure of the mutual action between the light in the body and the aether at its surface [83]. This greater strength of action between light and aether may be owing to the exceeding smallness of the particles of light, as the immense expansive force of the aether arises from the smallness of its particles. The special nature of light-aether interaction is now employed in explaining cohesion and fermentation, both requiring forces greater than that by which the aether causes gravity [114-15]. Acids acquire their power from the light they contain and are increased in strength by fire from the particles of light acquired from it [119-20].

One is first appalled, then bemused, intrigued, and finally almost seduced by this spate of argument. Does it matter that the very existence of the aether has not been demonstrated? Does it make any difference that the sections do not cohere, that the function and distribution of the aether in one instance would confuse and disturb that in another? What if the fine appearances of equations, measurements, and calculations are mostly irrelevant, giving a specious appearance of exactitude? The situation here is not greatly different from that earlier developed by the mechanistic physiologists, and even by some of the dynamic corpuscularians; but they did not achieve the quantitative promise of Robinson. No other Newtonian, save Newton himself, came nearer to the appearance of a quantitative measure of parameters. The effect is still removed from a thoroughgoing materialism, for, excepting the unique role of light in which quantity of substance is involved, the action of that special "material cause," the aether, is defined in mechanistic terms which are familiar—and some of which Robinson has evaluated—the size of its particles, their density distributions, and the repelling force between its particles. A transformation is still to be made between Robinson's aether and that diversified into electric, magnetic, light, and heat fluids. But Robinson has provided the justification, at least for the non-mathematical experimenter impressed, as usual in a century of Newton, with mathematically framed argument, to take the aether seriously. Once the aether, as a special material cause, is adopted, its materiality can be merged with that of other causative substances as these emerge from other sources.

CHAPTER SIX

Newtonian Pagans and Heretics

FEW STYLES OF TASTE OR THOUGHT are so thoroughly assimilated that there do not exist, at the same time, undercurrents of dissenting opinion. Such opinions flow submerged and, on the whole, unnoticed; but should the dominant current weaken, they may surface at least to disturb consensus if not to start a contrary fashion. The existence of such currents, anti-Newtonian in an Age of Newton, is well-known in Continental natural philosophy, but their eddies in eighteenth-century Britain are often passed unobserved. The reason is fairly obvious. Examination of the British dissenters from Newtonianism reveals the essential sterility of their effort. As their twentieth-century counterparts opposed Einstein to support values they derived from a Newtonian world view, so these men of the eighteenth century opposed Newton in favor of systems derived to preserve values they found missing in the new mechanical philosophy. Variously constructed about the views of Aristotle, of Descartes, or of some eclectic mixture of these with any other available system—neo-Platonic, Paracelsian, Boyleian, etc.—the variant proposals were always ill-digested, frequently ill-tempered, and are best characterized by their invincible ignorance not simply of mathematics but more significantly of the use of mathematical deduction, from quantifiable hypotheses, confirmable by measurement.

Under the circumstances, an examination of British anti-Newtonians would seem an exercise in futility, particularly in a study of Newtonian influences in the eighteenth century, were it not that the views of one group, at least, find an echo in the transformation of Newtonianism from dynamic corpuscularity to aetherial materialism during the middle part of the century. The various members of this group approached their several Zions from as many roads as there were issues and with a fine sectarian contempt for the dissent of others. They seem hardly to have influenced one another, and their direct effect on more astute contemporaries must have been imperceptible. Yet in the end they agree in explaining the phenomena of nature in terms of active substances, and a survey of their opinions indicates that a pervasive, if naïve, materialism lay at hand ready for elaboration to replace the overly sophisticated theories of the mechanists.

An early example of this genre, *An Essay at the Mechanism of the Macrocosm* of 1705, clearly originates in the scholasticism its author, Conyers Purshall, was taught at a pre-Newtonian Oxford.[1] Purshall adopts a modified Tychonic astronomical system, with planetary orbits and periods in numerical harmony with Kepler's law. The orbits, however, are circles, and their sizes are determined by the levels, appropriate to their several densities, at which the planets float in a plenum of aetherial matter decreasing in density from the earth-center [87-88, 130]. An expansive force, emanating from the sun, would be incapable of retaining the planets at the aphelion of an elliptical path [106]. The principle of inertia does not hold; bodies fall naturally according to their weights, and accelerate in falling as the impending expansive air follows and beats upon them [311-14]. Matter is resolvable into eight different and unchangeable principles, ultimately existing as particles, so small no agent in nature can further divide them, and having dimension and impenetrability [22-23]. Six of the eight are active principles, ranging in order from Aether—expansive, springy, and subtle; through Air, Spirit, Oyl, and Water; to Nitre—the contrary equivalent to aether, contractive, springy, and subtle [24-26, 35, 48, 53-54, 66]. Salt and earth are the passive or inactive principles, combining with the active through the medium of the less active, and naturally moving only to the center [70].

Aether is the principle of heat when its agitating motion moves parts of bodies and produces the sensation of warmth; its particles add weight to calcined substances. When put into violent rotary motion, in right lines, by the vibrations of the sun, aether particles become solar substance or light [24-25]. These are homogeneous in all respects, but their irregular and confused reflections, in different angles and numbers, their refractions in the eye with different forces, produce the various colors and explain all the phenomena of Newton's treatise on light [200-03, 250]. Cold is not mere privation, but the result of a true congealing principle, the frigorific particles of nitre [37]. Magnetism is the result of effluvial streams of nitro-terrene particles moving through the pores of iron, variously disposed like veins and arteries to receive currents to or from the poles [265-70]. Elec-

[1] Conyers Purshall (Coniers Purshull) (c. 1657-17?), matriculated at Pembroke College, Oxford, 1675, at the age of 18. He calls himself M.D. in the second (1707) edition of his book: [Conyers Purshall], *An Essay at the Mechanism of the Macrocosm: or the Dependence of Effects upon their Causes. In a New Hypothesis, Accommodated to Our Modern and Experimental Philosophy, etc.* (London: Jeffery Wale, 1705). See Joseph Foster, *Alumni Oxonienses: The Members of the University of Oxford 1500-1714* (Oxford: James Parker and Co., 1891), III, 1,221.

tricity is caused when expansive air, agitated and rarefied by rubbing, emerges with violence from the pores of amber, driving with it loose unctuous particles. These, in contact with the contractive principle of nitrous or aqueous matter in the air, condense and are repelled, by the contrary principle of the air, "to the place from whence they came" carrying any light thing with them [208]. Chemical action, throughout, is explained as the combinations of principles or substances by and to their degrees of affinity or similarity.

Purshall's *Essay* reached a second edition in 1707, but it is hard to believe anything so obviously reactionary had much influence. The next example cannot be taken as lightly. Robert Greene served for more than a quarter of a century as a tutor at Clare Hall, Cambridge, inculcating in his students the anti-Newtonian ideas embodied in his two major works: *The Principles of Natural Philosophy* of 1712 and the monumental *Principles of the Philosophy of the Expansive and Contractive Forces,* more than 950 closely printed folio pages, of 1727.[2] Greene's vanity and self-esteem so permeate his works as to make them even more difficult than their singularity of opinion requires. He conceives all things explicable according to his principles and sees no reason that he may not acquire a reputation equaling that of an earlier Cambridge student and teacher, Isaac Newton. Newton's philosophy is constructed on that of Galileo, Descartes, and Kepler. Greene's own is truly English and completely original, not derived from popish countries or those abounding in atheism or superstition. A typical fanatic, Greene knew that the changes in the *Principia* and *Opticks* were a result of his book, the *Principles of Natural Philosophy,* though Newton would not acknowledge the fact. When Roger Cotes demonstrated that the whole was less than a part in Greene's squaring of the circle, Greene responded that, in his algebra of infinities and geometry of space, a part may be greater than the whole. He wondered if Cotes' refusal to enter into subsequent further controversy about that new algebra of infinities "Proceeded from His Exceeding Modesty . . . his Utter Averseness to Dispute . . . or from his Conviction that his

[2] Robert Greene (1678?-1730), educated at Clare Hall, Cambridge, B.A. 1699, M.A. 1703, D.D. 1728. Fellow and tutor of Clare Hall from 1703, he was the author of a number of works besides those described in the text, including a treatise on squaring the circle. Among its many (ignored) requests, his will asked that his articulated skeleton be hung in King's College library, that Clare Hall publish an edition of his writings, and that four monuments be erected in his honor, each carrying a self-composed inscription praising him and deploring the lack of contemporary recognition. If only for his personality, he seems to deserve the recognition he craved.

Demonstration was not Just . . . [*Expansive and Contractive,* 924-26]."

Fanaticism is not, however, the necessary equivalent of stupidity, and Greene was not stupid—he merely refused to adopt the criteria from which other people were arguing. His *Principles of Natural Philosophy* demonstrates the dialectical skill of a man well trained in scholastic disputation.[3] Its purpose is not to describe and defend his own philosophy, but to demolish the doctrine of atoms or corpuscular philosophy. Repeatedly he demonstrates that Newton's proofs for the existence of the void and the relation between mass and gravitation rest on the prior assumption of the ultimate sameness of all matter—which Greene denies [*e.g.,* 98, 115-16]. "There is no contradiction in saying that different Matter may have different proportions of Gravity analogous and correspondent to their several Natures." "Aptness of *Solution* can be no very powerful Argument for *Principles* that cannot be maintain'd on Reason [113]." Observations may disprove a theory, but cannot prove one, as "present theories tho' supported now by observations will in time be disavowed by them [71]." Experiments with falling bodies cannot prove a void, as they are not performed in a void [84]. If all matter did gravitate alike in appearance, it would not be an argument that it did so in reality [104]. Space, extension, absolute place and time, and solidity are abstractions of the mind only [53, 60, 113-23]. There is no vacuum, as the nonexistent cannot exist [48]. Heat is not motion, nor cold the privation of motion, as both are positive forces and a positive force cannot result from the negation of a positive force [166-70]. The fundamental error is the assumption that matter is passive in its own nature and ultimately homogeneous [84]. ". . . from the Motions observable in . . . [Material Substance] we evince, that all Nature is active, and, that Matter it self is so, we are ready to think, and are fully perswaded, that such a system is more agreeable to the Divine Mind, than any other that can be supos'd . . . [392]."

Greene's *Principles of the Philosophy of Expansive and Contractive Forces* continues his polemic against the corpuscular philosophy while attempting, this time, to define and apply his

[3] Robert Green[e], *The Principles of Natural Philosophy, In which is shewn the Insufficiency of the Present Systems, To give us any Just Account of that Science: And the Necessity there is of some New Principles, In order to furnish us with a True and Real Knowledge of Nature* (Cambridge: for Edm. Jeffery, and James Knapton and Benjamin Took, 1712).

own principles.[4] As these extend to explain all things, including metaphysics and logic, ethics and natural religion, mathematics and natural philosophy, the book is a tangle of rationalization and contention, in an enormous mass from which the sections relating to matter and its action must be extracted and ordered. His system of nature begins with dissimilar matter and a plenum [946]. In describing matter, all its properties, "without frivolous and idle distinctions" between so-called essential and accidental qualities must be considered [286]. These properties are not to be derived from modifications of homogeneous matter nor from figures and motions of corpuscles, but from certain innate, inherent expansive and contractive forces [1, 23], defined as similar, homogeneous, equable action from or to a center [410]. The existence of such forces in material beings is evident, as some have a greater disposition to motion, velocity, and expansion, and others to rest and the contrary of action or motion. "From the Various and Infinite Mixtures and Combinations of these Forces in different Proportions, all the several Forms and Species of material Beings seem to arise . . . [62]." Matter, in fact, is "nothing else but Action," and belief in the existence of solid substance to support this action and in which it might inhere proceeds only from custom and corpuscular prejudice [409]. This variety of expansive and contractive forces derives ultimately from God, but, for material being, the immediate source of expansive force is the sun, or fire, or heat, and of the contractive, the earth, moon and planets, salts, or cold. ". . . we may truly Affirm, that all the Various Portions of Matter are Compounded of Infinitely Various Degrees and Various Combinations, of Heat and Cold, Mixed and Tempered together [415]."

The doctrine, as thus extracted, has a deceptive air of rationality imposed by selection and sequential arrangement on what is otherwise a random maze. There are, in Greene's treatise, perceptive criticisms and intriguing proposals, but before seeing shadows of Kant in his dissolution of matter into interacting forces, before reading into his work a premature striving toward fields, or a primitive conceptualization of energeticism, it is well to consider his application of these principles. There Greene demonstrates that his fundamental aim is toward infinite diversity

[4] Robert Greene, *The Principles of the Philosophy of the Expansive and Contractive Forces or An Inquiry into the Principles of the Modern Philosophy: that is, into the Several Chief Rational Sciences, which are Extant* (Cambridge: Cornelius Crownfield, E. Jefferys, W. Thurlbourne, J. Knapton, R. Knaplock, W. and J. Innys, and B. Motte, 1727).

in quality, while science was developing toward unity in quantity. Here are his examples of force in action:

> . . . there are Light, Heat, Transparency, Fluidity, Levity, Elasticity; and in Relation to Colours Red, Orange, Yellow; to Tasts Sweet and Luscious; to Smells, Fragrant; to Sounds Harmonious and Soft which fall under the Expansive, in which yet there is some Mixture of the Contractive; whilst the Contrary Qualities, as Darkness from the Earth's Contraction of the Medium which the Sun Expands, Cold, Opacity, Solidity, Gravity, Unelastickness; and in Relation to Colours, Blue, Violet, and Black, (for Green Lies Intermediate to both the Orders); to Tasts Acid and Sower; to Smells, Foetid; to Sounds Harsh and Dull, seem to derive their Origin chiefly from the Contractive, tho' not without some Degree of the Expansive in their Composition [123].

Somehow, even when considering his forces in their quantity, Greene loses precision. Pairs of similar forces add as vectors, whatever their directions [100], but an expansive force opposing a superior contractive increases the contractive force, and a contractive against a superior expansive, increases the expansive, as fire increases the weight of lead and water grows colder from being boiled [415]. Applying his doctrines to mechanics, familiar equations are made impossible to solve by the introduction of "propensities." For his "New System of Mechanics," Greene must write thirteen different equations to relate the ratios of moments of two bodies to the ratios of their velocities, their expansive forces, their contractive forces, and the "sums and quantities of the forces" (*i.e.*, bulks)—*e.g.*, $M/m = V/v \cdot E/e \cdot C/c \cdot B/b$ or $M/m = V/v \cdot e/E \cdot C/c \cdot B/b$, etc.—and views this chaos with complacency as "a manifest Conviction of the wonderful extensiveness of the Doctrine [59]." Naturally Greene declares against inertia.

Reflection is the combination of the expansive force of light meeting a superior contractive force in solid bodies. The expansive Force mixt with Contractive in every solid Body assists the Recoil of the Expansive of light. Hence metals reflect better than wood, having a greater proportion of expansive force in them, and hence the angle of reflection cannot, in general, be affirmed equal to the angle of incidence [308]. To explain Hauksbee's electrical experiments, "which have been hitherto deservedly accounted very Surprising," it may be noted that glass, made of flint and sand fused in fire, has a great degree of Elastick or Expansive Force under the Restriction and Constraint of its great Contractive. The expansive, being excited by friction and exerting itself from the center, disturbs the equilibrium of the ambient medium, which is restored by the threads of yarn, exerting their contractive force from the center. "From hence it is Manifest that an Equilibrium of the Air is necessary to these Phaenomonons [310]." Vinegar dissolves lead and not quicksilver, aqua fortis dissolves quicksilver but not lead, because lead, having less expansive force, from its softness, ductility, and dull color and less contractive from its weight, required the less expansive and contractive vinegar to move it, while the more expansive and contractive aqua fortis is required to move quicksilver, with its greater expansive force, from its volatility and bright color and greater contractive from its weight [317]. And, as a final example combining all of the worst features of the "*GREENIAN*" philosophy:

> If the Moments of Bodies are Equal to a Direct Reason of the Expansive Force, to the Expansive, and of the Contractive, to the Contractive, i.e. if they are as the greatest Expansive Force, into the greatest Contractive, to the least Expansive Force, into the least Contractive, it will probably be the Case of Gold, which has the greatest Contractive by it's Weight, and the greatest Expansive by it's Colour, which comes the nearest to that of the Suns, which has an entirely Expansive Force . . . and of Black Ashes or Dirt, which has the least Contractive and Expansive, by it's Weight and it's Colour: And from this Theorem, several Years ago, I had an Opinion, that if to the Weight of Quicksilver or Mercury, there cou'd be communicated by Fire the Expansion of it, Gold might be made [60].

Any suspicion that Greene has merely renamed Newton's attractive and repulsive forces, perhaps dissolving the corpuscular cores into Boscovich-like points, vanishes in the profusion of his applications. Greene's forces turn into qualities, for only by differentiating all species and kinds of matter through all their primary and secondary qualities can one begin to discuss their conceivable proportions of expansive and contractive forces. Whatever his initial intentions, he has exploded the unity of mechanistic natural philosophy into an infinity of material substances. Greene was, after all, correct in his insistence that his philosophy is one of dissimilar matter, and, after the complexities of such a materialism, it is a relief to turn to the material simplicities of the Hutchinsonians, even though these simplicities are something other than those generally ascribed to them.

The name "Hutchinsonian" is given to the only school of eighteenth-century British anti-Newtonians and is taken from that of its founder, John Hutchinson. The school is generally described as holding that Newton's philosophy was atheistic at least in tendency, while a complete, true, and Christian system of natural philosophy could be derived from a study of the Old Testament, read in the "original Hebrew" without vowel points. Any study of Hutchinsonian literature will reveal, however, that this inadequately characterizes its members. John Hutchinson's theological animus against Newton, as against learned men in general, is obvious, but Samuel Pike and William Jones of Nayland, both described as Hutchinsonians, are properly respectful of Newton, and Jones, indeed, is admiring. Hutchinson "extracts" his system from an unpointed Hebrew Testament, but Pike appears to have obtained his from a pointed text, and Jones employs reason and experiment. What unites the Hutchinsonians is agreement on the same general system of natural philosophy, and, without entering into etymological distinctions, it may confidently be asserted that this system does not derive from the Old Testament. Its roots are not primitive and Semitic, but sophisticated and French. The hands may have been those of Esau, but the voice was that of Descartes.

The cosmological outlines of the system were established by John Hutchinson.[5] As he adopts the form of deriving everything from his reading of the Scriptures, it is not possible to say positively what books in natural philosophy he may have read. His search for the primitive, uncorrupted Mosaic text led him to follow "the *Latin interlineary* version, as the most literal and fittest to show the order of the Hebrew words [42]." The same "scholarship" in natural philosophy would, no doubt, have been

[5] John Hutchinson (1674-1737), educated at home to be a land steward. About 1700 he began to assist John Woodward in collecting fossils, presumed for a work to confirm the Mosaic deluge. Disturbed by Woodward's desultory work habits and unhappy with the condescension toward his efforts, Hutchinson commenced a course of study leading to his first tract, *Moses's Principia* (1724-27). Given a sinecure post to support his continued writing in defense of Scripture, he responded with a flood of other works, including *Moses's Sine Principia* (1729), *Treatise of Power* (1732), *Glory or Gravity* (1733), and *Glory Mechanical* (1738). I have used all those named, but only in [Bishop Horne, ed.?], *An Abstract from the Works of John Hutchinson, Esq.; being a Summary of his Discoveries in Philosophy and Divinity* (London: for E. Withers, 1755), 2nd edn. But see also an extended and favorable essay-review of early Hutchinson works, "A Letter to a Bishop concerning some Important Discoveries in Philosophy and Theology," first published in 1732 and included in *The Works of the Right Honourable Duncan Forbes, Late Lord President of the Court of Sessions in Scotland* (London: by J. Morton, for T. Hamilton, R., J., and M. Ogle, 1809).

content with secondary accounts and popularizations. Clearly he had no use for mathematics. "It is to no purpose to stun us with *mathematical principles of natural philosophy,* till the principles themselves are simply proved: for mathematics are applicable to any data, real or imaginary, true or false: they have nothing to do with the dispute, and ought to take the last place in science . . . [156]." Hutchinson may have read the *Opticks;* probably he read Pemberton's *View of Sir Isaac Newton's Philosophy,* for the *Treatise of Power* refers caustically to some events of Newton's life described by Pemberton. He refers to the "romances of Descartes and Kepler" and may have known both only through such a work as Pemberton's. He knew and particularly disliked the theology of Samuel Clarke, and it is tempting to suppose that he ventured into Clarke's edition of Rohault and adopted the Cartesianism of that work because Clarke opposed it.

Hutchinson's natural philosophy is not complete nor is it entirely consistent from one work to the next, as his ultimate concern was theological not scientific; but if one smooths some discrepancies there is a fairly complete cosmology. Chaos, created out of nothing, consisted of atoms, or indivisible particles, in a bounded, unadhering mass, absolutely full and inactive [46, 140-41]. Put into motion by God, some of these cohered in various combinations, of different forms and sizes, to form the "Earth" —solids and fluids inactive of themselves and moved only when moved by the remaining, active, substance, the "Heavens," which, in its triune state of fire, light, and spirit (sometimes also called air) became the "substance of the *names,* those mechanical representations of the *ALEIM* [2, 95, 140]." The "Heavens" began in darkness, or stagnant air, whose particles, put into motion, became spirit and, the motion continuing, were "ground smaller by the collision of its concerted parts amongst each other, and became light [11]." To establish the mechanism of the Heavens, whereby the universe operates, the particles of light were concentrated in the sun, which acts as the source of heat and light streaming outward to fill the universe to its circumference. As these particles get further from the sun, their agility decreases, they slow and cohere into grains of spirit, too large to penetrate the smaller pores of bodies, which thus becomes "an instrument of support; it at once bears up and impels. . . . Hence weight or pressure [54]." At the greatest distance possible, the boundary of the universe, the particles wholly congeal into gross spirit or air, which is driven, by the expansive pressure of light flowing outward, back to the center where, as "fire can neither subsist, nor send forth light, without fresh supplies of air [68]," the sun is replenished, the air "melted down" and sent out again as light in a perpetual circulatory motion [367-68].

> These perpetual fluxes or tides of matter outwards and inwards, in every point, from the centre to the circumference, mechanically and necessarily . . . produce that constant gyration in the earth and the planets round their own centres, and round the sun; and . . . the same principle, with some circumstance arising from the situation and fluxes of light coming from the other orbs, will account also for the motions of the moon. . . . the adverse motions of the light pushing towards the circumference, and the air [spirit] pushing towards the centre with immense force . . . binds together solids, keeps fluids as they were, causes the variation of times and seasons, the raising of water, the production of vegetables and animals, . . . in short, produces almost all the effects and phaenomena in nature [Forbes , 298].

"To say matter can act without means or contact . . . is to advance a doctrine more senseless . . . than transubstantiation [146-47]." Projectile motion is caused by the body moving a pillar of air before, and forming a vacancy behind into which the aether and air push to thrust the body forward [386]. The corpuscles of water are light, small, round, and smooth. Particles of earth or vegetable matter are heavy, flat, thin, or fibrous, fit to adhere and compose bodies [318-20]. Hutchinson suspects cold to consist of inactive, rough corpuscles, though he recognizes it may be only the absence of fire [320-23]. Fire is, at once, both the separation and agility of subtle, active, matter producing light and heat [56], and the subtle particles themselves, like small and sharp spikes or wedges moving irregularly and capable of dividing the corpuscles of solid bodies if impelled by some grosser fluid [375]. Light also seems a duality, being sometimes the smallest particles themselves, but mostly the motion of the aerial matter or heavens in straight lines outward from the sun. "The light, which presses through a hole, moves quickest in the centre, and weaklier near the sides, whereby it exhibits divers colours: this diverted Sir *Isaac* excessively [372]."

Having traced the outlines of the Hutchinsonian system in John Hutchinson himself, we can discuss more easily the characteristics of that system as it was developed by two of his followers, Samuel Pike and William Jones of Nayland. Both accept the distinction between the inert, moved, matter of the earth and the moving matter of the heavens. Both deny action-at-a-distance, inertial motion, and a vacuum. And both explain most phenomena by the efflux of aetherial substance, in the form of light, from the sun, filling all space, congealing with distance into aether, until, reaching the circumference of the universe, it circles back to the sun to be melted, or broken down, into light again to flood out in a perpetual circulation.

Samuel Pike is the less interesting of the two in his further development of the system.[6] He does not explicitly support Hutchinson, nor does he view "his" system as an attack on Newtonianism, which describes but does not explain such phenomena as gravitation, cohesion, magnetism, and electricity. "Surely, then, an attempt to explain the causes of these appearances from revelation in an intelligible and mechanical way, cannot be justly call'd an opposition to the philosophy of an age; but rather an agreement with it and an improvement upon it [viii]." Pike gives little attention to the mysticism of "the *names*," those triune, active, substances of the heavens. In a sense, he is the least materialistic and most mechanistic of the three Hutchinsonians considered here. All matter is extended, solid, figured, and entirely passive or dead. To give any part of matter special unexplained powers would be as bad as accepting gravity without explanation [133-47]. To him, the motion of aetherial matter defined only by size and shape is justified by Revelation, and the kinematic mechanisms which follow this need no further causes or explanations.

The loadstone is so framed as to prevent the free motion of the aether through it in the direction parallel to the poles. This produces an elastic imbalance in the aether that causes steel—a body "similar" to the loadstone—to move toward it [93]. The operation of the electrical machine is a microcosm of the grand system of nature. The rubbed globe is an imitation of the friction of the sun, grinding masses of spirit into light which flows from it, as spirit flows toward it. Small bodies caught in the afflux of spirit move to the globe, where they are filled with the electric light and move with it out again, until, "by degrees or by any other

[6] Samuel Pike (1717?-73), educated in dissenting academies and established another in London to train students for the dissenting ministry. Defended Hutchinsonian principles in his *Philosophia Sacra: or, the Principles of Natural Philosophy. Extracted from Divine Revelation* (London: for the author, and sold by J. Buckland, 1753), which he repudiated after 1757 (though it was republished in a second edition in 1815), upon adopting the theological principles of Robert Sandeman. Excluded from the Presbyterian connection after 1765, he became an elder and then minister of the Sandemanian sect.

accident," they lose their electricity and are driven in again [94-95]. The same principles are applied to explain cohesion [86-89] and elasticity [90], and might be used to explain fermentation, suction, the action of light and production of colors, nature of wind and sound, vegetation and animal life; but Pike forbears, as they would run him into too great a length [96-97].

William Jones was willing to go the length necessary to develop the system more completely.[7] In many ways the most interesting of the Hutchinsonians, he denies that he is a member of that school. Hutchinson had, it is true, already taken up the problem that Jones wanted to study—the reduction of phenomena "to one simple and universal law, the Natural Agency of the Elements"—but had done so in "a manner neither acceptable nor satisfactory . . . [ii-iii]." Yet Jones' conclusions and many of his attitudes were those of Hutchinson and Pike. Newton had left the existence of the void in suspense and the causes of attraction and repulsion unanswered. But true attraction is impossible, as motion must be in the direction of its cause, and, when a body moves, one must conclude it is driven by a flux of matter in that direction. If Newton did not believe gravity was caused by impulse of a subtle matter, why did he propose it "to supply the defects which had been objected to" in his former answer [ii, 42-43, viii]? Mathematicians incline to consider the elements only as quantity, whereas physics should consider them chiefly as qualities, for quality is the soul as quantity is the body of nature [xvi]. It is safer, on many occasions, to be guided by reason and the nature of things, at least in argumentation, than by diagrams "which are applicable to contradictions, and may, indeed, be accommodated to any thing [4-5]." Nature is a system of parts, and it answers to no purpose to consider any part of it, except in its connections and relations to the whole [30].

[7] William Jones (1726-1800), called Jones of Nayland, educated at Charterhouse and University College, Oxford; B.A. 1749. Ordained priest in 1751; after several minor posts, he accepted the perpetual curacy of Nayland, in Suffolk, which became his permanent residence and provided the identifying suffix to his name. One of the most prominent churchmen of his day, he links the non-Jurors and the later Oxford movement clergy of the established church. Friendly from college days with George Horne, afterwards Bishop of Norwich, both were students of Hutchinson's writings and wrote in defense of his ideas. Jones' *Essay on the first Principles of Natural Philosophy: wherein the Use of Natural Means, or second Causes in the Oeconomy of the Material World, is demonstrated from Reason, Experiments of various kinds, and the Testimony of Antiquity* (1762) is the more clearly Hutchinsonian of his two major works in natural philosophy. F.R.S. in 1775, his *Philosophical Disquisitions: or, Discourses on the Natural Philosophy of the Elements* (London: J. Rivington and Sons, G. Robinson, D. Prince, Mess. Merrils, W. Keymer, Mrs. Drummons, W. Watson, 1781)—the work described here—develops the same themes, but in a less scriptural fashion.

In his attitude toward matter, Jones is closer to Hutchinson (without the mysticism) than to Pike. The matter of the world is not homogeneous, some of it does not have weight, atoms differ by their figures, and many of the effects of bodies may naturally be derived from the configuration of their particles, as, for example, the spiculated are sharp and corrosive, the globular are insipid and balsamic [2, 20, 26]. In fact, though modern chemists often dispute it, the ancient doctrine of the four elements—earth, water, air, and fire—appears true, and, of these, two are passive and two active. Fire and air have the powers of motion, water and earth are capable only of receiving their impressions [65, 78-80]. The endless variety of nature arises from different combinations of these few principles [71], and the aetherial medium, which is a mixture of fire and air, is subject to all those different affections which produce gravity, magnetism, elasticity, cohesion, electricity, heat, and illumination. "How far these affections may arise from the different densities of the medium or the different magnitudes of its parts, should be considered [62]."

Perhaps because of the influence of Boerhaave, under whom Jones' Oxford chemistry professor, Nathan Alcock, had studied, Jones was most intrigued by the active element of fire. "Whenever an effect seems far superior to its cause, the element of fire is concerned in some shape or other [512-13]." It is commonly divided into three sorts, which nearly agree in properties and effects and can convert into one another but differ as to the places of their residence and nature of their motions. Solar fire resides in the sun, from which it moves in right lines as light. Culinary fire resides in fuel, can be kindled on earth by artificial means, and vibrates and tends naturally upwards. Elementary fire is the subtle fluid residing constantly in all gross bodies, not necessarily to be distinguished by its heat or light. When filling the pores of bodies it is called aether [81-83]. It has no weight, though through chemical operations it may be attached to parts of solid matter and increase the weight of the whole mass [21]. Heat is the motion of elementary fire and, since fire is everywhere, it can be asserted that motion produces heat, as it causes the compression and agitation of the matter of fire without which heat cannot result. Cold is the privation, not of the matter of fire, but of its motion [88, 178-81]. The natural motion of elementary fire is diffusion in all directions and it presses, with a shock, to restore an equilibrium.

Electricity appears to be a form of elementary fire. Like fire it seems to fill the world, to move in every direction indifferently, and to invigorate all other matter [21]. Like fire it is an active fluid which, nonetheless, can be present and be neither seen nor felt [48]. It restores equilibrium with a shock and, as Nollet has shown, has a simultaneous double motion to and from bodies, modeling that of fire and aether to and from the sun [85, 50]. It may be more agreeable to nature to class air with fire as of the same elementary nature. It is hard to determine where air ends and fire begins [67]. Air supports fire as a pablum, and Hales speaks of "aerial particles of fire," as if there were a substance common in some degree to both fluids [68]. It is true that Hales maintained that fire was no element but an affection of bodies, but his own experiments could have shown him a transformation of the substance of air into that of fire. "An experiment in nature, like a text in the Bible, is capable of different interpretations, according to the preconceptions of the interpreter." Had Hales not been so strongly "attached to the qualities of attraction and repulsion, by which he accounted for everything," he would have seen that decrease in elasticity of air in burning was a decrease in the quantity of air and an increase in that of fire [148-50]. And, finally, "light is the mediating substance between fire and air," and its spectrum shows the alliance, for the extremes are red, the color of fire, and blue, the color of air [220-21].

From 1724 to 1781 and beyond, variations on Hutchinsonian themes obviously provided speculative materialists with a pattern and rationale. These were not, however, the only directions that materialists could go—or perhaps one should say rather that there were materialists in mideighteenth-century Britain who were not Hutchinsonian. Thus, in 1741, Thomas Morgan published a *Physico-Theology* which shows an independent development of the stigmata of the religious, only semi-Newtonian, materialist.[8] Morgan divides matter between active and passive substances, denies the void and inertial motion. His active substance is the "visive element" or light, which fills all of space and in which motion is excited to produce a sensation of lumination [37-38]. This motion is that of the material particles out from the sun,

[8] Thomas Morgan (?-1743), educated by a dissenting minister. Ordained a Presbyterian minister in 1716, he was dismissed from his ministry after 1720 for his freethinking. He studied medicine and appears to have practiced as a provincial physician while, as a "Christian Deist," he wrote controversial theological and physiological works, in some of which he calls himself an M.D. See particularly, Thomas Morgan, *Physico-Theology: or, A Philosophico-Moral Disquisition concerning Human Nature, Free Agency, Moral Government, and Divine Providence* (London: T. Cox, 1741).

but the quantity of elementary light remains constant, everywhere at the same distance from the sun, because the "luminous Rays are in a continual vibrating Motion, going and returning to and from the resisting Medium, in exceeding short and imperceptible Intervals. . . . Any one but moderately acquainted with the *Newtonian* Theory of Light, must see the Reason and Necessity of what I have . . . advanced . . . [34-35]." It is this material substance of light, "not endued with . . . any mechanical Power or Property whatever" which mediates between inert, resisting, matter and the continued regular Will of God, acting upon bodies and determining their mechanical powers by a means Morgan will not "pretend to explain [v, 56-57]."

Within the materialist context, this is reasonably commonplace and would hardly earn a separate reference were it not that, some fifteen years earlier, Morgan had written a physiological treatise on the *Philosophical Principles of Medicine*, which gives no indication of an obsession with the active matter of light.[9] Indeed, this earlier book might be a journeyman's version of Pitcairne or James Keill. The first hundred pages are devoted almost entirely to elementary mechanics and, though there are a few later references to elementary light or fire, these are no more prominent than they are in his contemporary sources. There is, perhaps, an excess of clinical, empirical, protestation, but the bulk of the work, as its complete title attests, applies the concepts of attraction and repulsion, the texture, force, and velocity of the blood, and all the other indicators of a Newtonian mechanistic physiology.

Sometime between 1725 and 1741, Morgan changed his philosophical outlook. The causes of his change are as obscure as are those for the contemporary changes in other people, but one must, at least, recognize that a change was also occurring in Newtonian interpretations. The materialist reaction against Newton was not entirely a result of immediate philosophical or theological revulsion. Some convinced mechanists came to reject the explication of nature in terms of phenomenologically neutral matter and its forces. For some reason, they came to prefer the concept of inhomogeneous matter, variously defined by active properties. Whatever the reason was, it prompted critical reappraisals of New-

[9] Tho[mas] Morgan, *Philosophical Principles of Medicine . . . Containing . . . A Demonstration of the general Laws of Gravity . . . The more particular Laws which obtain in the Motion and Secretion of the vital Fluids . . . The Primary and chief Intentions of Medicine . . . mechanically resolv'd* (London: J. Osborne, T. Longman and J. Batley, F. Clay, E. Symon, S. Billingsley, and S. Chandler, 1725).

tonian dynamic corpuscularity; and its promptings acted on both sides of the Atlantic.

In 1746 Cadwallader Colden, sometime physician, Surveyor-General and member of the Governor's Council of the Colony of New York, and one of the few colonial speculators on matters of natural philosophy, published in New York *An Explication of the First Causes of Action in Matter; and of the Causes of Gravitation.*[10] Colden's *Explication* was republished in London in 1746, translated into German in 1748 and into French in 1751. By that time, he had expanded it into the *Principles of Action in Matter, the Gravitation of Bodies and the Motion of the Planets explained from those Principles* (London, 1751). For a time these works stirred a minor response. They were reviewed and extracted in the popular magazines, they elicited a discussion between Colden and an unnamed correspondent in the 1759 *Monthly Review*, the *Principles* even excited Leonhard Euler into a denunciatory letter to the Royal Society. Perhaps it was the novelty of a colonial philosopher, maybe it was the semblance of mathematical form to Colden's argument, which momentarily bemused the Europeans. In any event, the flurry soon died. Colden's concepts differ only in detail from those of other materialists attempting to find causes for Newtonian attraction and repulsion in the innate activity of substance.

In Colden's case, however, all matter was presumed active, matter and activity being divided into three varieties: one, represented by light, had the power of self-movement; another (the aether) had the power of transmitting movement; and the third (such as earth) the power of resisting movement, or *vis inertia.* Typical of Colden's reasons for assuming an active principle in matter is the example he gives of a spark setting an entire city ablaze. Surely the spark, by itself, does not possess the power

[10] Cadwallader Colden (1688-1776), educated at the University of Edinburgh, A.B. 1704, M.A. 1705. Destined for the church, he turned to medical study in London and, in 1710, went to Philadelphia, where he practiced medicine. He visited Scotland in 1715 to marry, but returned to the colonies, where he entered government service in New York, and was Lieutenant-Governor at his death. Correspondent of Linnaeus, Gronovius, Franklin, and many American and Scottish physicians and scholars. Author of a popular *History* of the Five Indian Nations, of several papers on medicine and on botanical observation published in the *Acta* of the Royal Scientific Society of Upsala (1743, 1744-50), and of speculations on philosophy, mathematics, and natural philosophy. My comments on Colden are largely derived from the unpublished Master's research essay, "Dr. Cadwallader Colden, Iatro-Mechanism to Vitalism in 18th Century Natural Philosophy," by Michael Massouh, of the History of Science and Technology Program, Case Western Reserve University, 1967. See also Brooke Hindle, "Cadwallader Colden's Extension of the Newtonian Principles," *William and Mary Quarterly,* 12 (3rd. ser., 1956), 459-75.

represented by this result; instead, the activity of the spark releases the active principle inherent in combustible matter. The motion of a planet requires two forces to maintain itself in its orbit. The outward pressure of light from the sun repels the planet to near-rest at aphelion, where, light pressure being less than restoring aether pressure, the planet is returned by aether impulse to perihelion, with increasing velocity.

The similarities between Colden and his contemporary English materialists are greater than the differences, and Pike and Jones acknowledge the affinities by favorable comments in their work. Pike has a long and, on the whole, approving commentary in his *Philosophia Sacra* on Colden's "demonstration" that aetherial pressure could produce an inverse-square gravitational "attraction," though he finds three unexplained activities worse than a single one [133-47]. Jones cites Colden's spark example with approval in his *Disquisitions* [45-46], though, of course, he prefers his own conceptualization of material activity. Yet it seems clear that Colden had not read the work of Hutchinson, and, indeed, it appears that his only preparation for these ideas was a reading of the *Principia* and the *Opticks*. His far-from-unique misreading of these texts has a particular interest, in addition to its colonial setting, for his correspondence and his subsequent unpublished writings give some background for a motivation to Colden's work. As a student at Edinburgh, he was taught to admire Newton; either at Edinburgh or in London he had learned Newtonian physiology. In America he had used those principles in explaining the causes and cures of fevers, believing the "Mechanic system" gave the only satisfactory account of the "Animal Oeconomy." In 1717 he wrote in praise of the work of John Freind; in 1720 he was confident that proportions in geometry and laws of motion might be applied to physic with as great a certainty and as much advantage as to astronomy. By 1745 he was disenchanted. The "Mechanic system was very defective . . . very little depend on Mechanical principles . . . even the first and principal mover in the Circulation is in no way Mechanical," he wrote to Dr. John Mitchell of Virginia. Here too the answer was an active principle (or principles) in matter.

Colden found an example of active principles in physiology in the circumstance of smallpox inoculation. Here the active principle in the small amount of matter, put in a scratch, lies dormant for six or seven days and then releases the activity of other parts of the body into the eruption of pustules in the inoculated person. Another instance is in the circulation of the blood, where the blood loses velocity as it passes through the arteries, but acquires new velocity in the veins returning to the heart from the active principle of chyle extracted from food. Surely the similarities of examples of spark and pustulant matter, of planetary motion and blood circulation are not coincidences. Massouh argues that Colden's primary interests were physiological and that his principles had their origin in physiological speculations. His fundamental mistake, so far as his reputation was concerned, was applying these principles to astronomy, in an inversion of usual iatro-mechanical thinking; had he published in a more physiological form, his ideas would have been generally acceptable to an age moving toward vitalism.

This view raises some interesting questions. Certainly it is true that Colden's unpublished papers contain several examples of the application of vital principles to the animal oeconomy and to vital motion. It is equally true that physiologists were freeing themselves from Newtonian reductionism, only to take refuge in vitalism. Perhaps they would have found Colden's active principles more to their taste, though, as has been seen, active principles were by no means foreign to the scene of the physical sciences. The more intriguing prospect is that which relates the origin of these ideas to medicine and medical men, for this may have more than individual significance. Physicians had carried much of the burden of extending dynamic corpuscularity into untried fields. Physicians were, for the next half-century, to carry much of the burden in Britain of developing a materialistic experimental natural philosophy. One of the earliest and most successful ventures in this direction was the elaboration of a matter theory of heat, carried out in Scotland and by physicians. Under the circumstances, it seems fitting to end this chapter with reference to a Scottish source in which a connection between this material theory of heat and the speculative materialists, indeed the Hutchinsonians, is explicitly drawn. In volume twelve of the third edition of the *Encyclopaedia Britannica* (1797), the article "Motion" is written by the editor, John Gleig, and James Tytler. This article is a frank avowal of the principles of William Jones, coupling lengthy excerpts from Jones' *Essay* of 1762 with pointed references to contradictions in Newton and the Newtonians on the subject of the aether and attractive and repulsive forces. Insofar as the authors reach a decision, it is that the aether and elementary fire are equivalent and are probably responsible for the motions of the planets.[11] These conclusions find support elsewhere in *Britannica* articles by Tytler on "Electricity" and on "Fire." Both articles adopt Jones' view that the fluid of electricity and that of elementary fire are the same, while that on "Fire" explicitly rejects the views of those "celebrated philosophers of the last century," Bacon, Boyle, and Newton, in favor of that "strenuously asserted" by the chemists and "particularly maintained" by Boerhaave, that fire is a distinct fluid. According to Arthur Hughes, the "Motion" article, at least, was reproduced quite unchanged in the subsequent four editions of the *Britannica* and made its last appearance in the seventh edition of 1830-1842.[12] But the primary interest here is not so much this almost-authoritative continuation of Hutchinsonian ideas into the nineteenth century as it is the assumption of complementarity between the naïvely materialistic speculations of William Jones of Nayland and the materialism of experimenters in electricity and heat. By 1797 the study of both appears to have gone beyond Tytler's understanding of it, but he grasps the materialist fundamentals of both and he links the development of heat theory to the activity of the "chemists," Cullen and Black, each a physician trained in the Leyden-Edinburgh tradition, and to the pivotal influence of Boerhaave. It now seems time, therefore, to add to the two impulses toward a scientific materialism already discussed —the new emphasis on Newton's aether and the continued tradition of naïvely materialistic speculation—a third element, the influence of Continental, especially Dutch, science.

11 *Encyclopaedia Britannica; or, A Dictionary of Arts, Sciences, and Miscellaneous Literature, etc.* (Edinburgh: A. Bell and C. Macfarquhar, 1797), 3rd edn. Colin Macfarquhar was the chief editor from "A" to "Mysteries," in volume 12, when he died and the Rev. Dr. Gleig succeeded him.

12 Arthur Hughes, "Science in English Encyclopaedias, 1704-1875—I," *Annals of Science*, 7 (1951), 340-70, esp. p. 368. See also Hughes' second article in this series, "II. Theories of the Elementary Composition of Matter," *Annals of Science*, 8 (1952), 323-67.

CHAPTER SEVEN

Early Continental Interactions

FOR THE GREATER PART of the eighteenth century British science was effectively insulated from the best work of Continental scientists. Early in the century the Newtonianism of Britain stood in opposition to the continuing influence of Descartes and the lively objections of Leibniz and his followers across the channel. The result was the kind of negative influence already seen in the pugnacious anti-Cartesianism of most early Newtonian texts, in Samuel Clarke's use of Rohault as a foil for introducing Newton's ideas, in Clarke's and Keill's and Freind's arguments with Leibniz, and in Desaguliers' experiments before the Royal Society defending Newton against his Continental critics. With the general acceptance of Newtonianism on the Continent, sometime during the early 1730s, the situation changed, but only slightly, for the acceptance was usually for the wrong reasons and specifically it was in the wrong sense and for the wrong book to have much interaction in Britain.

Popular Continental Newtonianism was, in the hands of the *philosophes*, a political weapon to be used, like anti-clericalism, to belabor the conservatism of officialdom. This was a view of Newton which the British rarely understood. Nor, it develops, did they much understand the view of Newton that emerged in the work of the most distinguished creative scientists of Europe. Using the analytical formulation of the calculus, introduced by Leibniz, in contrast to the geometrical fluxion methods favored by the Newtonian British, Continental scientists such as Clairaut, d'Alembert, Legendre, and Laplace in France and the Bernouillis and Euler in Switzerland (and also Germany and Russia) developed the theorems of the *Principia* into an analytical mechanics which, in spirit, was curiously like the Cartesianism they had discarded. For use of the theorems of the *Principia* did not mean the adoption of the natural philosophy which lay behind it, as Euler's persistent rejection of gravitational attraction, interparticulate forces, and the materiality of light will testify. Meanwhile, the Newtonian emphasis in Britain had transferred to the *Opticks*; Newton had solved the problems posed in the *Principia*, the continuing scientific inspiration lay in the optical queries.

The gulf separating British and Continental experimental and observational sciences was less broad than that for the theoretical sciences. British books and papers in the *Philosophical Transactions* show continuing evidence of having drawn on astronomical observations, mineralogical, medical, even electrical and chemical studies as well as natural history reports from various European sources. Yet even here there were some difficulties and reservations. Much of this work retained a non-Newtonian, if not anti-Newtonian, flavor which delayed effective interaction until after the British reaction against mechanism made them more palatable. The appearance, in British natural philosophy, of the animism of Stahl and von Haller in physiology, the experimental positivism of Stahl and Macquer in chemistry, and the neo-Cartesianism of Nollet in electricity occurs after the rise of materialism. Study of the influence of this work will, therefore, be postponed until the discussions of those areas particularly represented. It is not in these men, however, that one can find the Continental influences in natural philosophy which help to induce that new materialism. For these influences one must look to the works early and freely accepted by British natural philosophers as representing broadly the same kind of Newtonianism which they had adopted. One must look to the earliest enthusiastic Continental Newtonian scientists, the Dutch.

For more than a century before the publication of the *Principia,* an unusually close relationship existed between Britain and Holland, which was hardly interrupted even by the trade wars of the mid-seventeenth century. The installation of a Dutch king on the British thrones in 1688 marked only an increase in the exchange of ideas, political, religious, cultural, and scientific, between the two countries. Descartes corresponded from Holland with some of the Cambridge neo-Platonists; through Henry Oldenberg, Boyle corresponded with Spinoza. Huygens' father was an ambassador to England; Huygens himself visited England, was a fellow of the Royal Society, and had learned of Newton's work from the time of the first optical papers of the 1670s. John Locke lived for a time in Holland as a political refugee; he corresponded with Huygens about the *Principia.* Clearly Dutch scientists were early aware of British scientific ideas, including the Newtonian. Moreover, there appears to have been a particularly active Dutch interest in Baconian natural philosophy. During a period when the only publication of Bacon's philosophical works in Britain were a *Sylva Sylvarum* in 1685 and an *Opera Omnia* in 1730, there were four Amsterdam editions of the *Opera Omnia* and separate editions of the *Sapientia Veterum,* the *Historia*

Ventorum, the *Novum Organum*, and *de Augmentis Scientiarum*.[1]

The Dutch universities were no quicker in adopting Newtonianism than their British counterparts, and, when it was accepted, it had from the first a stronger empirical flavor than that found early in Britain. This must, to some extent, be credited to the influence of Bacon, but the prevailing scientific philosophy to be supplanted was that of Descartes. Dutch universities had been the earliest to teach Cartesianism, and the influence of Descartes in Holland can easily be seen in the work of Spinoza and of Huygens. There remained, however, a strong conservative element in the universities against which liberal Cartesianism was combined with a corpuscular empiricism derived from Boyle. One may note, for example, that Boerhaave's most influential teachers at Leyden, Wolferd Senguerd, professor of philosophy, and Burchard de Volder, professor of mathematics, were eclectics attempting to combine Boyleian experimentalism with Cartesian mechanistic rationalism.[2] That strong element of empiricism noted in Pierre Brunet's studies of Dutch Newtonians and their introduction of experimental methods into France must, in large measure, be credited to the pre-Newton influence in Holland of Bacon and Boyle.[3]

Archibald Pitcairne's one-year stay (1692-93) at Leyden as professor of medicine may have helped introduce there a Newtonian physiology, though, as we have seen, Pitcairne's mechanism was more Italian and anti-Cartesian than positively Newtonian. Boerhaave's favorable reference to Newton in his *de Usu Ratiocini Mechanici in Medicina* of 1702 may ultimately derive from his attendance at Pitcairne's lectures, but Boerhaave's, and Leyden's, enthusiastic approval of Newton's work does not appear until 1715. Henry Pemberton was later to claim that the new emphasis on Newton, represented in Boerhaave's 1715 oration, *de Comparando certo in Physicis,* was a result of his having called Boerhaave's attention to the *Opticks* during his student term at Leyden in 1714-15. Another, and more clearly marked, channel of influence was that of Willem Jacob 'sGravesande, who visited England in 1715, as secretary to the Dutch ambassador, met Newton, attended Desaguliers' lectures, and returned to Holland a convinced experimental Newtonian, to become professor of

1 Gibson, *Bacon Bibliography, passim.*

2 Scott, "Sources of Boerhaave's Medical Lectures," pp. 5, 9-15, and *passim.*

3 Pierre Brunet, *Les Physiciens Hollandais et la Méthode Expérimentale en France au XVIII Siècle* (Paris: Librairie Scientifique Albert Blanchard, 1926) and *L'Introduction des Théories de Newton en France au XVIII Siècle.* I: *Avant 1738* (Paris: Librairie Scientifique Albert Blanchard, 1931).

mathematics and astronomy at Leyden in 1717. In either case, the Newtonianism which appealed was that of the *Opticks* and that Newtonianism was superposed on a Boyleian empiricism and a Baconian experimental rationale.

Whatever the origins of Dutch Newtonianism and the causes of its particular emphasis, it was soon reflected back into Britain. Under the guidance of Boerhaave, the medical school at Leyden became one of the most popular universities for English and particularly Scottish students. Only after Boerhaave's death in 1738, and not significantly until the second half-century, were the Scottish Universities of Glasgow and particularly Edinburgh to replace Leyden as the dominant medical schools, and, by then, the majority of the science faculties had either studied at Leyden or under someone who had studied there. The reputation of the school and teachers extended to their books: to the medical, botanical, and chemical texts of Boerhaave, to the natural philosophy text of 'sGravesande, and even to that of Musschenbroek, professor of philosophy and mathematics at Utrecht, all of which were translated and widely used in eighteenth-century Britain. So great was Boerhaave's reputation that as late as 1781, more than forty years after his death, this "wisest of chemists" was still the standard authority in Oxford chemistry classes.

To this triumvirate of Dutch Newtonian authorities must be added an earlier, and decidedly anomalous, figure, Bernardus Nieuwentyt, whose "major" work, *The Religious Philosopher,* enjoyed an unaccountable success in Britain.[4] It is hard to discover the origins of Nieuwentyt's Newtonianism—if, indeed, he ever was Newtonian. His early biographers speak of his ardent Cartesianism and decry the obscurantism of his mathematical works. More recently his *Fundamentals of Certitude or the right Method of Mathematicians,* . . . (Amsterdam, 1720, in Dutch) has been described as presenting "a surprising degree of similarity to those theories of science which have developed during the past fifty years and, in particular, to the doctrine of neopositivism or logical empiricism." All agree that his "on the Right Use of Con-

[4] Bernardus (Bernard) Nieuwentyt (1654-1718), son of an Evangelical minister of Westgraafdyck who educated him for the ministry. He was, however, allowed to choose a different career, and educated himself in science following the principles of Descartes. Phlegmatic and unambitious, he served his community as counsellor and burgomaster while he wrote and studied on scientific topics. Author of four works in mathematics questioning the validity of the infinitesimal in the calculus, he engaged in controversy with Leibniz and Jean Bernouilli. His most famous work, discussed in the text above, was completed and published in Dutch the year before his death. See [Michaud], *Biographie Universelle,* xxx, and E. W. Beth, "Nieuwentijt's Signifiance for the Philosophy of Science," *Synthese,* 9 (1955), 447-64.

templating the Works of the Creator," translated into English with a primary title, *The Religious Philosopher*, shows a wide knowledge of natural phenomena—astronomical, physiological, biological, and physical—but that it is dry and prolix. It is also pompous and occasionally even silly, but it had an extraordinary vogue and surprising influence. Translated into English in 1718-19, with a prefatory letter by J. T. Desaguliers, it reached a fifth edition by 1745.

Desaguliers called Nieuwentyt the Dutch Ray or Derham, and the name indicates the nature of his book, if the title alone does not.[5] In three wearisome octavo volumes, the author rings all the changes on the classical argument-from-design using his own experimental observations and those of others with a surpassing naïveté. A typical argument is the declaration that the magnitudes and distances of the stars have deliberately been made unmeasurable in order that men be forced to confess the boundless power of the Creator and live in continual astonishment at something which exceeds their learning [817]. The natural-philosophy frame within which design is demonstrated is a mechanistic one, for which Nieuwentyt drew on Descartes, Boyle (who ought never to be named but with respect [419]), Newton, the great mathematician—though it is the *Opticks* and particularly its queries which are most cited—and John Keill, whose argument on the expansion of the smallest amount of matter to fill the largest space is discussed approvingly [348]. Matter exists in the world in an infinite number of particles, each moved according to particular laws [1,003]. Bodies dispose themselves by percussion, attraction, "and some are wont to add consequentially repulsion." Attraction and repulsion are not to be denied because the manner of their acting is not comprehended, for such an argument would lead also to the denial of many other things demonstrated by experience [886-88]. But he explains the elasticity of air by "an inherent Elastical Power," like the steel springs of watches, and expects that it can, therefore, be destroyed by long continued bending or excessive expansion or contraction [388-99].

Like Derham, Nieuwentyt's mechanism is frequently and obvi-

[5] Bernardus Nieuwentyt, *The Religious Philosopher: Or, the Right Use of Contemplating the Works of the Creator: I. In the wonderful Structure of Animal Bodies, and in particular, Man. II. In the no less wonderful and wise Formation of the Elements and their various Effects upon Animal and Vegetable Bodies. And, III. In the most amazing Structure of the Heavens, with all its Furniture. Designed for the Conviction of Atheists and Infidels* (London: for J. Senex and W. Taylor, 1719), 2nd edn.

ously inconsistent, but most of these inconsistencies were unimportant, for it was the theology and not the science which attracted his readers. On one subject, however, his nonmechanistic explanations were to have a scientific impact. Nieuwentyt's defense of the materiality of fire was the first detailed support of that view presented in the eighteenth century, and he was cited as an authority for the remainder of the century. Heat, he declares, is to be attributed to adhering particles of fire and not, as some philosophers think, to the swift motion of the small and fine parts of all bodies [475]. Were heat motion, then all particles of matter, of whatever nature, could be turned into fire, if they were moved swiftly enough or divided small enough. But all matter will not burn, nor do all substances lose solidity as they become hot. Cooling is achieved by the removal of fire particles, as, for example, by blowing on hot substances, which also increases the motion of particles and would, therefore, increase rather than decrease the heat were heat motion [486].

Nieuwentyt will not assert that earth, air, water, and fire are "the only Principle or Foundation of all things," yet it cannot be denied that all of these enter into the composition of many natural things [591]. All corporeal things, not excepting air and fire, have gravity or weight [571]. Fire is a particular fluid matter, consisting of particles "very Elastical and Expansive [478-79]." Like air and water, fire acts as a menstruum dividing and separating bodies or, as Boyle has demonstrated, combining and fixing itself in bodies, increasing their weight while adding something to their composition [475, 603-606]. Light either consists in fire or brings fire along with it [592]. It is probably the effluvia of fire [854], and its particulate nature (as demonstrated by Newton) and its ability to become, in phosphorus, a solid body are further arguments for the materiality of fire [769]. It cannot be "perfectly assured, that Light does not circulate like the Blood in Animals, and having performed its Course, comes back to the Sun again," as Descartes seems to have thought [777]!

Here is materialism with a vengeance. Many of Nieuwentyt's British contemporaries were vague about the relation between heat and fire. Having been confused by Boyle's experiments, Derham, Clarke, even Freind, admit a substantial fire while talking of the motion of particles which constitutes heat. Nieuwentyt is not vague. He recognizes the arguments for a motion theory of heat and explicitly rejects them. Heat is caused by the possession of a substance with weight, elasticity, and chemical properties; cooling is caused by removal of that substance. Already, in his discussion, one can see the confusion between temperature and heat [593-606], or intensity and quantity, which was to bedevil the materialist until Black resolved the problem. But that confusion was less disturbing to Nieuwentyt's contemporaries than the inadequacies of a motion theory. Nieuwentyt's avowal of a matter theory of heat was the beginning of that theory as a serious consideration in eighteenth-century natural philosophy. Musschenbroek was particularly to praise this view of Nieuwentyt in describing the *Religious Philosopher* as an "excellent work" in the preface of his *Essai de Physique* of 1739, but Musschenbroek was only the last of the three Dutch natural philosophers, most influential in Britain, to adopt the theory. Both 'sGravesande and Boerhaave had already declared themselves heat materialists—and partly because of the influence of Nieuwentyt's reasoning.

To write of 'sGravesande, Boerhaave, or Musschenbroek in anything but mechanistic and Newtonian terms is a departure from the way these men saw themselves, from the way most of their contemporaries saw them, and from the way that they have generally been described by students of their work. It must, therefore, be emphasized from the outset that the general nature of their work was in confirmation of Newtonianism and that, in occasionally departing from the *opinion* of Newton, as they agreed that they did, they felt the departure was required by their following the *methods* of Newton.[6] It is just this conviction of the essential Newtonianism of their work, confirmed by experiments, which made their departures from Newtonian mechanism so convincing to their readers.

Consider first the texts in natural philosophy of 'sGravesande and Musschenbroek. Separated in their appearance in English by nearly a quarter of a century, by the appearance of the chemistry text of Boerhaave and the natural philosophy text of Desaguliers, these two works show much the same progress toward an easy, working acceptance of Newtonian mechanism seen in Britain between the texts of Keill or Clarke and those of Desaguliers or Rowning—except in their treatment of the nature of heat. 'sGravesande's *Physices elementa mathematica, experimentis confirmata: sive Introductio ad Philosophiam Newtonianam* was first published in 1720-21; before the year was over it had been

[6] See, for example, the preface to the second edition of 'sGravesande's *Mathematical Elements of Natural Philosophy,* quoted by Cohen, *Franklin and Newton,* p. 238, where 'sGravesande claims to be a true Newtonian philosopher because he pursues the method though disagreeing with the opinion of Newton.

translated in two different English editions, one (falsely) under the name of John Keill and the other by 'sGravesande's good friend Desaguliers.[7] By 1737 a fifth edition of Desaguliers' translation had appeared, and a sixth, posthumous, edition revised by Desaguliers' son was published in 1747. Clearly the work was influential and, equally clearly, the bulk of that influence was Newtonian—as it all was intended to be. Preceding Desaguliers' text by some 12 years, this was the first general Newtonian text in natural philosophy which made an explicit effort to state Newton's principles in non-mathematical terms and then present an elaborate series of experimental demonstrations by which those principles might be thought to be confirmed.[8]

The *Elementa Physicae, conscripta in usus Academicos* (1734) of Pieter van Musschenbroek was considerably less important as a general text in Britain.[9] Translated as *The Elements of Natural Philosophy* in 1744, it appeared in only one edition, for there were many competing texts by that time. As Musschenbroek was a distinguished experimentalist as well as teacher and had published papers on electricity and magnetism in the *Philosophical Transactions* and the *Memoirs* of the Académie Royale

[7] Willem Jacob 'sGravesande (1688-1742), educated in law at Leyden; LL.D. 1707. He maintained a continuing interest in science, published, at nineteen, an essay on perspective, and was one of the founders of the *Journal littéraire* (which became ultimately the *Journal de la République des lettres*) for which he wrote essays and reviews on scientific subjects. Returned from a visit to England in 1715 a confirmed Newtonian, to become professor at Leyden, where he remained till his death, in spite of invitations from Peter the Great and Frederick the Great. Ultimately taught civil and military architecture, logic, metaphysics, and ethics, as well as natural philosophy, in which, with Musschenbroek, he may claim to have introduced experimental physics into Holland. See [Michaud], *Biographie Universelle*, XVII. His *Physices elementa mathematica* was reprinted twice in Latin and translated into French (1746), as well as appearing in six English editions and was condensed as *Philosophiae Newtonianae Institutiones, in usus academicos*, 1st edn. (Leyden, 1723), English translation, London, 1735.

[8] Willem Jacob 'sGravesande, *Mathematical Elements of Natural Philosophy Confirm'd by Experiments; or, an Introduction to Sir Isaac Newton's Philosophy* (London: J. Senex, W. and J. Innys, and J. Osborn and T. Longman, 1726), 3rd edn., 2 vols.

[9] Pieter van Musschenbroek (1692-1761), educated in medicine at Leyden, under Boerhaave and 'sGravesande; M.D. 1718, with a dissertation which reveals his taste and talents for experimental physics. Practiced medicine for four years, but in 1719 was named professor of philosophy and mathematics at the University of Duisbourg sur la Rhin. Called to Utrecht in 1723 as professor of philosophy and mathematics, he remained there until 1739, when he moved to Leyden, where he was a colleague of 'sGravesande until the latter's death in 1742. In spite of invitations from Berlin, St. Petersburg, Goettingen, Copenhagen, and Madrid, he remained at Leyden until his death. See [Michaud], *Biographie Universelle*, XXIX. His most popular text appeared in English as *The Elements of Natural Philosophy. Chiefly intended for the Use of Students in Universities*, John Colson, tr. (London: J. Nourse, 1744), 2 vols.

des Sciences, as well as a book on magnets and another describing the experiments of the Accademia del Cimento, his text might reasonably attract the attention of other experimentalists who would never read that of 'sGravesande.

Both texts are primarily concerned with the macroscopic motions of rigid bodies, fluid mechanics, and the usual treatment of simple machines. Each has some preliminary discussion of the nature of matter in general and each an extended treatment of fire and light, in which 'sGravesande includes electricity, while Musschenbroek, appropriate to the date of his work and his own particular interests, has separate sections on electricity and magnetism. In their discussions of the nature of matter, they each begin by saying much the same thing, but 'sGravesande is more cautious than Musschenbroek. To the former, "Physics does not meddle with the first Formation of things," it is an impiety to suggest that man can deduce how the world must be framed, we do not even know the essence of matter [I, ix-xi]. To Musschenbroek, these things, if not yet known, are knowable and are recommended to the studies of posterity [e.g., I, 202-204]. Neither man will declare all matter ultimately homogeneous, but 'sGravesande is fascinated with Keill's supposition of the infinite divisibility of matter [I, 10], while Musschenbroek declares that the divisibility of matter can be nothing but the overcoming of cohesion between parts till one reaches the ultimate, physically indivisible, solids [I, 21]. Whether the "derivative corpuscles" of bodies, representing their infinite variety, result from the sameness of these ultimate solids or some differences, he will not decide, it being possible by suitable combinations to produce the same results from either hypothesis [I, 32-34]. Musschenbroek also defines the necessary attributes of all bodies: extension, impenetrability, inactivity, inertia, mobility, shape, gravity, and cohesion [I, 10].

Both agree that the phenomena of attraction exist in gravitation and cohesion. 'sGravesande repeats the Newtonian cautions about causation and, like Newton in the *Principia,* which is his major guide, says that apparent attractive forces may really be the result of impulse [I, 12, 18, 207]. He adds, however, that if gravity be the result of the stroke of any sort of subtle matter, it must act according to laws different from any known to us [I, 215-16]. Musschenbroek is more positive. Repeatedly he insists that the principle of attraction is true. He had formerly thought it a fiction, but a multitude of experiments have convinced him otherwise [I, vi.]. Now he declares that a subtle aetherial matter is a chimera to which anyone, by the strength of his imagination, assigns properties at pleasure [I, 89-90]. In those instances where the assumption of an internal active principle of attraction permits the mathematical deduction of properties to flow, he will fix on such a principle [I, 106-107]. Both also agree that repulsion exists and give the same examples, derived from the *Opticks*, in the dissolution of salts, decomposition of acid and alkali mixtures, etc. 'sGravesande declares the elasticity of air is caused by repulsion between particles which do not touch, the force being inversely proportional to the distance between particle centers [I, 216], though again he denies that he is speaking of causes. Musschenbroek is less detailed about the repulsive forces of elasticity, but more positive that they are not the result of impulse or of the elasticity of some subtle pervasive aether. This supposition is an hypothesis grafted onto an hypothesis, and moreover, leaves unanswered the question of elasticity unless one assumes some still more subtle and elastic substance permeating the pores of the aether, etc. [I, 175]. Both agree that where attraction ends, there repulsion ensues; but Musschenbroek makes this into a "constant and universal law of nature; *when the parts of bodies go out of the sphere of attraction, then they repel one another with immense force* . . .[I, 319]." And finally, as we may expect of the honest Newtonian mechanist of the period, Musschenbroek confesses that though there may be several internal principles in bodies acting in different proportions at different distances, it is, as yet, impossible to determine them.

> It would be very difficult to decide this matter because no trials can be made upon first elements. . . . nor can we know after what manner these parts incumb upon one another, or how much solid they have, how much pore, what are their figures. Yet upon these will depend their different force of attraction [I, 198-204].

Their most serious departures from Newtonian mechanism toward materialism come in their treatment of those phenomena where Newton had had the least, explicitly, to say—*i.e.*, electricity, magnetism, and especially heat. Here again 'sGravesande is the least positive, but here also he is the most mechanistic. He does not clearly assert fire to be a substance, but the implication of what he says is substantial. Fire is subtle, fast moving, is contained in all bodies and attracted to them from a distance [II, 1-3]. Heat is the effect produced in the mind from the irregular motion of fire in bodies, which may be caused by agitating the parts of the bodies and the fire in them or by the addition of moving fire from without [II, 14-15]. The action of fire conveys a repellent force to the particles of a body causing it to dilate; fluidity ensues when a balance is obtained between the repelling and the cohesive forces, and, when the agitation is severe, some parts of the body may be separated and carried off. Neither the heat nor the dilation is in proportion to the quantity of fire; it is the motion which is significant [II, 15, 18-20]. Light, the Newtonian archetype for material fire, is the effect produced by rectilinear motion of fire and Newton's demonstration of the action, at a distance, of bodies on light is confirmation of such action on fire [II, 13-14, 24]. 'sGravesande does not exactly declare electricity and fire to be the same, but he points to relations between their actions. Substances such as glass contain in them and about their surfaces an atmosphere which is put into vibration when heated by attrition. The vibratory motion expands the atmosphere by which its action of attraction and repulsion on lighter bodies at a distance is exerted while, at the same time, the fire contained in the bodies is expelled or, at least, moved. How such an atmosphere is confined in bodies is not clear, though the pressure of ambient bodies can hardly be the cause, as fire so easily penetrates all bodies by its subtlety [II, 7-13].

Musschenbroek, having himself contributed to the growth of electrical knowledge (he was one of the discoverers of the Leyden jar), was less sure of the nature and action of electricity. Clearly the power of some bodies to attract others when rubbed is due to their emission of some subtle exhalation, but what is the nature of this electrical effluvia [I, 186]? Is it fire alone, or mixed with other bodies, and is the fire terrestrial or of another nature? Is the twofold nature of electrical virtue the result of different subtleties or of different motions of the effluvia? How, indeed, is the effluvia moved to have those "wonderful fits" sometimes approaching and sometimes receding from bodies [I, 187]? Probably it has a vortex motion, but how then does the effluvia "run along a string [I, 196-97]"? All these are questions he leaves unanswered. And for magnetism, where he also had done some original investigations, he has still fewer answers, for here it is not even clear that the materialistic mode of effluvial explanation will answer. If, for example, magnetism is caused by effluvia, why does it work only on iron substances, and why cannot its effluvia be screened or dispersed as electrical effluvia can? Magnetism is of a different nature than other attractions, a nature as yet unknown [I, 210].

On the subject of fire, however, Musschenbroek has even fewer doubts than 'sGravesande. Following Boerhaave, he declares fire to be a fluid substance, occupying space, moving but able to insinuate itself and adhere to bodies. It has weight, solidity, is extremely subtle and perhaps elastic [II, 19-22]. It moves spontaneously from hot bodies to those less hot, "till it remains in equal quantity in all near and ambient bodies [II, 23]." Heat is a quantity of fire in motion and "bodies are so much hotter, as they contain more fire in motion [II, 46]." Cold is the absence of fire and absolute cold the absence of all fire. Light and fire differ not in substance nor in magnitude of parts, but in the direction of their motion [II, 57]. Light is fire in rectilinear motion, acted upon, at a distance, by bodies in attracting and repelling to produce refraction and reflection [II, 89, 136].

With 'sGravesande, and more clearly with Musschenbroek, the dilemma of the early eighteenth-century Newtonian mechanists is posed with increasing precision. They had adopted a frame of dynamic corpuscularity, but there were phenomena it could not, apparently, comprehend. The quantitative determination of sizes, shapes, solidities, even the variety of ultimate particles, could not be determined to give to theories the quality required of them. Newton himself had, at least, equivocated about the forces of attraction and repulsion and provided an apparent alternative in a material aether. Who can blame his successors for increasingly adopting such a solution for the "new" phenomena of electricity or even for the old one of heat? In their atmosphere or effluvia of electricity, in their substance of fire, 'sGravesande and Musschenbroek abandoned the Newtonian hope of reducing all phenomena to a system of homogeneous matter, motion, and forces. But the new, more empirically satisfying, system in which the heterogeneous types of matter are differentiated by their actions, could become "Newtonian" in another way by making it quantifiably predictable. 'sGravesande does not, perhaps, fully see the nature of his choice. He cannot quite describe the substance of fire nor abandon the notion of defining its effects through its motion. Musschenbroek comes closer to grasping the nettle. His matter of fire is described in standard mechanistic terms, but is identified by the heat its presence conveys; and though it is "fire in motion" which conveys the heat, the concept of equilibrium and that of absolute cold require what the measure of heat implies—quantity is to be found in a measure of amount of substance.

The same dilemma is more clearly and impressively revealed in the tensions between mechanism and materialism developed by Herman Boerhaave in his chemical studies.[10] Boerhaave was not, in the usual sense, a great scientist, but he was a great teacher and showed his creative skill in the syntheses of diverse facts of medicine and chemistry he attempted for his students. Dynamic corpuscularity had, in the work of the Keills, John Freind, and Stephen Hales, apparently been extended with great fruitfulness into the areas of physiology and chemistry, but the infinite variety of appearances, yet to be included in that frame, was such as to defeat every would-be reductionist for more than a century. Nonetheless, Boerhaave was forced to grapple with that variety and subdue it to some systematic study, for medical students cannot be placated with the natural philosopher's *ignotum*; where theory fails to provide an answer, there experience must be supplied.

Boerhaave's attempts to combine mechanistic theory with diversity of experience were to have great influence in midcentury Britain. During the nearly thirty years of his tenure as professor at Leyden, 659 students in the school of medicine were British. The majority of these returned to Britain to practice the medicine he taught; others returned to teach and to apply Boerhaave's ideas as they led in the development of a natural philosophy turning from mathematics to experiment. Of the five subjects—botany, chemistry, electricity, heat, and physiology—that dominated the interests of eighteenth-century British experimentalists, only electricity was not taught by Boerhaave. Botany, at least in its taxonomic aspects, soon escaped the confines of Boerhaave's tutelage in the simplicities of his protégé, Linnaeus; physiology was, increasingly, the preserve of the professional physician and the medical school; but chemistry, with its sub-

[10] Herman Boerhaave (1668-1738) entered the University of Leyden in 1683 as a theological student. He continued with graduate study toward the Ph.D. (Leyden, 1689), but also obtained an M.D. (Hardrwyk, 1693) at the urging of Burchard de Volder, his professor of mathematics. Wrongly accused of Spinozism, he gave up a theological career and commenced private medical practice until 1701, when he became lecturer in the Institutes of Physic at Leyden. By 1709, he was professor of botany and medicine; in 1714, professor of the practice of physic and Rector of the University. He lectured informally in chemistry until 1718, when he succeeded James LeMort as professor of chemistry. He also lectured, 1705 to 1716, on mechanics until replaced by 'sGravesande. A prolific author of works in medicine, botanical materia medica, etc., as well as of a famous text in chemistry, he was elected F.R.S. and Foreign Associate, Académie Royale des Sciences, Paris. See Wilson L. Scott, "Sources of Boerhaave's Medical Lecture"; Hélène Metzger, *Newton, Stahl, Boerhaave et la Doctrine Chimique* (Paris: F. Alcan, 1930); and Milton Kerker, "Herman Boerhaave and the Development of Pneumatic Chemistry," *Isis*, 46 (1955), 36-49.

ordinate field of heat, became a topic of general interest as its application to the practical arts was demonstrated. Boerhaave's chemistry thus affected a larger audience than its original medical orientation had intended, and not only were his students (*e.g.*, Andrew Plummer at Edinburgh and Henry Pemberton at Gresham College) and, in turn, their students, to lecture on chemistry following his methods, but his text in chemistry was to achieve the greatest popularity of any single text until that of Lavoisier at the close of the century.

The text was first published in an unauthorized edition of student lecture notes, as the *Institutiones et experimenta Chemiae* in 1724, and translated into English by Peter Shaw and Ephraim Chambers as *A New Method of Chemistry; including the theory and practise of that Art: laid down on mechanical principles*, etc., in 1727. Boerhaave angrily replaced the student notes with his own for the approved version, the *Elementa Chemiae* in 1732. The approved version was twice translated into English, by Timothy Dallowe as *Elements of Chemistry* in 1735 and by Peter Shaw in 1741 in the guise of a second edition of the *New Method of Chemistry*, with critical notes.[11] The Shaw version reached a "third" edition by 1753, while the text itself was destined to appear in nearly thirty editions. Nor was its impact confined to the subject matter of chemistry (and heat), for Boerhaave's mode of argument and method of presentation appeared to have wider applicability. Musschenbroek was summarizing for an age when he wrote, "one should have continually before one's eyes these two perfect models that the two great men of the century have left us . . . the *Opticks* of Newton and the *Chemistry* of Boerhaave."[12]

These "two perfect models" were not really commensurate, though clearly Boerhaave did his best to make them so. From the beginning, his definitions and discussions are couched, insofar as was possible, in dynamic corpuscular terms. The truths demonstrated in physics, mechanics, hydrostatics, and hydraulics are to be used, "since the properties which belong to all bodies in common, must hold good in chemical ones too [I, 2]." Chemistry is

[11] See Tenney L. Davis, "Vicissitudes of Boerhaave's Textbook of Chemistry," *Isis*, 10 (1928), 33-46. I have used the Dallowe translation of Herman Boerhaave, *Elements of Chemistry: being the Annual Lectures* (London: for J. and J. Pemberton, J. Clarke, A. Millar, and J. Gray, 1735), but compared it and examined the notes to Peter Shaw's Herman Boerhaave, *A New Method of Chemistry* (London: T. and T. Longman, 1753), 3rd edn., corrected.

[12] Quoted by Metzger, *Newton, Stahl, Boerhaave*, p. 191, and cited by Cohen, *Franklin and Newton*, p. 222.

the art of transforming bodies by physical operations on sensible bodies to produce particular effects, and the determination, by those effects, of their causes [I, 19]. Such transformations as can be induced on bodies are owing entirely to changes in motion, addition, destruction, alteration in degree or direction [I, 44-45]. In changing motion, chemists employ six "Instruments"—fire, water, air, earth, the solvents called menstruums, and "the common furniture of the Elaboratory [I, 78]."

Boerhaave is at his most Newtonian in discussing these instrumental operations of chemistry. Fire is the principal cause of almost all sensible effects, for it penetrates, agitates, and expands all bodies. Most corporeal effects are caused by the action and reaction between the expansible force of fire and the contractive, or attractive, force between the corpuscles of bodies [I, 114]. Water too is very penetrating; it is attracted to the particles of bodies, adds to their weight, can separate the parts, but also acts as a "Gluten" consolidating and binding the corpuscles together [I, 341-43]. Earth is the constituent principle in the corporeal fabric of animal, vegetable, and some fossil bodies. It serves as a basis for the other principles of body, fixing and retaining those parts of themselves that are too volatile, and thus qualifies a body to continue as it is [I, 383].

About the function and action of air, Boerhaave is not clear. Hales' *Vegetable Staticks* appeared between the unauthorized and the authorized versions of the lectures, and Boerhaave attempts to incorporate Hales' new discoveries and interpretations, but retains some material now made inconsistent if not contradictory. Air is the grand, efficacious and necessary instrument for almost all of nature's operations—only those of fire, the loadstone, gravity, and the "particular attraction and repulsion of corpuscles of bodies" are to be excepted [I, 247]. Rightly conceived, air is a universal chaos, in which corpuscles of every kind are confounded together and pure air is never to be found [I, 289]. It is probable that in its elasticity alone is to be found the necessity of air to the life of animals and vegetables [I, 292]. Yet, of itself, air barely affects our senses; it is only as agitated by fire that its action as a "mechanical pestil" is manifest in the compression, attrition, compaction, depuration, and union of homogeneous particles [I, 248, 316]. Air has a tendency to self-attraction—as revealed by its spherical bubbles in liquids [I, 251]—but Newton and Hales have declared it to be elastic, and Boerhaave agrees. Air is not concreted, coagulated, or altered in bodies but only lies concealed in them [I, 250]. When in liquids or united to other particles, it does not retain its elasticity, which it has only in virtue of its repulsion for other particles of air [1, 307]. This elasticity is regained when several air particles are pressed together and "begin to come within the sphere of each other's activity [1, 263-64]."

The dynamic nature of chemical "instrumental" activity is most clearly revealed in Boerhaave's discussion of effervescencies and dissolutions. There is a "prodigious reciprocal attraction" between acids and alkalis by which, when "placed at a certain distance they rush together with a mighty force." As a result of this innate power in these bodies, the air that is in them is disengaged, forced out, and separated from them, resulting in effervescence [1, 310]. Some chemical dissolution is produced by purely "mechanical means"—by which Boerhaave means the size, hardness, shape, weight, motion, and impulse of the particles of the menstruum and of the pores and particles of the solvent, all assisted by the action of fire [1, 396-97, 409-10]. By far the more important are the effects of the "attraction and repulsion between the Particles of the Menstruum and those of the Body dissolved [1, 489]."

Thus far, by careful exercise of selectivity, there are no essential contradictions with Newtonian mechanism, but this is only for part and that a very small part of chemistry. Freind had limited his chemistry to these operations or, as Boerhaave called them, the "Instruments" of chemical action, but Freind was more interested in the application of dynamic corpuscularity than in chemistry. Boerhaave's *definition* of chemistry and his discussions centering on that definition are still more restricted, for he does not reach toward the measure of quantification that Freind had attempted. Even in his volume on theory, Boerhaave's treatment of operations remains general and whenever he discusses practice, as in his second volume, his approach to chemistry extends further and further from the confines of mechanism. For what Boerhaave and his students really wanted, what chemists then and since have attempted to achieve, was not simply knowledge of the general attributes of substances or the general operations of chemistry, but rather the specific differences between the various substances and the specific consequences of chemical operations. For these objectives, dynamic corpuscularity was completely inadequate. In his application of the corpuscular hypothesis, the best Boyle had been able to achieve was *ad hoc* suggestions of particle shape—sharpness for acids, round smoothness for liquids, spring coils for elastic vapors—which might be consistent with a kinematic, corpuscular chemistry. Newton's addition of forces of attraction and repulsion added parameters to corpuscular chemistry, but in eighteenth-century Britain, at least, these were seldom adopted to explain the specific differences between "derivative corpuscles" or the results of chemical operations. That they might have been so used is indicated in the applications of attractive and repulsive spheres begun with Hales, Rowning, and Desaguliers and continued so imaginatively by Boscovich, but even here the usage retains the *ad hoc* character of Boyle's corpuscularity, for the parameters remain unmeasured.[13] There is little that more clearly reveals the essential poverty of the premature sophistication of dynamic corpuscularity than this failure of chemical reductionism. Whenever Boerhaave approaches a specific problem of chemistry, he systematically disavows his current mechanism and falls back into a taxonomic materialism.

This is clearest in his second volume on the practice of chemistry, where it is mechanism that is the intrusion on a text arranged by empirical process, on classes of substances producing stated kinds of medicinal preparations. It is also evident, however, and in conflict with mechanistic theory, in the first volume. Chemists are not to trouble themselves with fruitless inquiries into ultimate causes, but to study present causes, not to concern themselves with substantial forms, but to discover by their effects the peculiar qualities and powers implanted by nature in every particular body [I, 52]. However skilled a person be in mechanics, he will never be able to discover the effects of bodies from general and universal properties of all bodies, for chemistry depends entirely upon the proper and particular nature of certain bodies only and these depend upon peculiar virtues [I, 51]. One cannot affirm that the parts into which a body is divided existed before that division, nor that the nature of a combination is revealed in the general nature of the bodies put into it, for the act of combination or of division produces alterations in the corpuscles not to be found in the bodies themselves or in their general nature [I, 2, 46]. Yet examination of bodies tells us that there exist immutable corpuscles, not one and homogeneous, but several. Single, simple bodies are composed of lesser bodies perfectly like the greater and these again in the same manner beyond any limit the mind can fix. There is a principle implanted in some corpuscles by which they are united in a way not possible for art to separate and these special unions are called the elements of bodies

13 Some of this work will be discussed in Part III of this book.

[1, 46, 92]. But if matter is ultimately heterogeneous and if one cannot tell by mechanical analysis what properties either the elements or their compounds will have, how is one to proceed?

Boerhaave does not declare the answer explicitly, but he demonstrates it implicitly throughout both volumes of his lectures. He classifies and distinguishes sensible bodies by their obvious physical properties, their chemical behavior, and, to a limited extent, by their presumptive mechanical characteristics. Rarely do these specific or generic properties involve details of forces, even when he mentions chemical affinities, for, like Stahl and Geoffroy (even like Newton with his secret principle of sociability of substances), affinities to Boerhaave seem to mean the readiness of like substances to unite—as mercury with the mercury base of all metals [1, 23]. In his definition of metals in general, he talks of their fusibility and the mutual attraction of particles by which the molten metal assumes a spherical shape [1, 27], but different metals are classed by color, hardness, ductility, and chemical activity. As for the medicinal properties of metallic salt solutions, each has a "singular vertue which is proper and peculiar to every particular Metal, and which is generally inimitable by any other [1, 337]." Indeed, the principal differences observed among salts ought chiefly to be ascribed to real differences in their constituent elements, for though these cannot be examined separately and alone, "each sort has always some distinct and proper vertue [1, 439]." Plants and animals, in their corporeal parts, are defined as hydraulic bodies drawing various fluids through them as necessary for growth and nourishment [1, 38], but every single animal and vegetable body has an *Aura* or vapor proper only to that particular body, expressing its true genius and accurately distinguishing it from all others [1, 47]. For vegetables this involves a "vital principle" which characterizes a family of substances [1, 38], while animals have particular concoctive powers which make their substances distinct from all others [1, 466].

Even in the "instrumental" substances, it develops that there are individual and irreducible differences. Effervescent bodies have "some peculiar innate power" in them which occasions their effervescent motions [1, 310]. For menstruums, whoever "ascribes more to a mechanical power than the all-wise Creator has allotted to it, is certainly in the wrong"—and though Boerhaave primarily means here the kinematic mechanistic properties, he goes on to distinguish the actions of various menstruums by their peculiar "vertues" and to divide them into "certain Classes, prefixing to each of them some distinguishing character, to which they may be reduced [I, 414]." For earth, water, and air, the descriptions are basically physical and, perhaps, have some mechanistic significance. Earth is a simple, hard, friable, fossil body, fixed but not melted in fire, not dissoluble in water, alcohol, oil, or air [I, 364]. Water is a fluid liquor, inodorous, insipid, pellucid, colorless, which freezes in a certain degree of cold. Water is impossible to obtain absolutely pure, always containing fire and usually other impurities of air, salt, and earths [I, 319]. It is perfectly immutable, as its recovery with constant density, weight, and fluidity from chemical operations attests, and it is sufficiently provable that its ultimate particles are exceedingly small, rigid, solid spheres of adamantine hardness, more penetrating than any other substance except fire, and magnetism if that be supposed a fluid, and light if that be different from fire and a fluid [I, 324, 328-29]. Generically air poses a problem to Boerhaave as it did instrumentally. A substance which cannot be deprived of its fluidity [I, 249], it is hardly to be perceived by the senses, but is manifest by the resistance it offers to motion. It seems heavy, elastic, dense in proportion to compression or the "intenseness of the fire that acts upon it [I, 248]," subtle, and permeated with various kinds of corpuscles. But it seems possible that if the fortuitous ponderous corpuscles were removed, purely elastic air would be found "intirely without gravity" and equally, with fire, distributed throughout the universe [I, 293].

The primary "Instrument" in Boerhaave's arsenal of chemical operators is fire, and to that instrument he devotes 170 pages, more than a third, of his volume on chemical theory. Because of the importance he assigns it, the influence of heat on other phenomena, and the apparent similarity between heat phenomena and those of light and electricity, Boerhaave's treatment of fire and its action does indeed become a model for the substantial treatment of other qualities of action throughout the remainder of the century. For there is no doubt in Boerhaave's mind that fire is a substance. Acknowledging the clear explanations and demonstrations by Nieuwentyt and Frederick Hoffman of the existence of eternal fire [I, 77], he proceeds to define its properties. It can be neither created nor destroyed and, under static conditions, is equably distributed through all space, regardless of whether space be empty or contain body [I, 121, 113]. It is impossible to discover the limits of cold or of fire, for no place is found without it nor can one extract all the fire from a place or find the greatest quantity it may contain [I, 93]. Fire is movable, either in vibrations to generate heat or in being abstracted from one place and increased in another [1, 171]. It is very subtle—more so than anything but gravity and magnetism, and if these be caused by the emanation of material corpuscles, how much more must fire be corpuscular, which takes time for its penetration of bodies as these do not [1, 226-34]. Fire does not, however, penetrate the elements of bodies, but only the pores between the particles, from which it acts with equal force on all the corpuscles [1, 78, 115]. It is probably not affected by gravity, but there is "in the solid mass of Bodies, something like an attractive power" for fire which, once communicated, may adhere to the body for a considerable time [1, 119-20, 152].

The "pablum of fire" consists merely of bodies in which fire is detained and from which it is released as the bodies' particles are dispersed through the action of fire from without [1, 168-69]. The role of air in supporting fire is that of supplying an external pressure which prevents the outward expansion and dispersal of the particles of fire [1, 206]. Heat is generated in the attrition of elastic bodies when the alternate compression and expansion of the parts of body necessarily compresses and relaxes the fire in its pores. The surrounding fire is likewise agitated, and the combined excitation results in heat [1, 117]. The intense heat generated at the focus of burning glasses and mirrors was a continued problem to Boerhaave, but some of it was resolved in his decision that the sun could not, in reality, emit igneous matter, for that would be creating fire. The action of the sun was the determination of pre-existent fire into parallel right lines, giving its expansive power a prodigious efficacy [1, 125-42, 152]. This suggests an identity of fire and light which Boerhaave will not deny (nor positively assert). Admittedly the true and absolute simplicity of fire is contrary to the doctrine of Newton—but the contrariety Boerhaave refers to is that of color diversity and polarization of light, both of which Boerhaave will accept as attributes for fire, not knowing what marvels remain to be discovered about its properties [1, 226-34].

The failure of fire to change the weights of bodies in which its quantity is varied eliminates the obvious method of measuring the quantity of substance. Boerhaave declares that deciding on some measure is very desirable, but not as simple as some people believe it to be [1, 154]. The physiological effects of heating are clearly an inadequate measure, and eventually Boerhaave declares for the use of changes in body dimension which fire is known to cause though even this is not a simple matter [1, 79, 106].

When left to itself, fire diffuses uniformly and equably in all directions, and its force, or power, in equilibrium will be proportional to the space in which it is contained. But it is the disturbance of that equilibrium which produces the "infinitely powerful" operations of fire, and, though increasing the quantity of fire or decreasing the space it occupies will always have an efficacious effect in enhancing its power, it is by no means sure that its power is purely an effect of its quantity in a given space. Here again the effects of the burning glass and those of the "concentration" of fire in combustible substances combine to confuse the issue in Boerhaave's mind. Will not the near approach of its elements to one another, the condensation of fire, increase the intensity of its action to a degree larger than that of mere number [I, 142-43, 154]? Newton and Grimaldi have shown that the elements of fire acquire new motions in approaching opaque, reflecting bodies. Why may not the particles affect one another in the same manner [I, 244-45]? In trying to prove too much, Boerhaave's theory of fire and heat ends by failing its chief test, that of quantification. But the problems of quantity and intensity, of number and density, of heat radiation, transmission, and chemical production were to be the primary puzzles for heat theorists for the next hundred years.

It was to be expected that the richly seasoned arguments of Boerhaave's chemistry would initially be suspect in a country notorious for the simplicity of its diet. John Freind is said to have opposed Boerhaave's election to the Royal Society; Edward Strother's abridgement of Boerhaave's *Chemistry*, first published in 1732 and reprinted in 1737, declares that Lemery and Freind together sufficiently account for all the phenomena produced by chemistry; and James Alleyne's *New English Dispensatory* of 1733 is derided in a pseudonymous note in the *Gentleman's Magazine* for its attempt to yoke Dr. Freind and Dr. Boerhaave, "whose Systems and way of Reasoning are as different as that of Alkali and Acid."[14] The assimilation of Boerhaave's chemical ideas is most directly illustrated in the work of Peter Shaw.[15] Shaw's

[14] Scott, "Sources of Boerhaave's Medical Lectures," p. 98n, Davis, "Vicissitudes of Boerhaave's Chemistry," p. 44, and "De Duobus, etc., Tricks of Booksellers," *Gentleman's Magazine*, 2 (1732), 1,099-1,100. An answer to the above, *Gentleman's Magazine*, 2 (1732), 1,117-18, declares that many Members of the Colleges, both of *London* and *Edinburgh*, educated under that Great Man [Boerhaave] would have thought it very practicable to reconcile Freind and Boerhaave.

[15] Peter Shaw (1694-1763), son of a master of the grammar school at Lichfield. It is not known where he acquired the M.D. which qualified him as a licentiate of the Royal College of Physicians, London, in 1740. A popular physician in Scarborough and later in London, he became one of George II's physicians and was

alternative translation of the *Elementa* is profusely annotated, primarily with references to the work of Boyle and Bacon, whose *Philosophical Works* he had edited in 1725 and 1733, respectively. Newton's *Opticks* is, however, cited frequently enough and in such a manner as sometimes to give this edition something of the argumentative flavor of Clarke's edition of *Rohault*. This is particularly true in connection with the nature of fire, where Shaw declares that Boerhaave's doctrine is new and extraordinary "among us, who have been used to consider fire in the light it is set by Lord *Bacon*, Mr. *Boyle* and Sir. *I. Newton*." Homberg, Boerhaave, Lemery, and 'sGravesande are noted as the principal supporters of the material theory of fire, and passages from these writers are juxtaposed with conflicting passages supporting a mechanical theory [I, 206-07]. Between the publication of the notes to the unauthorized edition and those to this authorized one, Shaw has, however, already moderated the extremes of his Newtonianism. In a note describing the creation of heterogeneous substances from homogeneous primitive particles, Shaw initially stated that Newton had shown this to be true; in the revised version, the note was changed to read "Newton suspects."[16] A long note in the revised edition discusses the powers of attraction between the *minima naturae*, which Newton has established and Keill and Freind have applied to solve the phenomena of cohesion, coagulation, and almost all the phenomena that chemistry presents. "But," Shaw continues, "this seems a little too precipitate; a principle, so fertile, should have been further exhausted; its particular laws, limits, etc., more industriously detected and laid down, before we had gone to application. Attraction, in the gross, is so complex a thing, that it may solve a thousand different things alike: the notion is but one degree more simple, and precise, than the action itself: and till more of its properties are ascertained, it were better to apply it less and study it more [156-57]." In his edition of Boerhaave, Shaw does not suggest what might replace mechanism as an explanatory principle, nor does he indicate that an intervening translation of Georg Ernst Stahl's *Principles of Universal Chemistry* (1730) may have encouraged this change of attitude between his first and second versions of Boerhaave's text. The implication is clear, however,

granted a Cambridge M.D. in 1752 by royal mandate. He lectured publicly in chemistry, but is best known for his editions and translations of the works of Bacon, Boyle, Stahl, and Boerhaave. See, particularly, Peter Shaw, tr., Herman Boerhaave, *A New Method of Chemistry*.

[16] Noted by Cohen, *Franklin and Newton*, p. 223n.

that materialistic empiricism provided at least a temporary refuge from the excesses of mechanization. There is no indication that Shaw adopted, as part of this materialism, the substantial fluid theory of fire, but other investigators, using Boerhaave's text as their authority, were to do so, and were to extend the same principle to the understanding of electricity.

CHAPTER EIGHT

The Imponderable Fluids

During the middle years of the eighteenth century, the material concepts described in the last three chapters were reified in the imponderable fluids created by midcentury British natural philosophers to explain the phenomena of electricity, magnetism, and heat. Unlike the subjects of physiology and chemistry, where an increasing multiplicity of phenomena came to dim the once bright hope of reductionism, these problems of "physics" had, by the inherent nature of the phenomena themselves, always resisted comprehension by the methods of dynamic corpuscularity. Materialism, in the form of Newtonian aether, of non-Newtonian active substance, of Dutch fire, or indeed of varying mixtures of the three (for their tendencies were all the same), now offered an opportunity of explaining these phenomena by the inherent nature of appropriate fluids.

Although these fluids retained some characteristics of Newtonian dynamism, their essential materiality is revealed in the way their activity was quantified, for the parameters used are not those of dynamic corpuscularity. Paradoxically, this is perhaps most evident in the acceptance of the notion of imponderability. It is true that some efforts were made to determine weights, particularly for the fluid element of fire, but in the end most of the materialists were satisfied with other measures. The concept of weight as the primary measure of substance is, after all, a post-Lavoisier-and-Dalton achievement. The kinematic corpuscularians had regarded extension as the operative parameter; weight implies force, and force was a concern primarily to Newtonian dynamic corpuscularians, whose influence had now waned. But if weight was not to be the measure of substance, how was variation in its quantity to be determined? The development of fluid-concept explanations is essentially the successful definition of alternative parameters.

It is appropriate that the first success should be achieved in electrical studies.[1] Electricity had regularly been used by dynamic

[1] Electricity has been well served with historical studies, but for my purposes, Joseph Priestley's *History and Present State of Electricity* (New York: Johnson Reprint Corporation, Sources of Science No. 18, reprint of the 3rd edn. of 1775, 1966) best serves the purpose of providing information on eighteenth-century British (prior to 1765) studies, while Cohen's *Franklin and Newton* is the standard analytical work on the subject at that time and place. My dependence on Cohen is obvious, though I have reread many of the sources he studied and may give conclusions with a different emphasis than his.

corpuscularians as a demonstration of the existence of forces, but they had been unable to explain the phenomena mechanistically in terms of those forces. Prior to 1746, electricity had developed no explanations significantly beyond the stage of Robert Boyle's material effluvium which issued from and returned to certain substances called electrics when these were rubbed. It is hard to find, in such proposals, anything but the first stage of a materialistic explanation—the substantial emanation of a peculiar effluvium from a particular class of substances—and there is no avenue here to quantification. Yet the Newtonian mechanists offered no viable alternative solution.

Two sets of electrical papers in the *Philosophical Transactions*, published early in 1746, illustrate, by contrast, the range from mechanism to materialism in British electrical studies. In one of these, an anonymous writer describes his abortive attempt to determine whether, as electrical "Effluvia may be felt, heard and seen, they may be weighed" with a chemical balance.[2] It is not entirely clear whether he was trying to weigh a substance contained in bodies, or to measure the strength of an attractive force, but the parameters he is concerned with—whether the figure, size, density, and color of electrified bodies relates to the amount of effluvia which can be collected in them—are, for all their substantial aspect, just those which would have interested the dynamic corpuscularians. The writer's inability to measure them is the last sign of such an effort for the next twenty years, as the vogue shifted to the kind of arguments suggested in the papers of the Rev. Dr. Henry Miles. Miles supposes that vitreous and resinous effluvia are to be distinguished by "invariable and inherent properties" and suggests that these effluvia and those of light are the same. Through Newton's "elastic *Materia subtilis*," electricity and light are to be connected with the phenomena of fire—whether that be, as the "English philosophers" believe, mechanically produced or, with the foreign philosophers, Boerhaave, Homberg, Lemery, 'sGravesande, etc., "it be equally diffused throughout the Universe by the Creator." Heat, light, electricity, and the aether, invariable and inherent properties, English versus foreign philosophers—the stage is prepared for a new set of causative agents in electricity.[3]

2 [Anon.], "A Letter . . . of weighing the Strength of Electrical Effluvia," *Philosophical Transactions*, 44 (1746-47), 96-99.

3 Henry Miles, "Extracts of Two Letters . . . concerning the Effects of a Cane of Black Sealing-Wax, and a Cane of Brimstone in electrical Experiments," and "Part of a Letter . . . Concerning Electrical Fire," *Philosophical Transactions*, 44 (1746-47), 27-32, 78-81. Miles (1698-1763) was educated for the dissenting ministry, prob-

Dr. Miles was but one of a group of British authors who in 1746 suddenly invoked this range of causative agents for electricity. Yet Newton had related the aether and electricity in 1717; 'sGravesande, electricity, fire, and light in 1720; and Benjamin Worster had even offered an explanation of aetherial-electrical action in 1730. For practical purposes these suggestions had been ignored. What had happened to produce a spate of pamphlets and articles on fire, aether, and electricity in 1746? It seems clear that they are immediately traceable to the recently published pamphlets reviving Newtonian aether ideas: the anonymous *Examination of the Newtonian Argument for the Emptiness of Space* of 1740; and Bryan Robinson's *Dissertation on the Æther of Sir Isaac Newton* of 1743 and his *Sir Isaac Newton's Account of the Æther* of 1745. Reflections of these views are to be seen in Peter Collinson's letter of March 1745 to Cadwallader Colden, in which it is supposed that investigations in electricity will enable us "to discover the nature of that subtile, elastic and aetherial medium, which Sir Isaac Newton queries on, at the end of his *Opticks*."[4] When so sober a man as the Quaker Collinson is inspired to such hopes, there is small wonder in the extravagant enthusiasms of more ardent spirits, especially when these were also fed by active-matter theorists.

The first of these extravaganzas to appear was that of John Freke, whose *Essay to Shew the Cause of Electricity* of 1746 was followed with an extension of the same ideas in a *Treatise on the Nature and Property of Fire* in 1752.[5] Together these works demonstrate a talent for dogmatic assertion, varied and followed by nonsequiturs; they also demonstrate a wide reading in the materialist literature of the period. Electricity is fire, collected by rubbing air between the hand and tube or globe [101]. Fire is a fluid universally dispersed throughout the universe; it is the active element of matter, the Soul of the World, and first Cause

ably in London, and held pastorates in Surrey while assisting at Old Jewry, London. F.R.S. 1743, D.D. (Aberdeen) 1744; between 1741 and 1753 he published papers on natural history, meteorology, and electricity in the *Philosophical Transactions*. He assisted Birch in the editing of Boyle's *Works*.

4 Quoted by Cohen, *Franklin and Newton*, p. 435.

5 John Freke (1688-1756), was the son of a surgeon and apprenticed to a surgeon. Assistant-surgeon (1727) and then surgeon (1729) at St. Bartholomew's Hospital, he was one of London's chief surgeons until he retired. He fancied himself a judge of art and music as well as of science; F.R.S. 1729, he wrote principally on surgical cases and observations for the *Philosophical Transactions*. His *Essay* on electricity was published in three editions; two in 1746 and the third was included in his *Treatise on the Nature and Property of Fire* (London: W. Innys and J. Richardson, 1752), which is the form in which I have used it.

of activity and motion, as Earth is the passive Body [iv-vi]. Newton "appears grievously to have mistaken the Nature of Fire," as is shown by Boerhaave, and while there may be a relation between fire and the aetherial medium [11], on the whole the aether appears to be "unintelligible jargon [146]." Fire cannot be altered, increased or diminished [17]. The sun is the fountain and only source of fire, which it receives and ever circulates from south to north, thus explaining gravity as well as magnetism [13, 22, 148]. When the electrician collects fire from the ambient medium, the fire pervades the electrified body and envelops it with a covering, about a half inch thick, of fiery particles [101], from which the excess breaks out like lightning in a diverging spark with a force proportioned to the number of particles brought together [81]. Electrical attraction is caused by the great cohesive attraction of these fiery particles for those in other bodies, when those bodies lie within the vortex or whirlpool of spirally driven fire [103-108].

Although the reviewer in the *Gentleman's Magazine* of 1746 [16 (1746), 521-22] thought Freke's *Essay* a fair bid to account for electricity, it understandably had a less favorable reception from other electricians, particularly those who had alternate proposals. One of these was Benjamin Martin, who carried on the experimental and educational tradition of his mentor, Desaguliers, with an *Essay on Electricity: Being an Enquiry into the Nature, Cause and Properties thereof, On the Principles of Sir Isaac Newton's Theory of Vibrating Motion, Light and Fire* also in 1746. Martin's descriptions of experiments are clever and easy to follow; his attempted explanations, though less extreme than Freke's, are heterodox and confused.[6] Martin had been incensed by Freke's anti-Newtonianism and undertook to demonstrate that Newton's philosophy, in its doctrine of heat and light, clearly explained the "nature, cause, properties and effects of electric virtue [4]." Heat is the sensation produced by the agitation of parts of bodies. Should the agitation be sufficiently large and vibrating, effluvia may also be emitted which shine and appear

[6] Benjamin Martin (1704-82) began life as a farm laborer, but educated himself in natural philosophy and used a legacy to establish himself as a mathematical instrument maker and lecturer in sciences. According to George Horne's *Theology and Philosophy of Cicero's Somnium Scriptionis* (1751), Martin for a time assisted Desaguliers in his lectures. The *D.N.B.* lists more than 34 different works published by Martin on aspects of mathematics, scientific instruments, and Newtonian philosophy, of which I have used *The Philosophical Grammar* (London: John Noon, 1755), 5th edn. (first 1735); and *An Essay on Electricity, etc.* (Bath: for the author and Leake, Frederick, Raches, Collins, and Newbury, 1746).

as light and fire [7-8]. If the body is an electric per se, the effluvium is of that subtle, elastic kind called electric virtue, which repels as it flies off and attracts in its reciprocal return, as the body, within its space of action, attracts it back [14].

There is no clear reference to the aether here, as there had been none in Martin's previously published *Philosophical Grammar,* but the aether appears in a *Supplement* which he published the same year as the *Essay.* In this, Martin continues his attack on Freke, the "worst writer on electricity," and praises William Watson, the "best," and Watson's recent aetherial explanation of electricity.[7] Watson had, however, unnecessarily modified the word "aether" with "electrical," as Newton's aether was plainly the cause of light and fire as well as electricity. No one had shown fire to be an element, and its effects could be shown to follow from the activity of an aether pervading the pores and being emitted from them by attrition. Attraction is the result of an inward flux of aether to restore equilibrium, repulsion that of the elastic forces between bodies "impregnated" with this electric (aetherial) matter, which is retained there by an attracting power between the bodies and the aether.

The third of the electrical visionaries to write in 1746 must be taken more seriously. Benjamin Wilson's contributions to the study of Leyden-jar capacities and the pyro-electrical action of the tourmalin were positive achievements, though these and his pugnacious support of blunt-ended lightning rods in opposition to Franklinian points demonstrate both the confusion and the self-assurance revealed in his 1746 *Essay towards an Explication of the Phaenomena of Electricity Deduced from the Æther of Sir Isaac Newton.*[8] Greatly impressed by Bryan Robinson's trea-

[7] The 1755 edition of the *Philosophical Grammar* hardly mentions the aether and uses it not at all, but I have not seen the 1735 and 1738 editions and am dependent upon Arnold Thackray's "Newtonian Tradition and 18th century Chemistry," p. 4-34n, for the information that these do not refer to the aether. Nor have I seen Martin's *Supplement,* which is described in Cohen's *Franklin and Newton,* pp. 415-16.

[8] Benjamin Wilson (1721-88) was briefly educated at Leeds grammar school and then apprenticed to an artist. Going to London, he worked as a clerk and continued his art studies with the friendly assistance of such contemporaries as Hogarth, Hudson, and Lambert. He achieved success as a portrait painter after 1750, becoming a protégé of the Duke of York, and later of George III, who appointed him painter to the Board of Ordnance. F.R.S. 1756, he had educated himself in science, reading Locke, Euclid, Helsham, Desaguliers, Bacon, Boyle, and Newton. In addition to his *Essay . . . of Electricity* (London: C. Davis, 1746), he wrote many articles for the *Philosophical Transactions,* a joint treatise of electrical experiments and observations with Benjamin Hoadly, a treatise on tourmalin experiments, for which he received the Copley Medal, a work on phosphori, and, finally, a popular treatise on electricity in 1780. He engaged in disputes with Franklin, Æpinus, and Euler,

tises on the aether, Wilson visited Robinson in Dublin, and together they performed electrical experiments. Wilson's *Essay*, written in the "proof by experiment" form of the *Opticks*, is supposedly an account of these experiments and the conclusions necessarily drawn from them, with no framing of hypotheses except as questions to be disputed [Preface]. In fact, his work is an extended attempt to identify Newton's aether with electricity. Wilson repeats some of Robinson's propositions on the aether and includes major extracts from Newton's letter to Boyle and the aether queries [29-50, 52-57]. As electric matter, like the aether, is contained in all bodies in reciprocal proportion to their densities (except that it is in greater amount in sulphurous and unctuous bodies), electric matter is probably the same as the aether [4]. A body becomes electrified when it acquires a quantity of electric matter, in excess of its natural quantity, from the vibrating aether thrown off from some other body rubbed against it [8, 13]. Dense bodies may acquire an elastic atmosphere of electric fluid around them, which will repel an equally electric atmosphere surrounding another body; by inference apparent electrical attraction results from aether repulsion. The forces moving the bodies are equal if the aether densities are equal [18-19], but the acting force of electric matter in bodies is proportional to the volumes of the bodies containing the matter [20-21].

Wilson's statements of force relation are incomplete and contradictory—is it aether density which determines force or is it the volume of bodies, thus providing, with density, some measure of quantity of electric matter? Yet, for all his vagueness, Wilson anticipates that the solution of the cause of gravitation, muscular motion, and the action of light will be found in his "explanation" of electricity. It was time for more modest theorizers, more careful experimenters, and more consistent reasoners to take over the study of electricity. Such people were already at work in William Watson and John Ellicott; Benjamin Franklin would shortly integrate the work of these men and their predecessors into the first satisfactory theory of electricity, but these efforts did not end the ambitions of intellectual adventurers aspiring to illegitimate eminence. Speculators continued to dream of explaining electricity by all varieties of activity and of explaining, by electricity, all varieties of phenomena. In 1750 William Stukely, among

but some of his observations were of value, including that on the relation of glass thickness and covered area to Leyden-jar capacity. See his autobiographical sketch in Herbert Randolph, ed., *Life of General Sir Robert Wilson* (London: John Murray, 1862), I.

others, hoped to find in electricity the cause of earthquakes and saw electricity, in turn, as the same thing as aethereal fire, or the subtle fluid of Newton, disseminated through all matter. This aethereal, or electric, fire was one of the four elements of the ancients, it gave to vapors and exhalations that elasticity which made them air, in animals it was "Animal Spirit," the author of life and emotion. It was "the Life and Soul of Action and Re-action, in the Universe," the preservation "against the native Sluggishness of Matter," the "very soul of the Material world."[9]

From the many other possibilities, a final example will be made of Richard Lovett, whose *Electrical Philosopher* of 1774 continues the claims, made in his previous *Subtile Medium Proved* (1756) and *Philosophical Essays* (1766), to have solved by the action of the electrical fluid (which is identical with Newton's aether and elementary fire) the "mechanical Cause that we breath, live, and move; the efficient Cause of all motion; the physical Cause of Gravitation, Cohesion, Magnetism, the ebbing and flowing of the Sea, and of all the other the most abstruse Phenomena of Nature [19]."[10] Most previous books of philosophy have been obscure and perplexing, with matter inert, but possessing attraction and repulsion beginning where attraction ends, and other, similar, "arbitrary maxims, which never were, or can be proved or reconciled [32]." Newton himself had fluctuated between opinions respecting cause, between forces and the aether, though he alone has studied the nature of the subtle medium "so far as to investigate the laws by which it is governed [v]." Now that the electrical, i.e. rational, philosophy has dawned, these things can be explained without contradiction. The system of nature is replete with aetherial matter, or pure elementary fire, an element and permanent principle, taking the form of exceedingly fine air or aether, and discovered by electricity, to exist in the pores of all gross bodies [7-8]. It is "either hot or cold, according to the temperature of the Body in which it exists [8]." There are only two elements, fire and gross matter, one active and the other passive [75-76]. The active matter may be excited to heat by

[9] William Stukely, "The Philosophy of Earthquakes," *Philosophical Transactions* 46 (1750), 731-50.

[10] Richard Lovett (1692-1780), education unknown, was admitted to lay clerkship in Worcester Cathedral in 1722 and remained there all his life. Self-taught in electrical studies, with a particular interest in medical applications, with which he advertised himself able, in 1752, to effect cures. I have used his *Electrical Philosopher: Containing A New System of Physics founded upon the Principle of an universal Plenum of elementary Fire*, etc. (Worcester: for the author, and sold by R. Lewis, Bew in London, Fletcher in Oxford, etc., 1774).

the mutual attrition of its particles contained in bodies [9]; it is indefinitely elastic and rare, and acts by impulse on all other bodies, whence it is the cause of their centripetal tendency toward the sun or earth [75-76].

After nearly thirty years, here is another writer nearly Frekeish in his speculations; but Lovett has a major distinction. His is the only work which frankly acknowledges the authority of one of the materialists, in its citations to William Jones' *Essay on the First Principles of Natural Philosophy* (1762). Lovett's arguments do not differ in their nature from those of Freke or Stukely, they differ only in quantity from those of Martin or Wilson, and, because of this, there is revealed for all of these men, not necessarily the influence of a Hutchinsonian, but the pervasiveness of those impulses which gave rise to the quality of Hutchinsonian arguments. The aether and elementary fire join to become, with electricity, an active principle or substance to oppose to the passivity of ordinary matter. Such impulses are buried also in the work of those men, Watson, Ellicott, and Franklin, who make of these substantial arguments what their less gifted contemporaries were unable to achieve, a scientific theory possessing quantitative and predictive power. It is time now to turn to the accomplishments of these men and to examine how, together, they reformed fiery aetherial speculations into a science.

William Watson's earliest work, about 1745-46, differs little in concept from that of Desaguliers, whom he praises, and from those of his immediate associates, Martin and Wilson.[11] In two papers, illustrating "the Nature and Property of Electricity," he adopts the term "electrical fire" as an equivalent of electrical virtue or effluvia, believing the ignition of volatile fumes by electricity shows it truly to be flame. He also enumerates some similarities between electricity, magnetism, and light, and already shows his concern for quantification, asking "whether or

[11] William Watson (1715-87), educated at the Merchant Taylor's School, London, and, in 1730, apprenticed to an apothecary. By 1738 he had established himself independently and already demonstrated his interest in botany. In 1757 he was awarded the M.D. by the university of Halle and also by that of Wittemberg, and commenced medical practice. Licentiate of the Royal College of Physicians, 1759, he became Fellow in 1784, and was censor in 1785 and 1786. F.R.S. 1741, Copley medalist 1745, for his work on electricity, once vice-president of the society. He contributed more than 58 original papers, and summaries of the work of others, to the *Philosophical Transactions* on natural history, electricity, and medicine. His electrical papers of 1745, 1746, and 1748 were reissued as separate pamphlets, but I have used the following originals from the *Philosophical Transactions*: 43 (1744-45), 481-502; 44 (1746-47), 41-50, 704-709; 45 (1747-48), 93-120; 47 (1751-52), 202-11, 362-76; 48 (1753-54), 201-16; 52 (1761-62), 336-43; and 54 (1764), 201-27.

no the Fire emitted [when an excited nonelectric is touched] would be greater or less in proportion to the volume of the electrified body [43 (1744-45), 491]." In the "Sequel" to these two papers, read to the Royal Society in October 1746, Watson presents the first consistently developed concept of the electric fluid and its action. Watson inclines to the opinion of Homberg, Lemery the younger, and Boerhaave, that fire is an original, distinct principle, formed by the Creator, as opposed to the view of Bacon, Boyle, and Newton, that it is mechanically producible from other bodies. This fire can appear under different conditions as air, lambent flame, or heat; it is contained in all bodies, at all times, and is subtle and elastic [743-46]. Elementary fire is also the same as electrical aether, more subtle than common air, able by its elasticity to overcome the attraction of cohesion, and to pass, to certain and variable depths, through all known bodies [732]. The electrical aether surrounds excited bodies (both electrics and nonelectrics) in an atmosphere which extends to a considerable distance from their surface, and its "accumulation" is to be determined by the power of the body to attract at greater or lesser distances [729-30]. Electrical aether normally exists in all bodies in a determined quantity [734], but the quantity may be varied by using a tube or machine, which acts as a "fire-pump." Attraction and repulsion is an indication of the simultaneous afflux and efflux of electrical aether until the determined quantity is restored, and equilibrium is re-established [734-38].

Watson here makes the familiar identification of electricity with the element of fire, but he uses the concept primarily to lead, by analogy, to the idea that electricity is a substance contained in some determined amount in all bodies. The notion of afflux and efflux—which he apparently derives in part from the Abbé Nollet [see his footnote, 747]—will continue to be a stumbling block to his ideas, but the notion of measuring the amount of electrical substance by the distance through which the body containing it will attract, is a major step toward quantification. Unfortunately, his next paper (of 1748) is a step backward, from which he never recovered. Apparently influenced by analogy with Newton's aether, Watson here defines the natural quantity of the subtle, elastic, electric fluid that a body can contain in terms of that fluid's density—the natural state in all bodies being one of the same degree of density [95]. There is no exchange of fluids between bodies possessing it in the same density, and any discharge between them is in proportion to the different densities, which become the same after the exchange. Possibly Watson was reaching toward some notion of electrical intensity, but his discussion mentions only quantity, and the measure of discharge between bodies is a poor substitute for that of distance of attraction, while density considerations lead to problems of different sized bodies, with the same original density, acquiring different densities from equal transfers of electrical fluid. In this paper Watson refers to some early work by Benjamin Franklin, but fails to recognize that positive and negative electrification by variation in quantity of fluid is not the same as variation in density [98-100].

This blindness continued throughout the remainder of Watson's electrical work. In 1751 he favorably reviewed, for the *Philosophical Transactions,* Franklin's book on electricity, asserting, however, that he had previously and independently achieved many of the same conclusions. In a paper of 1752, on electrical discharge in vacuo, he returned to his speculations on the nature of the substance of electricity, joining to the notion of "Theophrastus, Boerhaave, Niewentyt, 'sGravesande, and other philosophers" of elementary fire, that of Homberg and the chemists on the chemical sulphurous principle [phlogiston?] and showing the similarities with electrical fluid. In 1753, and again in 1761, he reviewed editions of Nollet's *Letters concerning Electricity* for the *Philosophical Transactions.* He notes Nollet's disagreement with Franklin, reasserts his own claim "to have laid the foundation for the doctrine of plus and minus," and does not (quite) agree with Nollet's hypothesis of the simultaneous affluence and effluence of electrical matter, which he reports fully and without disapproval. His final electrical paper, on the effects of lightning, in 1764, still insisting on his prior development of the concept of plus and minus, shows that he retains the view of electrical action as the restoration of a state of equilibrium between densities of electrical aether. He had reached for and missed, without quite knowing why, the key to electrical quantification.

John Ellicott, by way of contrast, grasped the key—though after Franklin had done so—and then, apparently unable to develop, prove, or use it, let it drop.[12] In a single gem of a paper, pub-

[12] John Ellicott (1706?-72), a clockmaker, like his father, gained considerable reputation for the beauty and excellence of his workmanship. Clockmaker to George III, he was also a mathematician of some ability, F.R.S. 1738. In addition to his paper, "Several Essays towards discovering the laws of Electricity," *Philosophical Transactions,* 45 (1748), 195-224, he published papers on pyrometers, clocks, temperature compensating devices, the specific gravity of diamonds, transits of Venus, and the height to which rockets might ascend.

lished in the *Philosophical Transactions* for 1748, Ellicott outlines nearly every one of the basic principles Franklin was, independently, to establish and apply. Electrical phenomena are produced by means of effluvia, composed of particles so exceedingly small as to pervade the pores of all substances [208], which strongly repel each other and are strongly attracted by most if not all other bodies [196]. These effluvia, though replete in bodies, produce phenomena only when separated from the electrified body and put into motion by excitation, all bodies not being capable of equal excitation [207, 217-18]. Attraction is the attempted transfer of electric fluid from a body with a greater amount to one with a lesser, attraction ceasing when quantity equilibrium between them is established [222]. ". . . the greater Difference there is in the Quantity of Electrical *Effluvia* in any two bodies, the stronger will be their Attraction. . . . if the effluvia in each are equal, instead of attracting, they will repel . . . [208]." ". . . the Attraction decreases, as the Squares of the Distances increases . . . [215]." Ellicott even explains the effect of points in favoring discharge, by the increase of particle density at a point, whence the repulsive force acting on each particle is increased [220].

Saved (possibly by the influence of Bryan Robinson) from diversion into aether, fire, and light, Ellicott concentrated his attentions on electrical phenomena and produced the beginnings of an electrical theory possessing real possibilities—even including a speculation on the form of the law of electrical attractive force.[13] As events were to show, however, an electrical force law was a prematurely sophisticated intrusion into electrical quantification, and Ellicott missed the essential ingredient of Franklin's theory—the concept of natural quantity of electricity, with its plus and minus consequences—which made that operative in the null determination of conservation of electrical matter or charge.

Thanks to the labors of Professor Cohen, more is known of Benjamin Franklin's odyssey toward the creation of a theory than is, perhaps, known for any other great scientist.[14] Frank-

[13] In a letter of 25 January 1746 (Wilson Papers, British Museum, Add MSS 30094), Robinson requests Benjamin Wilson to transmit to "Mr. Elicot," further identified as an instrument maker, a caution against regarding the Aether, light, and fire as the same thing!

[14] Benjamin Franklin (1706-90), briefly educated (2 years) in Boston grammar schools. He was apprenticed as a tallow chandler to his father at ten and as a printer to his half-brother at 12. Franklin left Boston for Philadelphia in 1723, where he worked as a clerk and printer, eventually establishing a successful printing business. Active in community and national affairs, for which he visited London, 1757-62, 1762-75, and represented American interests in France, 1776-85. Largely

lin's reading of Hales, Desaguliers, Boerhaave, Newton's *Opticks,* Martin, Wilson, and Watson has been traced and these works analyzed for their influences on Franklin's single-fluid theory of electricity. That theory itself has also been described in such considerable detail that it is necessary here only to summarize from Cohen's studies with, however, a greater emphasis on the materiality of Franklin's views than was entirely consistent with the purpose of Cohen's work.

Franklin's electrical researches essentially began when the Library Company of Philadelphia received a gift of apparatus, with directions for its use, from Peter Collinson, late in 1745 or early in 1746. Franklin's first report on his work was contained in a letter to Collinson of May 1747 (published in extract by Watson in January 1748). Outlining his progress in subsequent letters, he presented the first complete statement of his single-fluid theory in July 1750. Collinson had communicated the early letters to the Royal Society, which, however, did not publish them, and by February 1750 he commenced arrangements for their publication separately. The *Gentleman's Magazine* for January and for May 1750 contained reports on Franklin's work, and, in April 1751, Edward Cave, its publisher, issued what was to be the first edition of Franklin's *Experiments and Observations on Electricity, made at Philadelphia in America.* Cave issued supplements to this edition in 1753 and again in 1754, and later editions were steadily enlarged. Not until the fourth edition, of 1769, was Franklin personally to oversee an edition of his work. In all, the book appeared in five English editions, three French editions, in two different translations, and an Italian and a German edition, while it was included in collections of Franklin's *Works,* of which there were at least two English editions and one American by 1808.

From the beginning, Franklin's electrical papers were distinguished by the clarity of description of the experiments and the simplicity and directness of the theoretical ideas he used to explain them. The concepts he used were fewer than those of most of his contemporaries and were made more immediately graphic

---

self-educated, he learned to read Latin, French, Italian, and Spanish, starting in 1733, and taught himself science. M.A. Harvard and Yale, 1753, William and Mary, 1756; LL.D. St. Andrews, 1759; D.C.L. Oxford, 1762; F.R.S., 1756; Copley medalist, 1753. Foreign Associate, Académie Royale des Sciences, Paris. See Cohen, *Franklin and Newton,* esp. pp. 363-467. I have used Cohen's edition of the 5th (1774) edition of Franklin's book, *Benjamin Franklin's Experiments* (Cambridge: Harvard University Press, 1941), from which bracketed references are taken.

by the terms (plus, minus, conductor, etc.) in which they were described. Franklin modified his ideas and developed them over the years, but his last major contribution to electricity was made in 1755, roughly ten years from the date he commenced his work. The following description, with one major exception to be discussed later, is of that theory as it existed in 1755.

All common matter, says Franklin, normally possesses, evenly diffused through its substance, a natural quantity of electrical fluid, which may differ in amount for different kinds of matter and, for the same matter, in different circumstances. This electrical fluid is a substance *sui generis*, subtle, elastic and particulate; it can neither be created nor destroyed. Particles of electrical matter are mutually repulsive, but are strongly attracted to those of all other matter. Whenever a body, of whatever kind, possesses its natural quantity of electrical fluid, in even distribution, it is in a neutral state and will not attract other neutral bodies. The process of electrification, whether by friction, communication, or induction, is one of removing electrical fluid from one body, or a place in a body, and adding it to another. The earth is an inexhaustible reservoir for electrical fluid, which it can emit or receive without losing its neutrality, but if a body, from which fluid is removed, is left with less than its natural quantity, it is negatively charged, while electrical fluid added to a body in excess of its natural quantity, spreads over its surface forming an atmosphere and giving that body a positive charge. The electrical fluid moves freely and from place to place in and through the substance of non-electrics while, in electrics per se, it will be attracted and held fixed more strongly and in greater quantity. These are, therefore, more properly called conductors and non-conductors, respectively.

A positively charged body attracts a negatively charged one, a neutral body will be attracted to both positively and negatively charged ones, because of the attraction between the (relatively) excess electrical fluid and the (relatively) unsaturated common matter. Under the attraction of a plus or minus body, the electrical fluid in a proximate "insulated" conductor (*i.e.*, one placed so that fluid may not escape to ground) may be redistributed, concentrated in one place and equally deficient in another, so that it is temporarily polarized with one end plus and the other equally minus. If the conductor is "grounded," electrical fluid may pass to or from ground to balance the charge equilibrium between conductor and proximate charged body. Removing the connection to ground will leave the conductor with a deficiency or an excess of fluid. This is charging minus or plus by induction.

The concept of electrical atmospheres had been invented primarily to explain the action across space of electrical force, for Franklin, like his contemporaries (*e.g.*, Wilson and Watson), disliked the notion of action-at-a-distance. The atmospheres surrounding positively charged bodies did not mix when brought together, but elastically pushed against one another, producing repulsion. Atmospheres could also be used to explain the easier discharge of electricity from points or acute angles, as the surface area of common matter, which attracted and retained the atmosphere of electric fluid, is less for a point than for a flat surface, and the particles of fluid can therefore more easily be withdrawn. Unfortunately, atmospheres became something of an embarrassment in Franklin's notable explanation of the action of the Leyden jar, in which the repelling force of electrical fluid passed through glass, while the substance, or fluid itself, did not; and atmospheric explanations for the mutual repulsion of negatively charged bodies, presumably devoid of atmospheres, were difficult to conceive.

Franklin and his fellow American experimenters had early noted this repulsion of negative bodies and, wondering, passed it by. In 1754, John Canton forced Franklin's attention back to the subject with an elaborately argued paper in the *Philosophical Transactions*, "Electrical Experiments, with an Attempt to account for the several Phaenomena . . . [48 (1753-54), 350-58]." Here Canton reintroduces the concept of atmosphere fluid density to explain charging by induction and the repulsion of negatively charged bodies. His proposals were ingenious and complicated, involving a double motion of electrical fluid—away from the bodies along their supports, and to the bodies from the surrounding atmosphere of a proximate, positively charged body—and they invoked that variation of fluid density which Franklin always avoided when possible. Franklin responded with his explanation of induction, outlined above, and a frank, if uneasy, avowal of the repulsion of bodies deficient in electrical fluid, without attempting an explanation. This curious reluctance to accept repulsive forces, previously noted with the dynamic corpuscularians prior to Stephen Hales, turns up again in 1761, when Ebenezer Kinnersley, Franklin's American colleague, wrote to him with an attempt to resolve apparent repulsion into attraction. Franklin responded that he too had been tempted in the same way, but there were many other examples of repulsion in nature, *e.g.*, the elasticity of steam, exploded gunpowder, and magnets, and he had concluded that two causes were more convenient than one alone. A different solution for the repulsion of negative charges, also involving two causes but abandoning the easy application of plus, minus, and neutral concepts, was attempted in the development of two-fluid theories of electricity, proposed, for example, by Robert Symmer in his "New Experiments and Observations concerning Electricity," in the *Philosophical Transactions* for 1759 [51 (1759), 340-89], and two-fluid theories became increasingly popular toward the end of the eighteenth century and in the nineteenth. Franklin, however, preferred the proposal contained in Franz Ulrich Theodor Æpinus' *Tentamen theoria electricitis et magnetismi,* published in St. Petersburg in 1759. According to Æpinus, the particles of ordinary matter normally repel one another while attracting the particles of electrical matter, which also mutually repel. In bodies containing their natural quantity of electrical fluid, the natural repulsion of each kind of matter is reduced to zero, while bodies with an excess of fluid or a deficiency will repel according to the mutual repulsion of fluid or matter and attract according to the attraction of unlike substances. As thus modified, Franklin's single-fluid theory lost its emphasis on electrical atmospheres—which developed into the sphere of action of electrical charge—and this is the form in which the theory was adopted and argued during the last half of the century.[15]

There was a material decrease of graphic simplicity in Æpinus' suggestion, accompanied by an increased attention to problems of mechanism, for Æpinus' *Tentamen* also contained an elaborate mathematical theory involving the forces of attraction and repulsion between these various particles, though the precise nature of these forces was left unspecified. This development of Æpinus' theory was ignored, on the whole, as were Joseph Priestley's deduction, in 1767, of the inverse-square law of electrical forces and Henry Cavendish's beautiful theoretical development of the force-law and its consequences, in 1771 (these will be discussed in Section III), because Franklin's theory was not primarily a mechanical one, for all its description of electrical fluid in the dynamical mechanistic terms of particles and forces. When he began his work, electricity was confused in supposed similarities with the matter of fire, of light, of Newton's aether, or of all three. His ideas were developed in this material-

[15] For a brief discussion of Æpinus and his theories of electricity and magnetism, see Cohen, *Franklin and Newton*, pp. 538-43, 546.

ist atmosphere and his fluid of electricity has its qualities. He managed to escape, however, the pitfalls implicit in the necessity of joining heat, light, gravity, etc., with electricity in some related modes of action. Part of his escape may be due to the model existing, in Stephen Hales' elastic air, for his own elastic fluid of electricity, fixable in common matter. But his fluid also has characteristics similar to Boerhaave's elastic fluid of fire: in its indestructability, subtlety, pervasiveness, and expansive tendency to form atmospheres about bodies already replete with it. In his early letters, Franklin wrote of "electrical fire," and went so far as to suggest, in 1749, that electrical fire and common fire "may be different modifications of the same thing [210]," but the same letter treats them as different elements and he was shortly to drop the term "fire" in relation to electricity. At least from 1750, Franklin carefully distinguished between the action of fire and that of his electrical fluid.

One reason for his discrimination may lie in an early, mistaken belief that electricity can fuse metals without heat, but a contributing factor may also have been the influence on Franklin of the writings of Cadwallader Colden, who also saved him from confusing electrical action with the aether. In 1745 Franklin had offered to print Colden's "piece on gravitation" at his own expense and risk, as something "valuable to the world [18]." Writing to Kinnersley, in 1762, on the nature and existence of fire as an active element diffused through the universe, Franklin paraphrases Colden's example of the spark setting a city ablaze, with a typical Franklinian graphic variation to a spark from a flint exploding a magazine of gunpowder [371]. A letter from Colden, distinguishing fire from the aether—"for aether properly speaking is neither fluid nor elastic; its power consists in re-acting any action communicated to it, with the same force it receives the action"—is printed in the edition Franklin prepared of his own book. Not only are there no arguments offered by Franklin against this view of the aether, the book contains an index citation to "Aether, what" with a page reference to Colden's statement.[16] In an early letter, Franklin had also speculated on a relationship between his "electrical fire" and the common properties of ordinary matter: perhaps removal of enough of it from glass might change the glass, it might lose its transparency, or its brittleness, or its elasticity [198]. But this type of reasoning soon disappeared. As he pared his theory to an electrical simplicity, he

[16] Benjamin Franklin, *Experiments and Observations on Electricity, etc.* (London: Francis Newbery, 1769), 4th edn., p. 277 and first index item.

came to see the nature of his fluid as distinctly different from that of any other substance; the electrical fluid becomes and remains an element in its own, and singular, right.

There is a similar development of increasing simplicity in Franklin's consideration of the action of that fluid. The confusion over the cause of electrification—with Ellicott, for example, as the increased activity of a subtle fluid, or, with Watson, as a variation in its density—vanishes in Franklin's simple transference of substance from one body to another. In the beginning, Franklin had discussed the relation between quantity of fluid and its effect. A test pith-ball "will be repelled . . . more or less according to the quantity of Electricity" in a positively charged body [172]. The attraction between positively charged bodies and negatively charged ones will be greater than of either to neutral bodies, the difference in the quantities of electric fluid being greater [185]. The "quantities of . . . fluid . . . being equal, their repelling action on each other is equal [230]," and the "greater the quantity of electricity . . . the farther it strikes or discharges its fire [221]." Continued discussions of Franklin's theory retain this materialist conclusion that attraction or repulsion is proportional to the quantity of matter in excess or deficiency in the charged body, but Franklin himself appears to have slighted these considerations in his development of the still more materialistic concept of conservation of charge. As the predictive quality of this conservation of substance concept is realized, particularly in explanation of the charging of the Leyden jar, any need for regarding relations between quantity of fluid and force seems to disappear. The theory continues to make some allowance for differences in total charge and for changes in density or distribution of that charge, but the major emphasis is on comparative changes in charge, which can ultimately be tested by a null determination. Whenever a quantity of fluid is thrown in at one surface of a jar, an equal quantity goes out the other [190], and the equivalence can be demonstrated. The explosion and shock from the outside to the inside is the same as that from inside to out, the discharge from a charged to an uncharged jar leaves each half-charged [188-89]—and again the statements are testable by ingenious variations of null experiments. The comment that there is an equal repulsion from equally charged phials or jars [189] becomes almost irrelevant, as one demonstrates the equality of charge, not by the equal repulsion, but by the total disappearance of charge when plus and minus sides of different jars are alternately connected. Any initial tendency to regard Franklin's single-fluid theory of electricity as an elaboration of dynamic corpuscular concepts, in its forcible particularity, disappears in the persistent disregard, by its adherents, of the essential aspects of the dynamics of such concepts. The electrical fluid is an element *sui generis* with no thought of its convertibility, no consideration of relating its particle size with those of other corpuscles, no concern with the measure of its forces, nor with any idea (apparently not even for Æpinus) of transitions from attraction to repulsion. The electrical fluid acts through the characteristics of subtlety and elasticity but is not identified with them. A body becomes electrical by its possession of fluid or its lack of it, and both are determined by Franklin's creation of the law of conservation of charge. There is a mark of greatness in this hardly-mathematical quantitative law, a greatness indicated by the fact that, one of the earliest of the conservation laws to be formulated, it still stands essentially unchallenged. But it marks a theory of substance not one of mechanics.

Although there continued to be some opposition, and alternative theories were proposed, Franklin's single-fluid theory of electricity (as modified by Æpinus), with its associated law of charge conservation, remained an assured success for eighteenth-century British natural philosophy.[17] Naturally such a success prompted attempts to apply a similar explanation involving a unique fluid to magnetism. These attempts were, however, not successful. Like electricity, magnetism had frustrated efforts by corpuscularians, kinematic and dynamic alike, to find explanations to fit their systems. Indeed, because of its polarity, its unique association with iron materials, and its ability to pass through all would-be screening substances except iron, magnetism offered even greater problems to the effluvialists than did electricity, problems not convincingly solved in Descartes' and Boyle's assumption of screw-form particles passing through appropriately shaped pores. Nor had it proved easy to measure the form of a magnetic force law; Helsham's assertion, in 1739, of its inverse-square nature was rightly submerged in the contradictory and inconclusive, but detailed, reports of experiments by Brook Taylor, Musschenbroek, and Desaguliers.

The state of magnetic studies, prior to 1746, is perhaps best

[17] For the general success of Franklin's theory see Cohen, *Franklin and Newton*, pp. 548-68, but additional text citations can easily be found, as, for example, in William Nicholson, *An Introduction to Natural Philosophy* (London: J. Johnson, 1805), 5th edn., one of the most popular texts of the late eighteenth and early nineteenth centuries, whose discussion of static electricity, II, 273-335, is a simplification of Franklin's theory for popular consumption.

illustrated in a *Philosophical Transactions* paper of 1730, by Servington Savery, "Magnetical Observations and Experiments [36 (1729-30), 295-340]," which offers no observations significantly in advance of those made by William Gilbert more than a century earlier. Some of the interpretations of this, otherwise unknown, Mr. Savery are, however, suggestive. The loadstone draws, "by an invisible force different from that of Gravitation, and also Electricity [300]." Though the "directive Vertue" of north and south poles is contrary, "yet the unknown Cause of the Attraction and Repulsion seems to be the same." Magnetic polarity is "constituted of Attraction and Repulsion"; these two powers are "equally strong in the same pole of every Magnet [301]." And, finally, Savery cannot quite believe that the loadstone loses none of its "vertue" by communicating it to Iron or Steel [306].

If this last observation implies a conservation of magnetic virtue, it was an idea that was not pursued. Nor could it easily have been, for a fundamental problem to materialist explanations of magnetism was the inability to find the kind of substance-transfer phenomena for magnetic fluid which had served Franklin so brilliantly for the electrical. Yet the supposition of such a transfer, without conservation notions, seemed to Gowin Knight, in 1747, to offer the possibility of an explanation of magnetic action.[18] In "A Collection of the magnetical Experiments communicated to the Royal Society," published in the *Philosophical Transactions* [44 (1746-47), 656-72], Knight supposes that "the magnetic Matter of a Loadstone moves in a Stream from one pole to the other internally and is then carried back in curved lines externally till it arrives again at the Pole where it first entered, to be emitted again [665]." Two or more magnetical bodies are supposed to attract as a consequence of the flux of a single stream of magnetic matter through them all, while repulsion is the "conflux and accumulation of magnetic matter [669, 671]." Attraction and repulsion vary with distance, as more

[18] Gowin Knight (1713-72), educated at Leeds Grammar School and Magdalen Hall, Oxford, B.A. 1736, M.A. 1739, M.D. 1742. He settled in London, where he is said to have practiced medicine for a time. Beginning magnetic researches early in the 1740's, by 1744 he had invented a method for making the artificial magnets which he sold for a living. F.R.S. 1745, Copley medalist, 1747, for his papers on magnetism (published separately in a collected edition of 1758), he was an unsuccessful rival of Thomas Birch for post of secretary to the Royal Society, 1752. In 1756 he became first principal librarian of the British Museum, where he remained till his death. Of an experimental and practical bent, his only speculative work was his *Attempt to demonstrate, that all the Phaenomena in Nature May be explained by Two simple active Principles, Attraction and Repulsion* (London: J. Nourse, 1754).

magnetic matter will pass from pole to pole, or converge between like poles, as the distance between poles diminishes. Concentrating his attention, Knight achieved a simple, graphic explanation of magnetic phenomena, including that of the curved lines assumed by a scattering of iron filings. But his explanation is scarcely more than a description. It fails to define the character of the magnetic matter which, in its action and quantitative nature, is little if any superior to the magnetic effluvia of his predecessors. Unfortunately, in his attempt seven years later to remedy these deficiencies, Knight deserted the directness of material explanations in the chaos of an elaborate, ingenious, but unworkable and essentially ignored, mechanistic theory.

Just as Franklin, in achieving operational simplicity for his fluid theory of electricity, had submerged the mechanical properties of the electrical fluid in its substance, so, inversely, Knight lost material simplicity in the mechanical properties of a new corpuscular hypothesis. His *Attempt to demonstrate . . . all Phaenomena . . . by Attraction and Repulsion* of 1754 is a curious work. Although he cites both Newton and Hales, Knight writes as though no one, previous to himself, had ever used repulsion as an essential principle in nature [*e.g.*, 2]. There are no references to the score or more mechanistic or aetherial works published, *e.g.*, by Keill, Desaguliers, Rowning, and Robinson, which had made an attempt similar to his, nor to the electrical works which had intervened since his papers of 1746-47, though his studies at Oxford must have introduced him to some of the early works. His preface retains the usual eighteenth-century British castigations of the scholastics and Cartesians for their departures from observed fact, but he denies the applicability of such a test to his own work, where a proposition might differ from observations "without lessening the Truth of the Proposition; the evidence of which is sufficiently clear, without any Reference to Facts, provided the Proposition referred to in the Demonstration be True [35]." Guided by two main principles of reason: nature always operates with the greatest simplicity possible [11], and "the more grand our Conceptions of the Works of the Creation, the more likely they are to be just [14]," Knight proceeds to formulate a theory of matter and its action and, in doing so, provides a fine illustration of the failure, by over-sophistication, of dynamic corpuscular hypotheses.

In its initial stages, Knight's theory differs little from those, abundantly familiar, of other dynamic corpuscularians, but it soon takes on peculiar distinction. The general properties of all matter are existence, extension (and therefore bulk and figure), impenetrability, mobility, and *vis inertiae*. By the rule of simplicity, all primary particles are originally round and of the same size (indefinitely small by the indefinite division of extension) [4, 11, 12]. Besides these general principles, others are needed, capable of producing and continuing motion. These are the principles of attraction and repulsion which, again by the rule of simplicity, act always inversely proportional to distance. Both are the result of the immediate and continuing will of God, both are attested by experience—attraction by gravity and cohesion, repulsion by electricity, magnetism, the elasticity of air, and some chemical preparations [5-7]—and both, as mutually opposing principles, are necessary, for, without repulsion, attraction would coalesce matter into spherical, inactive bodies, and, without attraction, repulsion would reduce matter into atoms and disperse it equably through every part of infinite space [10, 14]. Attraction and repulsion, being contraries, cannot both belong to the same individual substance at the same time. There must therefore be two kinds of matter, one mutually attracting, the other mutually repelling [10].

Every corpuscle of attracting matter condenses, about itself, repellent matter, with a particle density in proportion to the space from which it is excluded and decreasing with distance from the center [41]. This mutually repellent matter will be impelled toward the center with an *apparent* attractive force, reciprocally as the distance, and, following "the great Founder of our Modern Philosophy," it will also experience a *true* attraction to the mutually attracting particle [10, 16-17]. At contact, repellent particles will adhere to the surface of attracting matter with an indefinite cohesive force [19]. Every corpuscle of attracting matter will have about it (either adhering to its surface or as a surrounding atmosphere) as many repellent particles as balance the attracting force. This combination of attracting and repellent matter comprises the various primary corpuscles of ordinary bodies [28]. When the repellent matter, just balancing the attraction, is all united to the surface, the primary corpuscles are "neutral," and neither attract nor repel at sensible distances, though near the repelling surface a body will be repelled [23-24]. Larger corpuscles, in which the surface-to-volume ratio is such as to require the greater part of balancing repellent matter to be in a surrounding atmosphere, are attracting corpuscles. Within such an atmosphere, the particles increasingly attract as the atmospheres begin to coincide and recede behind the corpuscles.

Smaller corpuscles, with comparatively large surface-volume ratios, being unable to repel particles impinging against them before contact, as larger corpuscles with repellent atmospheres might, will have their surfaces "covered as thick as the larger with repellant particles," and thus become repelling corpuscles [21]. Neutral corpuscles will be repelled by repellent corpuscles and attracted to attracting ones, as the sum of repellent matter is greater in the first instance and of attracting matter in the second [26]. Attracting and repelling corpuscles will either attract, or repel, or be neutral with regard to each other, according as the sizes of each are greater or less, and according to their distances [27].

Neither the primary corpuscles (surrounded as they are with surface-adhering repellent matter) nor any bodies composed of them can ever actually come into real contact [25], hence all bodies contain more pores than solid matter, and these pores will contain more or less repellent matter, in proportion to their size [37]. Within a repellent atmosphere, the force of repulsion decreases directly as the distance to the center decreases; the force of attraction, on the other hand, will be in a reciprocal ratio of the squares of the distance, being compounded of the decrease of repulsive force with distance and the similar increase of attracting force [42-43]. This explains the apparent inverse-square gravitational attraction, when attracting forces are actually an inverse function of distance. If a force separates attracting, cohering bodies, resistance to separation will increase with distance, to a certain point, after which it will decrease, for, at near approach, surface repulsion counteracts attraction; at a greater distance equal and opposite forces mutually destroy, the attracting corpuscles are at rest and may be said to be in contact. At greater distances, attraction falls off with the increasing amount of repelling atmosphere intervening [35-36]. This "law of cohesion" explains the phenomena of elasticity, with repulsion at close distances, equilibrium at mean, and attraction at larger, until the corpuscles are drawn beyond the range of attraction and separation (breakage) ensues [36-37].

Knight points out that he cannot be expected to demonstrate the application of his theory to all the phenomena of nature [35], but as he has declared that it can be so applied, he is constrained to show its relevancy to some problems, in addition to those of gravitation and elasticity mentioned above. Since space is filled with repellent particles (with vacuum everywhere interspersed) [15-16], Knight conceives light to be this repellent matter in vibration [58]. He finds it easier to conceive the propagation of light as a progressive movement of a tremor or wave through a series of mutually repellent particles, than to suppose that particles of light can traverse, with motion and direction unaltered, a vast space filled with thousands of repellent particles [53]. Admitting that such a wave would spread from a right line, he declares the force of any lateral vibrations would be too weak to affect our sight, while colors will result, not from differently sized particles, as Newton had supposed (for, by the rule of simplicity, all repellent matter is of equal size), but from the quickness or slowness of the vibrations [54-55]. Heat also is the vibration of repellent particles, but differs from light in the strength of vibration and number of particles vibrating; light being the brisk vibration of a few particles in a line, while heat is the weaker, diffused vibration of many particles. A dense body is slower to be heated and retains heat longer as the heat is not caused by repellent fluid entering the body but by the communication of its vibrations from without [59]. Evaporation is produced because heating expands fluids "to such a Degree, that their Component Corpuscles will be carried out of the sphere of each other's Attraction, and then they will begin to repel each other"—and Knight cites Newton's statement that where attraction ends, there repulsion begins [61].

The smallest of the primary corpuscles of bodies (those compounded of attracting and repelling matter) are the repelling corpuscles, which, brought together in sufficient numbers, constitute pure elementary air. The smallest neutral corpuscles are those of water [29], while larger neutral corpuscles comprise other fluids. The smallest attracting corpuscles "seem to be what Stahl and other Chemists have called the Phlogiston [33]," while still larger attracting corpuscles constitute gums, resins, salts, and earths; and an infinite variety may be produced of various magnitudes of attracting corpuscles and their mixtures with air and water [30]. Chemical operations may now be explained as the progressive substitution of one kind of corpuscle by another, more attractive, one. Air will replace the repellent matter in pores, and may, in turn, be replaced by water, and this by such fluids as contain still more attracting matter, allowing always for the tenacity of the substances and the size of the pores [39-40].

The real test of Knight's theory is, however, its explanation of magnetism, for that was the occasion of its inception, and a third of the book is devoted to magnetic phenomena. Knight supposes iron to be a dense body so constituted that its pores contain more repellent fluid than would belong to an equivalent space without. If that fluid, for whatever reason, were to be set into motion, it would flow in a stream through the iron, the deficiency thus produced would be supplied from without, while the emerging matter would be carried, in the direction of least resistance, around to the side from which the entering fluid had been taken [66-67]. Once begun, such a circulation, being unopposed, would be perpetual and a permanent magnet would result; we may call the pole from which the fluid ever emerges the south pole and that, where it enters, the north. If a north pole be applied to, or near, a south, the stream of matter will pass through both magnets before it is carried back. As the excess repellent matter in the pores must be that which normally would serve to balance the attracting matter and, moving, can no longer do so, the repellent force is diminished in proportion to the amount of fluid passing from one magnet to another, and attraction will prevail in the same proportion [68]. The mutual repulsion of south poles is the opposition of accumulated confluent streams of repellent matter; of north poles, the double flux of repellent matter crowding between them as it enters the intervening space from contrary south poles [78-79]. As the attractive (or repulsive) force is a compound ratio directly of the quantities of relative magnetism (the amount of fluid entering both magnets) and reciprocally of the distances, and as, in different pairs of magnets, the quantities of relative magnetism will differ and vary differently at different distances, there can be no certain law of attraction peculiar to magnets [69-70]. The reason no bodies but iron are magnetic is clear. "No other bodies contain more repellent Matter in their Pores than what belongs to the Space occupied by the Pores; for it follows . . . that they would be magnetical if they did [81]." Knight observes, finally, that explaining electricity is too great a task at present, but the consequences of putting the repellent matter, adhering to the surfaces of bodies, into motion by rubbing should be sufficiently obvious [95].

The ingenuity of Knight's theory is manifest, as is the similitude of some of his explanations of phenomena with those, then or subsequently, adopted by others. The impossibility of precise application of the theory is equally manifest. That it is a mechanistic theory and, at least, proto-scientific is easily seen by a simple comparison with that hardly scientific, semi-materialistic theory of Robert Greene, who had earlier proposed the application of expansive and contractive forces to the explanation of all things. But Knight's parameters are, finally, no more determinable than Greene's infinitely varied mixtures of contractive and expansive forces producing all the qualities of the sensory world. The progression of variously sized corpuscles compounded of various proportions of attracting matter and repellent matter, variously adherent to or surrounding the surfaces of the attracting particles, with attraction and repulsion variously a function of distance and degree of penetration into repellent atmospheres of varying densities and dimensions—how could one hope to assign numerical magnitude to this confusion of varying parameters? Even for magnetism, the original impulse to all his invention, Knight had ultimately to declare the impossibility of finding any determinant law of forces. The consequence of his efforts is easy to judge. Although Joseph Priestley contemplated a history of magnetism, as one of the earliest of his projected histories of all the experimental philosophies, he does not list Knight's book either among those to which he already had, or wanted to have, access. When William Nicholson came to discuss magnetism, in his popular *Introduction to Natural Philosophy,* he declared "that the cause is entirely unknown," while Tiberius Cavallo's *Elements of Natural or Experimental Philosophy* prefers that theory "proposed by Mr. Æpinus . . . which is similar to the Franklinian Theory of Electricity."[19]

In Æpinus' theory, ferruginous bodies are supposed to contain a subtle, elastic magnetic fluid, equally diffused through their substance under normal conditions. When the fluid is concentrated in one part of a body and equally deficient in another, the body is polarized and said to be magnetized, though it contains no more fluid than before magnetization. As the particles of magnetic fluid are mutually repelling but attracted to those of iron, poles with an excess of fluid repel similar poles while attracting those with a deficiency; mutual repulsion of deficient poles is again explained as that of the particles of ordinary matter. Only magnetic force (including that of the earth) can directionally move magnetic fluid to produce the polar imbalance, but heat, hammering, electricity, etc., can aid the motion of fluid by increasing pore dimensions or disturbing the attraction between fluid and matter. If the iron substance is hard, or steel, resistance

[19] Robert E. Schofield, ed., *Scientific Autobiography of Joseph Priestley* (Cambridge: The M.I.T. Press, 1966), item 28a, p. 76; and Priestley, *History and Present State of Discoveries relating to Vision, Light, and Colours* (London: J. Johnson, 1772), appended list of books. Also, William Nicholson, *Introduction*, p. 260; and Cavallo, *Elements of Natural or Experimental Philosophy* (London: T. Cadell and W. Davies, 1803), III, 557-60. See also Benjamin Franklin, *Works* (Philadelphia: William Duane, 1808), III, 223-35, for Franklin's outline of Æpinus' theory and his expressed approval of it, in 1773.

to fluid motion will restrain a return to equilibrium, though this, in turn, may be promoted as well by heating, etc., in a bar not aligned with magnetic polar forces. As with his modification of Franklin's electrical theory, Æpinus defined his magnetic theory in the context of mathematical equations involving the (unknown) forces of attraction and repulsion—and again this part of his system was generally ignored. The inability to separate magnetic fluid from iron and transport it to the increase of magnetism elsewhere blocked further development of null-quantity determinations and the magnetic fluid theory therefore lacks the empirical, quantitative character possessed by that for electrical fluid. Nonetheless, the similarity of the magnetic to the popular and satisfactory electrical theory (a similarity leading even to an alternative, two-fluid magnetic theory), and possibly its graphic contrast to the chaos of Knight, prompted its wide acceptance.

Eighteenth-century progress toward the explanation of heat phenomena somewhat inverted the process seen in electricity and magnetism, for heat began in mechanistic solutions and ended in nearly total materiality. There is, it is true, an exaggeration in the common assumption that the seventeenth century generally regarded heat purely as motion. Bacon had declared heat to be the motion of constituent particles of matter and cold the deprivation of that motion, but the apparent agreement by such scientists as Galileo, Isaac Beekman, and Boyle was anything but consistent and complete. Beekman asserts the existence of "fire particles," small, subtle, pervasive, and apt to motion, which heat bodies by setting their atoms in motion; Galileo also postulates particles, of such size, number, motion, and penetration as to pervade bodies and heat them, the effect being "in proportion to the number and velocity" of these fire corpuscles. "Since the presence of fire corpuscles alone does not suffice to excite heat, but their motion is needed also, it seems to me that one may very reasonably say that motion is the cause of heat. . . . But I hold it to be silly to accept that proposition in the ordinary way, as if a stone or piece of iron or a stick must heat up when moved."[20] Robert Boyle, while explicitly declaring his belief that heat is the local motion of fundamental particles of matter, in his *Experi-*

[20] For Beekman, see Marie Boas, "Matter in Seventeenth Century Science," in Ernan McMullin, ed., *Concept of Matter*, p. 348; for Galileo, Stillman Drake, trans., *Discoveries and Opinions of Galileo* (New York: Doubleday Anchor Books, 1959), "The Assayer," pp. 277-78.

ments, Notes, etc. about the Mechanical Origine or Production Of divers particular Qualities, "Heat and Cold," of 1675, also introduces the concept of ponderable fire particles, in his *Essays of . . . Effluviums . . . New Experiments to make Fire and Flame Ponderable,* of 1673, to explain the increased weight of tin and lead calcined in sealed glass retorts. Even Newton confuses the issue, for his Query 20 (of the 1706 Latin *Opticks*) supposes heat to be the vibrating motion of the parts of bodies, while Query 18 (of the 1717 *Opticks*) proposes an intimate relationship between a vibrating aether and the heating of bodies. So far as authorities were concerned, there was ample justification for the ambiguities of such confirmed dynamic corpuscularians as Clarke and Freind, or of such experimental Continental Newtonians as 'sGravesande, Boerhaave, and Musschenbroek, in holding that, if heat is motion, it must be the peculiar motion of particular particles, those of a substance of fire.

Authority aside, however, there was need of a special faith in the mechanical philosophy to enable such men as Hales, Desaguliers, and Rowning to persist in viewing heat as the motion of constituent particles of matter, for a motion theory of heat had many obstinate and contradictory observations to reconcile. One of these, the production of heat or cold by chemical dissolution, was to be troublesome at least as early as 1701. In that year, the French chemist Etienne Geoffroy sent "Observations upon the Dissolutions and Fermentations which we may call Cold . . . ," to be published in the *Philosophical Transactions* [22 (1700-01), 951-62]. Geoffroy accepts "with all physicians," the principle that cold is the diminution of motion, and he proposes a conservation of motion explanation for the cold produced when salts dissolve in water—"the Salt Particles being without motion, and dividing that [of the] Liquor, diminishes it so much the more. This . . . produces the Cold greater or less in the same Liquor [955]." But what then produces the heat of solution of some salts in water? ". . . these Salts . . . are loaded with many Fiery Particles which they hold, as it were in Prison, in their Pores. These Igneous Particles regain their Liberty by . . . Dissolution . . . and being very active, do augment the Agitation of the Watry Particles till they make it very hot . . . [956]." Another apparent contradiction between theory and observation appeared in the relation between heating and the mechanical properties of the bodies heated. George Martine notes these in his essay of 1740, "On the heating and Cooling of Bodies."[21] Oil, being more

[21] George Martine, Essay Five, in *Essays, Medical and Philosophical.*

tenacious than water, should be slower to heat; yet, consisting as it does of "sulphureous particles," it should be more apt to attract heat. On either account it should be slower to lose heat, but this is not generally true [262-63]. From the doctrine of *vis inertiae*, Boerhaave and Musschenbroek adopt, as a principle, that the faculty of being heated, or of cooling other bodies, is in proportion to the density of the medium, and this also is not generally true [263, 271]. In accordance with his general attitude, Martine recommends ignoring speculations concerning the nature of things and a return to the solider footing of empiricism.

Martine's was the counsel not simply of ignorance but of irredeemable defeat, and it naturally was unacceptable to a generation of natural philosophers well aware of the use of hypotheses for all their praise of Bacon and their parroting of Newton's "hypotheses non fingo." Yet the situation seemed inescapable. If an hypothesis was to be made Newtonian, in an acceptable experimental sense, it must yield quantities capable of some confirmation, and neither the purely mechanistic theory nor the semi-mechanistic, material-fluid theory had afforded such possibilities. Both Desaguliers and Musschenbroek, for example, had confessed the impossibility of deciding the mechanical properties (size, shape, motion, and forces) of bodies and their relationship to heat, but the alternative fluid-of-fire hypothesis involved these properties as well, and also, possibly, the state of condensation of the fluid. With confusion rampant among the disciplined natural philosophers, it is not surprising that the speculations of the visionaries ran uncontrolled. Who could prove that fire was not the active matter, soul of the world, and equivalent to electricity, aether, and air, rarefied and abraded?

Nor, in this instance, was the example of the men who had tamed the fluid of electricity of much direct value. Benjamin Franklin, for example, naturally adopted a material theory of heat. Fire exists, diffused through all matter, as a subtle, elastic fluid incapable of being created or destroyed. Normally quiescent, by suitable operations it can be excited, disengaged, and brought into action. Active fire permeates bodies, according to their conductivities, and there overcomes the normal cohesion of body particles, causing expansion, fusion, evaporation, or the appearance of sensible heat, flame, and light. When two equally good conductors, "one heated, the other in its common state, are brought into contact . . . the body with the most fire readily communicates of it to that which had the least . . . [which] readily receives it, till an equilibrium is produced."[22] Perhaps Franklin's assumption of the fluid nature of fire, coupled with the evident successes of his theory of electricity, influenced others to a further development of the concept, but Franklin's own development of it, violating as it does nearly every principle by which his electrical fluid had achieved success, can not have provided much guidance. For this is no substance whose activity is a function of its departure from some natural quantity. This fluid of fire must be excited into action and that action may exhibit a range of consequences, only one of which, the heat equilibrium of conducting bodies in contact, has quantitative implications. And without the standard of "natural quantity," what does heat equilibrium imply? Is it, Franklin forbid, an equal density of the fluid of fire? Or is it an equivalent amount of fluid per unit mass, or the same temperature for each body? Franklin does not say, and though, operationally, one may assume he meant temperature equalization, it is by no means clear just what, in relation to heat, temperature measures. These were the unanswered problems whose solutions were comprehended in the first successful theory of heat, by the Scottish chemist and physician, Joseph Black.[23]

Details of the development of Black's theory of heat and its influence during the eighteenth century are, as yet, imperfectly known. Black himself never published a written account of his theory and that published by John Robison, in the posthumous edition of Black's *Lectures on the Elements of Chemistry*, did not appear until 1803.[24] Black began lecturing on his theory as

[22] Franklin, *Works*, III, 295-96; also 180-81. See also Cohen, *Franklin and Newton*, pp. 323-40.

[23] Joseph Black (1728-99), born in Bordeaux of Scottish parents, was educated at home and in Belfast, before he entered the University of Glasgow in 1744. Though he began there his study of medicine and chemistry, under Cullen, he migrated, in 1752, to the University of Edinburgh, where he completed his degree, M.D. 1754. In 1756 he became professor of anatomy and chemistry at Glasgow; in 1766 he commenced his life's career as professor of medicine and chemistry at the University of Edinburgh. Member of the Royal Society of Edinburgh; foreign associate, Académie Royale des Sciences, Paris, and of the St. Petersburg Academy, he was distinguished for his work on heat described above and for pioneering work in pneumatic chemistry to be discussed in the next chapter. There exists no adequate biography of Black, but see the excellent article, Henry Guerlac, "Joseph Black and Fixed Air: A Bicentenary Retrospective, with some New or Little Known Material," *Isis*, 48 (1957), 124-51, 433-56; and the classic study by Douglas McKie and Niels H. deV. Heathcote, *The Discovery of Specific and Latent Heats* (London: Edward Arnold and Co., 1935).

[24] John Robison, ed., Joseph Black, *Lectures on the Elements of Chemistry* (Edinburgh: by Mundell and Son, for Longman and Rees, London, and William Creech, Edinburgh, 1803). Bracketed page references, unless otherwise identified, will be to this work.

early as 1762, however, and copies of student notes of these lectures were freely circulating as early as 1767. A short, anonymous, account of his ideas was published in 1770, as *An Enquiry into the General Effects of Heat, with Observations on the Theories of Mixture,* and subsequent publications, by such of his students as William Cleghorn and Adair Crawford, contained statements of Black's theory and spread knowledge of it well before the end of the century.[25] Although Black's ideas must have undergone some evolution in time, and the secondhand reports available surely include the interpretations of others, there is sufficient agreement among these versions on which to base an account of his theory of the nature of heat.[26]

That theory must, however, be extracted, frequently by inference, from Black's explanations of heat phenomena. Early in his career, Black disavowed "theories which pretend to investigate a cause [Enquiry, 2]." After the publication of Cleghorn's "de Igne" in 1779, he is supposed to have declared Cleghorn's "idea of the nature of heat . . . the most probable of any that I know [34]," and his *Lectures* are permeated with that "idea" of heat as a fluid with certain dynamic properties. Cleghorn surely derived some of his notion of heat from Black, but John Robison's penchant for dynamic interpretations of nature makes suspect any extraction, from his version of Black's *Lectures,* of a theory involving forces. Nevertheless, the *Enquiry* had earlier dismissed, as inadequate, both the view that heat is the motion or agitation of the parts of bodies and the idea that it is the agitation of that subtle, elastic fluid, known as Newton's aether—this latter view being further described as fanciful and uncertain [*Enquiry,* 1-2]. Discussing, in the *Lectures,* the belief of "the greater number of English philosophers" in the mechanical origin of heat, Black declares, "I acknowledge that I cannot form to myself a conception of this internal tremor, that has any tendency to explain, even the more simple effects of heat, or those phenomena which indicate its presence [32-33]." He

[25] [Joseph Black], *An Enquiry into the General Effects of Heat, etc.* (London: J. Nourse, 1770). The potter, Josiah Wedgwood, possessed manuscript extracts of Black's lectures, dated 1766-67 (now in the Wedgwood Museum, Josiah Wedgwood and Sons, Barlaston, Stoke-on-Trent), and Matthew Boulton received a copy, from Erasmus Darwin's son, Charles, of notes on Black's theory in 1777 (now in the Assay Office, Birmingham). The work of Cleghorn will be discussed in Part III.

[26] Studies by the late Dr. Douglas McKie and David Kennedy, "On Some Letters of Joseph Black and Others," *Annals of Science,* 16 (1960), 129-70, have shown that Robison expanded and modified the text of Black's *Lectures* in publishing them. A "variorum" edition of these *Lectures,* collating a selection of students' notes over the years, is badly needed.

also criticizes the view of "the greater number of French and German philosophers, and Dr. Boerhaave" that the motion of which heat consists is that of "the particles of a subtile, highly elastic, and penetrating fluid matter, which is contained in the pores of hot bodies." The main burden of his complaint about this theory is, however, less the inadequacy of the concept than the failure of its authors fully to consider or apply it to the whole of the facts and phenomena [33].

After eliminating, from the *Lectures*, the discussions directly associated with Cleghorn's mechanization of the fluid of heat, and keeping only material consistent with views expressed in his lectures prior to 1779, the conclusion is still clear. From his repeated recommendation of Boerhaave's text, especially the section on the subject of heat; from his lecture references to the idea "which I lately mentioned as most plausible," that heat depends on the abundance of a subtle matter, highly elastic, which easily enters the pores of bodies, an idea so plausible that notwithstanding heat's seeming imponderability, he imagines his students would find themselves "more and more impressed with the belief that . . . [it] is the effect of a peculiar substance [49]"; but most of all, from the internal consistency of his own interpretations of phenomena, it is apparent that Black, from very early, held a material-fluid theory.

The extent of his genius is shown in the way that Black combined the operations of the laboratory with the simplest concepts of substance transfer to discover and explain heat phenomena, all without conscious articulation of the nature of the substance whose existence his work implies. He adopts, from Boerhaave, the conviction that expansion, a uniform gradual effect of heat as registered on a thermometer, is the only objective indicator of the presence of heat; but, unlike Boerhaave, Black insists that this is a measure of general strength or intensity, not of quantity. From Boerhaave, he also adopts the concept that heat seeks an equilibrium, diffusing uniformly and equably in all directions until that is achieved; but he denies that this equilibrium is one in which the quantity of heat has become equal in all space. The equilibrium is one of temperature, not necessarily of quantity. Finally, he employs the technique of mixtures, combining substances of varying natures, weights, and temperatures with great facility and a firm conviction in the conservation of heat.[27] In

[27] See McKie and Heathcote, *Discovery*, pp. 59-63, for a model exposition of the difference between Black's experiments, guided by a theoretical conviction in heat conservation, and the empirical experiments of Georg Wolfgang Kraft.

effect, Black substitutes temperature, equal diffusion of heat, a standard substance (usually water), and conservation of heat for the forces, natural quantity, and conservation of charge of the electrical-fluid theorists.

Inevitably Black is most famous today for his "experimental discovery" of specific and latent heats, but these were dependent upon his prior invention of the technique of heat measurement, now called calorimetry, and this, in turn, required the establishment of those operational-theoretical concepts just listed. Boerhaave and Musschenbroek had supposed that equal volumes of space contained, at equilibrium, equal quantities of heat, because a thermometer registered equal temperatures in whatever substance filled that space. Black demonstrated that equivalence of temperature seldom meant equal quantities of heat. Exposing mercury and water, in equal quantities by weight and at equal temperatures, to the same degree of heat, the mercury reached equilibrium temperature for less heat absorbed, *i.e.*, at a faster rate than the water. Moreover, when mercury and water were mixed together, water raised the temperature of the mixture more than did an equal weight of mercury, at the same temperature. Clearly mercury required less heat to raise its temperature than was required to raise that of an equal weight of water by the same amount; water had, in effect, a greater "capacity" for heat than mercury. Nor was this "capacity" a function of the densities of the materials involved. In a series of mixing experiments, Black showed that tin, lead, oil of turpentine, iron, and copper each had a different capacity for heat which was a unique function of the material. He regarded this independence of heat capacity and density as "totally inconsistent" with a motion theory of heat, for the laws of motion should require a relationship [78].

The discovery of heat capacities was the first fruit of Black's theory of heat, and, once the capacity of any substance is determined, it permits the prediction of the consequences of almost any mixing experiment. Heat capacity, or specific heat, related originally, however, only to substances which retained their physical state—*i.e.*, which remained solid, liquid, or gaseous throughout the mixing process. By a brilliant extension of his heat absorption and mixing techniques, Black also arrived at a quantitative heat conception of change of state. Convinced that "Coldness is only the absence or deficiency of heat" and that "frigorific particles," necessary to concretion, were "a groundless work of imagination [26, 30]," Black began an investigation of the processes of freezing and melting. Exposing to uniform high, or low, temperatures two flasks of equal size, form, and weight, containing equal weights of various fluids, Black observed the resulting changes of state. When the temperature was high, and the flasks contained ice and a snow-water mixture, though both received the same amount of heat in the same time, the snow-water temperature was raised by seven degrees in 30 minutes while the ice did not thaw for 10 hours. When the temperature was well below freezing and the flasks contained water and a water-salt mixture, though both lost the same amount of heat in the same time and cooled at equal rates until they reached freezing temperature, the salt water continued to cool to eight degrees below freezing in 30 minutes while the water remained at the freezing point for a long time, until it was all converted to ice. Black argues that there is an amount of heat, not apparent in sensible form but latent in a fluid, whose repulsive power prevents the attraction of cohesion called freezing unless it is released and converted into that sensible heat which a thermometer registers. "When we perceive that what we call heat disappears in the liquefaction of ice, and reappears in the congelation of water . . . we can hardly avoid thinking it a substance, which may be united with the particles of water and may be separated . . . [192]."

Having determined that fluidity depends upon some amount of heat latent in fluid, Black went on to find a similar latency of heat in vaporization. Water at 54 degrees reached a constant boiling temperature in four minutes, but did not completely evaporate for another 20 minutes. During that time, it must have received five times the amount of heat required to raise the temperature to boiling, hence 790 degrees of heat entered the composition of water to make it steam.

Black's determinations of the latent heats of fusion and vaporization were subject to change, as were his evaluations of the capacities for heat of various substances, but the concepts themselves resisted reinterpretation for nearly a century. By their use, a range of previously unrelated phenomena were joined and explained. In the change of heat from latency to sensibility, the delay of freezing was, for example, to be understood. Each particle of water become ice released heat to warm the surrounding water. Similarly, each particle of water become vapor took latent with it heat formerly sensible, leaving the remaining water cooler for its departure. Problems remained—the heat of chemical combination, the transference of heat through vacuo, the apparently inexhaustible production of heat by friction—and these, not easily explained by a material theory, prompted subsequent renewed attacks from mechanists. By 1805, the cautious William Nicholson refused to decide between a substance and a mechanical theory; but Cavallo, at nearly the same time, was still prepared to accept, as the most satisfactory hypothesis, the "modern theory" in which the existence of a "subtile and elastic, fluid . . . called *elementary heat*, or simply, the *caloric*," is assumed.[28] Particularly for chemists, Black's substance of heat, called by Lavoisier the *élément calorique*, remained a viable explanatory concept well into the nineteenth century.

Avoiding speculation on the modes of its action, Black nonetheless created a substance of heat, the variation in whose simplest property, its quantity, was alone sufficient to explain the most pressing problems of heat phenomena. Here was pure materialism in its most successful effort. It is hard to blame Black for his one flight of speculative exuberance, the declaration that "heat is the general material principle of all motion, activity and life, in this globe [243]." And if that statement seems an echo of Hutchinsonian faith in the active matter of fire, surely that is appropriate. For Black's fluid of heat, like Franklin's fluid of electricity, represents the product of an experimental policy based on a materialist creed.

[28] Nicholson, *Introduction*, II, 114; Cavallo, *Elements*, III, 8-9.

CHAPTER NINE

Vital Physiology and Elementary Chemistry

DURING THE ASCENDANCY of dynamic corpuscularity few areas were more prolific in ingenious mechanistic speculation than physiology and chemistry. With the shift in intellectual climate, in none was the flight from mechanism more explicit and complete. When the transition began, about 1740, the standard authorities cited in British works on physiology were Borelli, Bellini, Pitcairne, and James Keill, and those on chemistry were John Keill, Freind, and Boerhaave.[1] The progress of change, by 1760, is indicated by the nearly unanimous substitution, for these authorities, of Stahl, Hoffmann, and von Haller in physiology and Stahl, Geoffroy, and Macquer in chemistry. The end of the evolution is illustrated, for physiology, in William Heberden's declaration, published posthumously in 1802: ". . . to living bodies belong many additional powers, the operations of which can never be accounted for by the laws of lifeless matter. The art of healing, therefore, has scarcely hitherto had any guide but the slow one of experience . . . nor will it . . . till . . . some superior genius . . . [appear] capable . . . of discovering that great principle of life, upon which its existence depends. . . ." The empiricist, anti-mechanist, equivalent in chemistry is the assertion of Lavoisier: "We must trust to nothing but facts: These are presented to us by Nature, and cannot deceive. . . . All that can be said upon the number and nature of elements is . . . confined to discussions entirely of a metaphysical nature. The subject only furnishes us with indefinite problems, which may be solved in a thousand different ways, not one of which, in all probability, is consistent with nature . . . if, by the term *elements*, we mean to express those simple and indivisible atoms of which matter is composed, it is extremely probable we know nothing at all about them. . . ."[2]

For both physiology and chemistry, the development during the second part of the eighteenth century is one of escape from mechanical reductionism, in which causation is sought in un-

[1] Newton and Hales also belong to the panoply of early authority, but they are retained in the later lists, and the reinterpretation of their ideas, to retain their prestige for newer views, is part of the transitional process.

[2] Heberden is quoted by Robert T. Gunther, *Early Science in Cambridge* (Oxford: for the Author, 1937), p. 284; Antoine Lavoisier, *Elements of Chemistry*, trans. by Robert Kerr, with a new introduction by Douglas McKie (New York: Dover Publications, reprint of the 1790 edition, 1965), pp. xviii-xxiv.

differentiated matter, motion, and forces, and the achievement of independent positions as autonomous disciplines. Physiology attained its goal by developing an empirical nosology—*i.e.*, a taxonomy of diseases—and by erecting a barrier of vitalism behind which it defined its own problems and modes of investigation. The relation between these and the work in other branches of natural philosophy thus becomes increasingly obscure, and the task of extracting theory from apparently empirical descriptions is a job for the specialist. This study will, therefore, confine itself to investigating those aspects of the transition which seem to show peripheral connections, at least, with general changes in eighteenth-century natural philosophy.[3] In chemistry, the autonomy was less sharply divergent from the other physical sciences, but here a rampant materialism expressed in element-taxonomy supplied generations of chemists with sufficient labors and successes to justify their resistance to frequently concurrent efforts at the reestablishment of a physical chemistry. For the most part these efforts lie outside eighteenth-century British studies, though some of the mechanists, discussed in the next section, seem to have been working toward such a goal.

The transition from mechanism to a vitalistic autonomy in British physiology is perhaps most easily surveyed by examining changes in attitude toward explanation of some single set of phenomena. In the contrast between the early Croonian lectures on muscular motion, delivered before the Royal Society between 1738 and 1747, and the concepts developed by the Scottish physicians Robert Whytt and William Cullen, there is a convenient introduction to just such a set of phenomena on which opinion underwent a critical and rapid development. In the first set of Croonian lectures, by Alexander Stuart in 1738, a somewhat naïve application of dynamic physiology can still be seen. All matter consists of Newtonian *minima,* small, hard, and impenetrable, which form fluids and solids by joining in various combinations of varying sizes and shapes. Matter possesses powers of attraction, which act reciprocally as the squares of the

[3] This restriction is an attempt to avoid the detailed analysis of professional physiological works of the period after 1750, for which I have no competence. Many of these works I have read for the more obvious relationships to trends in contemporary natural philosophy, but I am here particularly dependent upon accounts in the secondary literature. Much of the writing in medical history seems unconcerned with developments in non-medical subjects. Happily this is not true of the work of Dr. Lester S. King, on whose *Medical World of the Eighteenth Century* (Chicago: The University of Chicago Press, 1958), and *Growth of Medical Thought* (Chicago: The University of Chicago Press, 1963), I am clearly dependent, though my interpretations occasionally depart from his.

distances from the center; apparent repulsion, as in elasticity, is actually the result of attractive forces restoring the equilibrium between *minima* in combination. These central forces of attraction, of which God is the immediate, acting, and ubiquitary cause, are, in turn, the cause of gravity, cohesion, elasticity, hydrostatic principles, and of all motion in the animal "oeconomy." In muscular motion, the originating cause is the immaterial power of the mind, which impels an aqueous fluid called animal spirits from the brain or spinal marrow through the tubes of the nerves into the vesicles of muscular fibers. Being there in greater than normal amounts, the animal spirits distend the vesicles, and thus also the associated capillary vessels, expanding them to lessen resistance to blood flow and open them to still more nervous fluid. Hence the muscles are enlarged and moved, until the mind ceases to send fluid, when, by their elasticity, the veins and blood globules contract, shortening the muscle and expelling excess fluid through the veins toward the heart.[4]

Stuart's views are probably a reflection of those held and taught by Boerhaave, whose medical theory, like his chemistry, contained strong elements of mechanism. Though Boerhaave did not deny the existence of vital and animal powers, he rarely employed them in explanation and, like most of his mechanist colleagues, he had to defend himself against accusations of eliminating the soul from his considerations. Boerhaave insisted that his medical system was not purely mechanistic, "but if some portions of the human body correspond in their structure with mechanical instruments, they must be governed by the same laws. For all the power of these parts is in the motion which they produce; and motion, by whatever body it is performed, takes place according to the universal laws of mechanics."[5] In its details, however, Boerhaave's medical theory, as revealed in his most influential

[4] Alexander Stuart, "Three Lectures on Muscular Motion," supplement to the *Philosophical Transactions*, 40 (1739). Stuart had earlier read an "Explanation of an Essay on the Use of the Bile in the Animal Oeconomy," *Philosophical Transactions*, 38 (1733), 5-25, in which he invokes the standard mechanistic parameters of velocity of circulation, secretion by passage in lateral branches of the circulating system, particle size, etc., in explaining the cause of sleep. He is not memorialized in the *D.N.B.*, but the *British Museum Catalogue of Printed Books* lists other publications by Stuart, including a doctoral dissertation on muscular motion published in Holland in 1711. From this, and some praise of Boerhaave, we may assume Stuart to be a product of the Medical School of Leyden and guess that he was in his mid-thirties at the time of his lectures.

[5] Quoted by John Thomson, *An Account of the Life, Lectures, and Writings of William Cullen* (Edinburgh and London: William Blackwood and Sons, 1859), I, 208-09; see also King, *Medical World*, pp. 59-121.

text, the *Institutiones Medicae* of 1708, appears to contain the same eclectic mixture of mechanical hypotheses and empirical materialism previously described for his chemistry. Where he could, he used an early, Pitcairnian, version of iatro-hydrodynamics, but he supplied the deficiencies with an admixture of empirical practice and of iatro-chemistry, in which disease-producing substance was specialized in species categories of acid, alkali, purulent, ichorous, and putrid matter. The influence of his teaching was confused by later variations of explanation, as expressed, for example, in the composite *Academic Lectures on the Theory of Physic* of 1743, where James Keill is named as the proponent of the attraction theory of muscular motion and Newton becomes the defender of an aetherial-action theory.

The existence of a reactionary element in physiological explanation is revealed in the set of Croonian lectures delivered by James Parsons in 1745. Surveying previous attempts to discover the cause of muscular motion, Parsons finds all equally wrong. Descartes lacked even anatomical justification for his ideas; iatrochemical ferments were all fanciful, whether they were, with Croone, supposed to be a mercurial liquor in the blood impregnated with volatile salt and sulphur, or, with John Mayow, nitro-aereal particles from the brain combined with salino-sulphurous particles from the blood. Nor were aetherial explanations any better. Bernouilli's proposal of spiculae of the nervous fluid penetrating blood globules to release their elastic aether failed to demonstrate the existence either of spiculae or of aether, while Bryan Robinson's suggestion that the muscles contained Newtonian aether which was set into vibrations at the power of the will did not explain how such vibrations caused muscular contraction. Of dynamic corpuscular concepts Parsons had no understanding. James Keill's preferential attraction by the small particles of nervous fluid for the particles of blood, allowing expansion of entrapped air, differed, for Parsons, in no understandable way from Bernouilli's theory. The blood did not contain air and, besides, how could Keill suppose rarefaction without heat, or condensation without cold? Parsons, instead, supposes the existence of an "aura" (sometimes also described as an air) whose essential property is "to inflate." This aura normally fills the hollow cells (he has discovered to exist) in muscle fibers and the tubular nerves to just that amount which maintains an equilibrium between antagonist muscles and between these and interstitial air or aura. By impulse of the will "an additional Inflation is made to the Cells," by which their force becomes superior to opposing forces, and an expansion ensues until the impulse ceases or is given alternately to antagonist muscles, and equilibrium is restored.[6]

The conservative origin of Parsons' reaction is suggested by his book, *On the Analogy between the Propagation of Animals and Vegetables* (1752). As the title indicates, the work does not relate primarily to questions of muscular motion, and any physiology in it is confused in its curious combination of religious moralizing and mystic speculations about life. It is clear, however, that Parsons rejects mechanism as a solution to biological problems. ". . . the different Particles, that go to constitute the most simple Bodies . . . are innumerable, and never to be comprehended by any human Power [152n-154n]." In vain have authors of secretion theories (*e.g.*, Keill) "sought for Particles and Pores of different Configurations, in vain had Recourse to the *Momentum* of the Blood, and in vain endeavour'd to reconcile the Doctrine of Secretion to Mathematical Calculation [169n]." Plants and animals are characterized by their possession, in such different degrees as God has determined, of an animating principle and a principle of organization. Neither of these in the least resembles the "plastic Power" of such philosophers as Henry More, which is "destitute of any Possibility of being made manifest to our Understanding" or of agreeing with the "surer Testimony and Assistance of experimental Philosophy [17]." The animating principle has the regulating power which renders the action of organization compatible with "social Harmony . . . prevents its Interruption, corrects Movements of all its springs, makes the distinction between use and abuse, and propagates virtue and Truth, instead of vice and confusion by its guidance of the organization [253-54]." The organizing principle, which may lie inactive for a time in seed or egg, is that which makes it possible for living beings to grow [20-21]. The different parts of bodies select from the general fluid, by a Galenic process of attraction by similitude, kindred particles necessary for their growth, till the whole organization arrives at full

[6] James Parsons, "The Crounian Lectures, on Muscular Motion," supplement to the *Philosophical Transactions*, 43 (1744-45). Parsons (1705-70) was educated in Ireland and studied medicine for several years in Paris before receiving the M.D. at Rheims, 1736. Returning to England, he was a licentiate of the Royal College of Physicians, 1751; a member of the Society of Antiquaries, Society of Arts, and Spalding Society. F.R.S. 1741, foreign secretary of the Royal Society, 1750, he wrote extensively on medical problems, attacked the lithontriptic properties of Mrs. Stephens' "remedy" for the stone, and published *Philosophical Observations on the Analogy between the Propagation of Animals and Vegetables* (London: C. Davis, 1752).

growth [140n-42n, 150-54]. One may wonder at Parsons' ability to discriminate between plastic nature and animating and organizing principles, as earlier his "aura" with its inflating property is distinguished from Keill's elastic air. Nonetheless, Parsons finds, in Hales' *Vegetable Staticks,* proof that plants rid themselves by perspiration of just those particles which are not matched anywhere within it [157], and such varied observations as those of Newton on heterogeneous light and of Henry Baker on the polyp provide "proof" for Parsons' theorizing.

In the Croonian lectures of 1747, Browne Langrish makes a valiant attempt to return to unprincipled mechanism, but his efforts also reveal the contemporary revival of aetherial ideas, with the presumed relationship to fire, light, and electricity.[7] Langrish dedicates his lectures to Stephen Hales, from whom he had received "many personal particular Favours," and notes that he has not altered his sentiments on the cause of muscular contractions since the publication of his essay on that subject in 1733. The first of his three lectures outlines a theory of matter which appears completely to support that assertion. The *minima* or primary particles of matter are perfectly hard, solid, and inseparable. Elasticity can proceed only from aggregates of *minima.* The elasticity of solids proceeds primarily from the attraction and repulsion of every least corpuscle of matter which, Langrish believes, is individually polarized with an attractive virtue on one side and a repulsive power on the other. In muscular fibers, the corpuscles are united at particular points corresponding to their attractive virtues. In distention, some of these are disunited while others are turned so their repulsive poles are in opposition. The combination of attraction and repulsion acts to restore the original shape. Newton's Query 31, with particular reference to crystallization and the polarization of light, is cited in support of polarized primary corpuscles and, not surprisingly, Langrish notes that Desaguliers had approved "what I had formerly published beyond all other accounts of muscular motion."

[7] Browne Langrish, "The Croonian Lectures on Muscular Motion," supplement to *Philosophical Transactions,* 44 (1747-48). Langrish (?-1759) was educated as a surgeon and, before 1733, practiced as such in Hampshire. In 1735 he was an extra-licentiate of the Royal College of Physicians and began medical practice. He obtained an M.D. in 1748, having been F.R.S. since 1734. In 1733 he published an *Essay on Muscular Motion* of which his 1747 Croonian lecture was an extension. He also published a work on *Modern Theory and Practice of Physic* (1735, 2nd edn. 1764), showing original clinical research; a vivisectionist study on dissolving the bladder stone (1746), and a clinical account of smallpox (1748).

In his 1733 essay Langrish had, however, ascribed to "animal spirits" the variation in corpuscular attraction which resulted in muscular motion. Now (with his contemporaries and, perhaps, following Hoffmann), he has identified these spirits with the aether or, because he always calls it "nervous aether," perhaps some particular modification of that substance. Noting that "nature seems to delight in Transmutations," he observes, from Query 30, that Newton believed light entered the composition of bodies and was the source of much of their activity, while Homberg declares that fire also is part of material composition, increasing the weight of bodies exposed to it. So was the aether, the most refined matter in nature, a part of bodies, perpetually secreted from the blood by the glands of the brain and flying into the nerves for the uses of muscular motion. The muscle fibers are normally in a state of tension, balanced between their attractive forces and the opposing impulse and pressure of circulating fluids and the repulsion of fiber corpuscles if distended. The will has the power to direct the aetherial medium contained in the tubular nerves into the cavities of muscular fibers, where it increases the attractive virtue of fiber corpuscles, causing them to overcome the opposing forces. When the will withdraws the impetus given the nervous aether, the muscle relaxes and is extended again under the tension of antagonist muscles, the impulse of blood, and the elasticity of its particles. The well-known effects of magnetical and electrical effluvia illustrate the instantaneous influence of nervous aether on muscle fibers, as electrical attraction carried down a string may be very similar to the motion and action of nervous aether.

In his third lecture Langrish turns to involuntary muscular action, and here he is least confident. Adopting the form of queries, he asks if the heart is not a compound organ, containing within itself antagonist muscles such that some were always stretched as others were contracted. Then, if the nervous aether were transmitted to the heart in a pulsating manner (a suggestion previously made by Boerhaave), possibly in part as the nerves were compressed by an out-of-phase systole and diastole of the dura mater, the heart would alternately contract and relax as the tension in the fibers was abated, the overstretched muscles pulled back, and blood fluxed into the heart. The weakness in this particular argument is suggested by the query form in which it is proposed, but a more general weakness to all his arguments had earlier been revealed. For Langrish, like his contemporaries Desaguliers, Hales, and Musschenbroek, was forced to recognize the lack of experimental proofs for his elaborations of mechanism. "It must be confess'd indeed, that these *Intima Naturae,* or secret operations in the Animal Oeconomy are all skreen'd from our Knowledge, the Agents being too subtile ever to become the Objects of our Senses . . . so that we can only . . . deduce our Argument from such collateral Proofs, or from such *Data* as we are pretty sure are true."

Failure to find empirical justification did not, however, restrain another aetherial theorist from developing a physiological system which obtained an anomalous longevity by its connection with that extension of Lockeian philosophy called associationist psychology. In 1746, David Hartley published his "conjecturae Quaedam de Sensu, Motu, et Idearum Generatione" as an appendix to *de Lithontriptico a Joanna Stephens,* and in it founded what has been called the first systematic physiological psychology.[8] Hartley expanded the "Conjecturae" three years later in his *Observations on Man* (1749) and further developed a physiological doctrine of necessity (or determinism) and of the association of ideas. For a time Hartley's views were enormously popular among liberal intellectuals of eighteenth-century Britain; Coleridge named a son Hartley, and Joseph Priestley twice republished Hartley's *Theory of the Human Mind* (1775, 1790), shorn, however, of most of the physiological sections. The entire text was translated into German, with notes, in 1772 and into French in 1802, while the German notes were translated in 1791 for a new English edition, which was republished through a sixth edition in 1834. As the associationist elements could be somewhat separated from the aetherial physiology, it is not easy to judge the importance of the latter in the development of British physiology. Apparently it was small, as the professional physiologists cite Hartley only to criticize him; but surely there

[8] By Benjamin Rand, "Early Development of Hartley's Doctrine of Association," *Psychological Review,* 30 (1923), 306-20, especially p. 313. The "Conjecturae" were translated into French, were republished by Samuel Parr in 1837, and have recently been published in English translation as *Various Conjectures on the Perception, Motion, and Generation of Ideas,* translated by R.E.A. Palmer, with notes by M. Kallich (Los Angeles: William Andres Clark Memorial Library, University of California, Augustan Reprint Society, Publication 77-78, 1959), which is the form I have used. David Hartley (1705-57), was educated at Jesus College, Cambridge; B.A. 1726, Fellow 1727, M.A. 1729. He declined taking Holy Orders from scruples about the thirty-nine articles and became a physician, though without a medical degree. F.R.S. and friend of Stephen Hales, he wrote in support of Mrs. Stephens' lithontriptic remedy and the strongly religious rationalist treatises on physiology and psychology embodied in the *Various Conjectures* and the *Observations on Man, his Fame, his Duty and his Expectations* (London: first published 1749, reprinted by J. Johnson, 1791).

was enough association between the prestige of the psychology and the physiology attached to it to justify a brief summary of Hartley's ideas.

The fundamental principle of Hartley's system is that of small vibrations ("vibrunticles") impressed upon the aether in the solid filaments of the nerves by external objects. These sensations are transmitted by aetherial vibration to the infinitesimal particles of the soft, but solid, medullary substance. By their differences in degree, kind, and place, these vibrations represent different primary sensations, or "simple ideas" in the brain, which becomes increasingly disposed to vibrate in any particular mode by each repetition of the sensation. Other vibrations, particularly if they arrive at the brain simultaneously, may therefore induce this mode of vibration and become regularly associated with it, modifying one another, causing recollection of sensation, and building chains of induced vibrations called ideas, or more complex concepts. Such, briefly, is the physiology of associationist psychology as created by Hartley, and the system obviously has physiological connotations other than just those for sensation and concept formation.

Hartley declares that he was initially led to his idea of vibrations by reading Newton's *Principia* (the General Scholium), the aether queries, and Newton's letter to Boyle [*Conj.* xiv, *Obs.* 13-14]. He admits that he does not fully understand the nature of the aether, but he knows that it is involved in the attractions of gravity and cohesion, attractions and repulsions of electrical bodies, mutual influence of light and bodies, effects and communication of heat, and the performance of animal sensation and motion. Certainly the aether is as easy an explanation of accretion, secretion, sensation, and motion as that offered by "a glandular secretion called nervous fluid, animal spirits, etc. [*Obs.* 18]." To the effects of the aether, one adds the active properties of the particles of the medullary substance and muscular fibers—powers of attraction and repulsion at different distances—and, as the existence and importance of reciprocal motion in nature is attested by examples of heat, sound, elasticity, electricity, etc., it is possible that the vibrations of the aether may merely modify and maintain the vibrations of these particles rather than set them in motion [*Conj.* 6-8]. The vibrations of medullary particles will differ in degree in proportion as the aether is more or less condensed in the mid-points of the pulse. If the vibrations are moderate, the particles return to their customary situations and a pleasurable association ensues; if the agitation carries the particles beyond their spheres of attraction, they run into new cohesions and pain arises [*Conj.* 10-11, 14]. Muscular contraction is agreeable to the doctrine of vibrations, as these, descending along the nerves (resembling the motion of electric effluvia along hempen strings), stimulate the small particles of muscular fibers and put into action "an attractive virtue, perhaps of the electrical kind" (resembling electrical attraction at the end of the strings) [*Conj.* 34-35, *Obs.* 28]. Hales is cited as having shown that blood globules may contain an electric virtue, and Hartley supposes that the use of air in respiration is as an *electric per se,* restoring the electric virtue lost by the blood in sustaining the movement of the heart [*Conj.* 38].

To the empiricist complaining of the conjectural nature of these speculations, Hartley provides the classical response of the rationalist. ". . . Let us suppose the existence of the aether, with these its properties, to be destitute of all direct evidence, still, if it serves to explain and account for a great variety of phenomena, it will have an indirect evidence in its favour by this means [*Obs.* 15-16]." Like the decipherer who recognizes the truth of his key as the meaning of a message unfolds, so, Hartley claims, the natural philosopher recognizes the validity of aetherial explanations as they advance the understanding of nature. For many of his contemporaries, and most of his successors, Hartley's defense was, however, inadequate. As the Germanic physiological theories of Stahl, Hoffmann, and von Haller were introduced, it seemed increasingly clear that the language of the cipher was not that of the Newtonian aether.

Historians of medicine are able to draw nice distinctions, which need not entirely concern us, between the concepts of Stahl and Hoffmann, and between these and the work of von Haller. Nonetheless, as these differences are involved in the variations from mechanism employed by the British, it is appropriate briefly to characterize them. Georg Ernst Stahl (1660-1734) was a pietist professor of medicine and chemistry at Halle and in both subjects maintained a practical empiricism coupled, in medicine, with a belief in an Anima, a conscious, rational, immaterial principle in living substance responsible for the unique properties of life. The Anima directs the substantial (*i.e.*, real, in the sense of substantial qualities), but immaterial motions of inert matter in the goal-oriented functions of adaptation, growth, resistance to corruption, etc., and, as such, exhibits intention and denies determinism. As the Anima excites such motions in the body as are suited to obviate illness, the Stahlian physician is relatively unconcerned with detailed anatomy, but is intent to observe the body and its motions as instruments of the Anima and to leave cures to the *vis medicatrix naturae*.[9]

Friedrich Hoffmann (1660-1742), Stahl's colleague in medicine and chemistry at Halle, was much closer in spirit to the British aetherialists. One of the great systematists of the eighteenth century, he opposed Stahl's views of the passivity of matter, which, he believed, was provided with innate motive force and powers of resistance. Living matter has, in addition, a primitive *virtus organans plastica* diffused through all its parts, according to their number, size, configuration, measure, site, and position. This organizing force acts particularly through a subtle spirit or "aetherial fluid," in whose smallest portion it inheres, to order the motions, resist decay, and achieve the functions of life according to a determined compulsion, the necessity of law.[10]

The third and greatest of the "Germanic" influences of mid-century British physiology was Albrecht von Haller (1707-77) who, in fact, was a Swiss and had studied under Boerhaave at Leyden, but who taught anatomy at Göttingen for 17 years before returning to Berne to serve as a minor public official and continue his research and writing. The author of more than 650 books and articles, including romantic poetry glorifying the Alps and pietist religious works, his major contribution to physiology was his eight-volume *Elementa Physiologiae corporis humani* (1757-65), which established a pattern for all future investigation. His earlier text, *Primae Lineae Physiologiae* (1747, 2nd edn. 1751, 3rd 1767), served as a link between Boerhaave's teaching and Haller's own *Elementa,* but it already begins the departure from mechanistic reductionism to be seen, for example, in his classic *De partibus corporis humani sensibilibus et irritabilibus* of 1753 or the *Dissertation on the Motion of the Blood* of 1757, where he severely criticizes the "geometrical physicians," those members of the "medico-mathematical sect."[11] Like Hoffmann, and in opposition to Boerhaave, Haller tended increasingly to emphasize the solid rather than the fluid parts of the body. Unlike both, he was very much an empiricist and

[9] See Lester R. King, "Stahl and Hoffmann: a Study in Eighteenth Century Animism," *Journal of the History of Medicine,* 19 (1964), 118-30.

[10] King, "Stahl and Hoffmann"; King, *Growth of Medical Thought,* pp. 159-74.

[11] Albrecht von Haller, *Dissertation on the Motion of the Blood . . . to which are added, Observations on the Heart, proving that Irritability is the primary Cause of its Motion* (London: J. Whiston and B. White, 1757), and his *First Lines of Physiology* (New York and London: Johnson Reprint Corporation, from the 1786 Edinburgh edn., Sources of Science No. 32, 1966), with an introduction by Lester S. King.

adopted the typical naïve materialism of the empiricist. Thus he describes the solid parts of the body as: sensible, when they react to stimuli by means of a fluid in the nerves, which is an element in its own right, capable of transmitting sensation to the brain; or irritable, when they react to stimuli by contracting; or neither. ". . . The theory, why some parts . . . are endowed with these properties, while others are not, I shall not at all meddle with. For I am persuaded that the source of both lies concealed beyond the reach of the knife and microscope, beyond which I do not chuse to hazard many conjectures. . . ."[12] But, of course, he does hazard conjectures—and they are essentially materialistic ones. Nerves are the source of sensibility which resides in their medullary part, a production of the internal substance of the brain [674]. Irritability is independent of nerves, soul, and will; nothing is irritable in the animal body but the muscular fiber, whose power of producing motion is different from all other properties of bodies [675, 690]. Irritability is to be accepted as a property of animal gluten, doubtless owing to some arrangement of the ultimate particles of animal bodies, but no more explainable than attraction and gravity, as properties of matter in general [691-92]. Hence this greatest of his anatomical discoveries remains finally only a way of categorizing different species of animal fibers, and his "explanations" of muscular and nervous response to stimulation, through these fibers and the nervous substance, become a new materialistic base for the development of vitalism.

Such were the concepts respecting muscular motion when Robert Whytt published, in his *Essay on the Vital and other Involuntary Motions in Animals* of 1751, a critique of previous theories and an exposition of his own.[13] Whytt taught medical

[12] Albrecht von Haller, "A Dissertation on the Sensible and Irritable Parts of Animals," with an introduction by Owsei Temkin, *Bulletin of the Institute of Medicine*, 4 (1936), 651-99, esp. 657-58.

[13] Robert Whytt (1714-66) was educated at St. Andrews, M.A. 1730; Edinburgh, in London hospitals, Paris, and Leyden before receiving the M.D. at Rheims, 1736, and St. Andrews, 1737. He was Professor of the Institutes of Medicine at Edinburgh from 1747 until the end of his life. F.R.S. 1752, he wrote on lime water as a cure for the stone; on nervous, hypochondriac, and hysteric diseases; and many short treatises included in the *Works of Robert Whytt* (Edinburgh: T. Becket and P. A. deHondt, and J. Balfour, 1768). The edition of the *Essay* reprinted in the *Works*, pp. 1-329, was revised after the appearance of Haller on sensitive and irritable parts and refers to it. William Seller, "Memoir of the Life and Writings of Robert Whytt . . . ," *Transactions of the Royal Society of Edinburgh*, 23 (1864), 99-131, is the most extensive biographical study; but for a correction of Seller's overly defensive attitude, see also Leonard Carmichael, "Robert Whytt: a Contribution to the History of Physiological Psychology," *Psychological Review*, 34 (1927), 287-304.

theory for nearly 20 years at one of the world's most famous medical schools. He was a scholar—the *Essay* refers to the work of Newton, James Keill, Hales, Bryan Robinson, Alexander Stuart, Browne Langrish, Stahl, Hoffmann, and Haller—and a skilled anatomist. His researches on the central nervous apparatus which led to the discovery of the reflex arc are classic; for nearly seventy years he was the chief authority on the individuation of nerve filaments. His *Essay,* therefore, is a useful source for contemporary and accepted views on muscular action.

Whytt begins with the ubiquitous British statement of experimental principles. Like Newton, whose philosophy was founded on plain fact, he has not indulged his fancy "in wantonly framing hypotheses, but . . . rather proceeded upon the surer foundation of experiment and observation [Preface]." He then sets out the certain and accepted facts which are to be explained. The immediate cause of muscular contraction is the power or influence of the nerves [6]. Though we have generally called it animal or vital spirits, we know nothing of the nature of the substance in the nerves, nor do we know that of the muscular fibers [1-2]. Muscular fibers have the power to act, but a stimulus is needed to excite that power. The stimulus producing voluntary motions originates in the mind, but the way in which this stimulus acts must be the same as that in which other stimuli act to produce the vital and involuntary motions [2-3]. An irritation on the bare muscles of a living animal produces a contraction in proportion to the vigor of the stimulus, though any given stimulus may be weak or vigorous depending upon the peculiar constitution of particular nerves and muscle fibers [9-11]. An irritated muscle alternately contracts and relaxes while the stimulation remains constant, and a muscle may continue to act for some time after the stimulus has been removed [12-13].

Matter itself is incapable of sensation or thought. It invariably acts according to laws prescribed to it, and these laws do not involve feeling, inclination, choice, or self-movement in the most refined and subtle parts any more than in the most gross and sluggish. We might as well pretend that the eye sees or the ear hears by virtue of material organs as to suppose that animal fibers move owing to their mechanical structure or the peculiar disposition of their parts [128]. How, then, are the muscular fibers framed to contract when a proper cause is applied to them? To ascribe this to some general elastic "power of resilition" is to give to dead, inactive matter the power of generating motion [122-23]. It is no answer to talk of the "peculiar energy" or mode of action of some hypothetical, subtle animal spirit, secerned from the blood and lodged in the brain, nerves, and cavities of muscular fibers [123]. Such a matter can no more be the vital principle or source of animal life than the material blood from which it is derived [147]. Some have imagined muscular contractions to be occasioned by effervescences of nervous and arterial fluid, or by a subtle, aetherial, or electric matter residing in the nerves and brought into action by the will. These, however, all require some other agent to excite them, and, when excited, electricity and effervescences do not act by alternate efforts as the muscles do, while the oscillations of the elastic aether follow the laws of other elastic bodies and are inconsistent with the contractions of muscular fibers [124-27]. Finally, some (Haller) suppose a latent power or property in the muscular fibers, to which their motions, in consequence of irritation, are to be referred, but they have been able to prove the existence of such unknown active powers only by appearances which can be explained without them. It is, besides and again, "improper to attribute active powers to that, which . . . is yet no more than a system of mere matter . . . [127]."

In fact, "the sensibility of our fibers is owing to their being animated by a living principle different from matter and of powers superior to it [200]." Stimuli can only excite our muscles as they are animated by this sentient principle [143]. It is through the sentient principle that "sympathy between the nerves" has meaning, for only thus can an irritation of the extremities of the nerves of one organ or part of an organ initiate a more than ordinary derivation of spirits into nerves having their origin in a different organ or part [98]. This explanation is not to be confused with that of the Stahlians, who by extending the influence of the soul as a rational agent a great deal too far, have made it a subject of ridicule. The mind does not act by previous conviction, rationally and consciously to preside over vital motions, but as a sentient principle automatically exciting such motions, in consequence of stimuli, as may be most proper to remove the irritating cause [150-52]. And if we do not understand the manner of its operation, neither do those philosophers who make use of the power of attraction between parts of matter know its cause [144-45].

For all his reservations concerning Stahl's animism, Whytt clearly preferred the tendencies of that explanation to the materialism he found embodied in ferment, aether, electrical, or "irritable fiber" explanations. Not everyone entirely agreed.

Whytt's successor to the chair of the Institutes of Medicine (though he would have preferred that of the Practice of Medicine) was William Cullen, who thought Stahl's medical system "fanciful," and deplored the Stahlians' dependence on the wisdom of nature and their reservations in the use of such general remedies as bleeding and vomiting. Only when the impotence of "our art is very considerable and manifest, ought we to admit the *vis medicatrix naturae* in practice."[14] Cullen believed that the distinction of the genera of diseases, of their species, and often even of the varieties was "a necessary foundation of every plan of physic," and, in his *Synopsis Nosologiae Methodicae* (1769), he provided such a classification, endeavoring to apply the facts to an investigation of proximate causes [*Lines*, I, xxv-xxvi]. The disposition to classify nature, which is endemic in the more empirical sciences, became epidemic in the eighteenth century through the botanical taxonomy of Linnaeus. For medical nosology, Linnaeus also provided an example, with his *Genera Morborum* of 1763, and in this, as in his botanical work, he adopted external characteristics, *i.e.*, signs or symptoms, as the basis for classification. Now any consistent system of groupings has the advantage of imposing some order on a multitude of data, and it further encourages precise observations for the purposes of distinction, but such systems differ in their usefulness, and there is a tendency to regard classification as saying something profound about the objects classed. There is always an implied hope that the "right" groupings will somehow relate to, or reveal, a natural affinity between the objects grouped. The system of Linnaeus did not provide such a natural classification, nor did that of his major rival Sauvages, whose *Nosologia Me-*

[14] William Cullen, *First Lines of the Practice of Physic* (Philadelphia: Parry Hall, 1792), pp. xii, xvi-xvii. William Cullen (1710-90) was educated at Glasgow University and privately by a physician. After brief experiences as a ship's surgeon and apothecary's assistant, he commenced medical practice while attending winter sessions of the medical school at Edinburgh. He received the M.D. from Glasgow, 1740, having already become a private teacher in medicine (one of his students was the famous surgeon and anatomist William Hunter). In 1744 he settled in Glasgow to practice and assist in the development of the medical school there. He lectured on chemistry from 1746, and was named professor of medicine at Glasgow University in 1751; in 1756 he moved to Edinburgh as professor of chemistry. At the death of Robert Whytt, he became professor of the Institutes of Medicine, which he alternated with John Gregory, professor of the Practice of Medicine, until he succeeded Gregory at the latter's death. F.R.S. 1777, in 1783 he helped secure the charter for the Royal Society of Edinburgh. Distinguished for teaching in chemistry and medicine, rather than for originality, his most notable works were the *Synopsis Nosologiae Methodicae* (1769) and the *First Lines of Physic*, 1st edn. 1776. See particularly John Thomson, *Life of Cullen*, and Lester S. King, *Medical World*, pp. 193-226.

thodica, of 1763, maintains an explicit nominalism while concealing a wistful conviction in natural classes. Cullen, with a strong primary interest in therapy, hoped to find a better system.

He insists that he has "avoided hypotheses, and what have been called *theories*" and that his general doctrines, physiological and pathological, are but generalizations of facts or conclusions from "cautious and full induction [*Lines,* I, xxix]." As always, these words are a signal to look for some concealed theoretical structure, and, in Cullen, this is not hard to find. Although he thought it would be easy to show that the mechanical philosophy "never could, nor ever can be applied to any great extent in explaining the animal economy [xii]," and rejected hypotheses of nervous vibrations or vital fluids secreted by the brain, he was impressed with Bryan Robinson's physiological extension of Newtonian aetherial concepts and, with Hoffmann, postulated the existence of an aetherial medium in the brain, spinal marrow, and nerves. As electrical matter may exist in bodies without being sensible, and may, by artifice, be excited through accumulation or depletion, so the aether in the nervous system exists and is excited by heat, circulation of the blood, tension in the vessels of the brain, etc. The excited state of the nervous fluid, or aether, *is* life; when the fluid is no longer excitable, a state of death has ensued, and illness results from abnormal degrees of excitement or collapse.[15] All the action of muscular fibers depends upon the degree of excitement, or nervous energy, of the aether fluid in brain and nerves.

When this theory was applied to nosology and practice, it took on the protective coloration of clinical empiricism, for Cullen was an observer of the first order, and his teaching combined method with clarity and precision. His *First Lines of the Practice of Physic,* based on his nosology and that, in turn, on his doctrine of nervous energy, illustrates the application of this method. Fevers, for example, constitute one of the Orders of the Class Pyrexiae, characterized by a unique succession of symptoms and a "diminution of strength in animal functions [I, 35]." The proximate cause of fevers is a state of debility in which the energy of the brain is diminished. As it is a "general law of the animal economy" to obviate the effects of noxious powers by exciting appropriate motions, the *vis medicatrix naturae* induces a spasm of the extreme vessels, irritating and constricting the heart and larger arteries. This stimulates motions in the sanguiniferous system to restore the energy of the brain and produces

[15] Thomson, *Life of Cullen,* I, 315-17.

the requisite changes in the state of the moving powers of the animal system. Various genera of fevers are to be distinguished by the degrees of resistance of the initiating cause and the levels of energy exhibited by the system in their expulsion [I, 55-59]. Another class of diseases, and that one most commensurate with our previous concentration on muscular motion, is that of "Neuroses or Nervous Diseases," within which are to be found the Orders of Comata, Adynamiae, Spasmi, and Vesaniae, characterized by the interruption or debility of nervous power from the brain to the muscles or from the sentient extremities to the brain [II, 67ff]. Here Cullen makes it clear that, though the body fibers are so constituted as to contract (irritable) or feel (sensitive), they do so only through the intervention of nervous fluid, or aether, in a state of excitement.

In his lectures, Cullen cautioned his students to regard the nature of nervous fluid and its conditions of greater or lesser mobility merely as a representation of the physiological facts—the increased or diminished force of the energy of the brain and the different states of the nervous system.[16] To maintain a position of such unstable equilibrium, intermediate between clear materialism and vitalism, was, however, not possible for British physiologists. Cullen's colleague at Edinburgh, John Gregory, cautioned him against the use of subtle fluids in explaining life and later wrote that the phenomena of the world were not all explicable in terms of attracting and repelling powers of ultimate particles. Newton was deceived; ". . . even in the unorganized kingdom, the powers by which salts . . . concrete into regular forms can never be accounted for by attraction and repulsion . . . ; and in the vegetable and animal kingdom there are evident indications of powers of a different nature from those of unorganized bodies . . . certain effects are produced which the laws of matter are not able to explain."[17] Suspicions, like Gregory's, of Cullen's materialism were not usual, however, and more people found and approved a vitalistic tendency in Cullen's definition of life in terms of nervous energy, for that was the tendency of British physiology in the second half of the eighteenth century. Mechanists were reinterpreted into materialists, materialistic concepts substituted for mechanistic actions, and finally both

[16] *Ibid.*, p. 310.

[17] John Gregory, *Lectures on the Duties and Qualifications of a Physician* (1770), quoted by Philip C. Ritterbush, *Overtures to Biology: the Speculations of Eighteenth-Century Naturalists* (New Haven: Yale University Press, 1964), p. 187. Gregory was a second cousin of David Gregory, who had inspired Pitcairne to Newtonian mechanistic physiology.

were superseded by vital force, vital energy, or simply "life." John Pringle, for example, expressed his love for his teacher, Boerhaave, and his admiration for Stephen Hales by making them both iatro-chemists. In his *Observations on the Nature and Cure of Hospitals and Jayl Fevers* of 1750, Hales' mechanistic instrumentalism of the role of air's elasticity in the animal economy is transmuted into air poisoned by putrid effluvia, while Pringle notes, in his *Observations on Diseases of the Army* of 1752, that Boerhaave had observed the error of the mechanical writers in too sparingly admitting chemistry into their explanations. ". . . tho' he retained the use of mechanics, yet [he] revised and reformed the *doctrine of acids and alkalies*; and under these last comprehended all that he thought *septic* or *putrid.* But as my celebrated master had not time to ascertain every part of his doctrine from experiments of his own, it was no wonder some mistakes were made, and that the extent of these principles were not fully understood."[18]

Even in a problem of physiology most closely related to physical phenomena, the same development is to be seen. Professor Everett Mendelsohn has traced the British explanations for animal heat from the early mechanism of heat as motion and the attrition of blood in the lungs and arteries, through a revived iatro-chemistry and heat by fermentation, to the final ingenuities of Adair Crawford and heat as a fluid, with Black's latent and specific heat theories invoked. The story does not end there, however, for other investigators were introducing the complications of animal cold (*i.e.*, maintenance of low temperatures in heated environments), and a flurry of *Philosophical Transactions* papers in 1775 agree with Charles Blagden's ". . . no attrition, no fermentation, or whatever else the mechanical and chemical physicians have devised, can explain a power capable of producing or destroying heat, just as the circumstances require. . . . it can only be referred to the principle of life itself. . . ."[19] The best, and most characteristic, example of this conclusion is that reached by John Hunter, one of the most brilliant British anatomists and surgeons of all time. Hunter repeatedly invokes the "living principle" to explain nearly all animal and vegetable physiological phenomena —of growth, heating, cooling, healing, resistance to putrefaction,

[18] Quoted and discussed by Dorothea Waley Singer, "Sir John Pringle and his Circle—Part II, Public Health," *Annals of Science,* 6 (1950), 229-61.

[19] Charles Blagden, "Experiments and Observations in an heated Room," *Philosophical Transactions,* 65 (1775), 111-23. See Everett Mendelsohn, *Heat and Life, the Development of the Theory of Animal Heat* (Cambridge: Harvard University Press, 1964).

digestion, etc.—and, through the influence of his teaching and writing, passed to the next century, cloaked in the infinite respectability of his persistent empiricism, a heritage of vitalism.[20]

Clearly this progression in physiology from dynamic corpuscularity to fluid spirits, the aether, electricity, fire, and irritable or sensitive fibers parallels the changes of mechanism to materialism in other branches of natural philosophy. There is, however, an incongruity in any declaration that the continuing change from these to a vital, sentient, or living principle continues that parallel, particularly considering the vehement anti-materialism expressed by the vitalists. There is at least as little incongruence in calling materialistic an anti-reductionist "substantial quality of life" as there is in the vitalists' calling materialistic a mechanistic explanation in which none of the essential causes of phenomena inhere in matter as such, but rather result from the geometry of size and shape and the dynamics of force. Nonetheless, it is possible that the greater parallel lies between vitalism and the extension of dynamic corpuscularity into a dynamism independent of corpuscles—to be discussed in Part III.

The revolution in eighteenth-century British chemistry, which ended in the substitution of the element-taxonomy of Lavoisier for the physical reductionism of Keill and Freind, also had its beginning in Continental resistance to mechanism. Some part of this stemmed from a long alchemical tradition, allied to German vitalism, and was represented in seventeenth-century Europe by van Helmont and, to some extent, by Becher. This reaction against chemical mechanism can be seen even in persons otherwise sympathetic to the corpuscular philosophy. Fontenelle, though an ardent Cartesian, is said, for example, to have complained that "Boyle was far too rational and too little mystic ever to be a proper chemist."[21]

In the long run, except for organic chemical problems where chemistry joined physiology in flights of vital fancy, this was not, however, a primary objection against dynamic corpuscular chemistry. Of far greater importance was the failure of mechanism to come to grips with a major part of chemical phenomena—the persistent identity of certain substances and the regularity

[20] See, for example, John Hunter, "On the Digestion of the Stomach after Death," and "Experiments on Animals and Vegetables with respect to the Power of Producing Heat," *Philosophical Transactions*, 62 (1772), 447-54, and 65 (1775), 446-58, respectively; and Ritterbush, *Overtures to Biology*, pp. 187-88.

[21] Marie Boas Hall, *Robert Boyle on Natural Philosophy* (Bloomington: Indiana University Press, 1965), pp. 70-71.

of the varying combinations. Newton's addition of forces to the parameters of chemical corpuscularity increased the potential of mechanism to explain the processes of chemical and physical change, but, as the works of Keill and Freind show, forces were of little use in distinguishing between substances. Indeed, with Newton's dynamic corpuscularity as with the kinematic corpuscularity of Boyle, there was rather the continued implication of intra-convertibility, the transmutation of substance, than any serious concern with permanence of identity. Even hopefully mechanistic chemists, such as Boerhaave, had to recognize the impasse, and most of them superposed on some mechanistic theory of matter a chemical theory of principles, forms or qualities, or elements carrying with them inalienable properties by which they were identified. In short, chemists deal essentially with matter, and when the mechanists failed adequately to explain the forms in which their matter was important to them, they adopted a less sophisticated, but much more operative, materialization of the chemists' qualities.

An early example of British reactions to these problems of mechanization is to be found in the *Course of Chymistry* of Henry Pemberton, first published posthumously in 1771 but based on lectures he delivered as Gresham Professor of Physic about 1731.[22] The work is primarily a collection of pharmacological processes and recipes. The frequent and admiring references to Newton, expected of a recognized Newtonian, are there, but most of these are methodological—the inevitable shunning of hypotheses—or to the "de Natura Acidorum"; surprisingly few of them clearly derive from the "chemical" and corpuscular queries. The operations of chemistry are described in terms of the motions, division, reunion, and variation, of the parts of matter achieved through the material instrument of heat, communicating vibrations by the action of aetherial substance [202, 28]. The regular forms of salt crystals are ascribed to particles whose different sides had different powers of acting, in analogy to the different sidedness Newton had demonstrated for the particles of light [93-94]. At the same time, however, "Sir Isaac Newton seems to think, that the acid principle is the great agent in nature [72]," while metals are known to contain an inflammable substance. Hence the bulk of all bodies consists of inactive particles of matter, actuated by two principles, the acid and the inflammable, or sulfurous [244]. Pemberton notes that the increase in weight

[22] Henry Pemberton, *A Course of Chymistry*, edited by James Wilson (London: J. Nourse, 1771). Pemberton, as has been noted, was a student of Boerhaave.

of metals calcined and the concurrent consumption of air are causally related. "Without doubt while the air by acting on the inflammable substance either in metals or other bodies expels it from them, it unites itself (in part at least) to the remains of the body [244-45]." The dubious pleasure in finding another person to set beside Jean Rey as a questionable precursor of Lavoisier merely emphasizes the elements of materialism in Pemberton's chemistry. Here is the characteristic contrast between inactive and active matter, with specific active qualities or principles defined by their chemical properties, and the substantial as well as instrumental role of air in chemistry is a part of that materialism.

In fact, the entire question of elements, their role, instrumental or constituent, in chemical change, and the nature of chemical composition itself had been reopened for British chemists just before Pemberton's lectures and in a way which explicitly disavowed mechanism as a fundamental approach for chemists. In 1730 Peter Shaw published a translation of Georg Ernst Stahl's *Fundamenta Chymiae Dogmaticae et Experimentalis* (1723) as *The Principles of Universal Chemistry: or, the Foundation of a Scientifical Manner of Inquiring into and Preparing . . . Natural and Artificial Bodies . . . Design'd as a General Introduction to the Knowledge and Practise of Artificial Philosophy.*[23] The influence on Shaw of Stahl's work is immediately apparent, in the title of his next book, *Three Essays in Artificial, or Universal Chemistry* (1731), and in Shaw's changed view of chemistry. In Shaw's 1727 translation of Boerhaave, the notes had praised Newton, Keill, and Freind, as had the preface to his earlier edition of the works of Boyle; but in his 1734 *Chemical Lectures*, Shaw was already beginning to wonder if an increased concern for the chemists' elements: water, earth, salt, mercury, and air, was not called for, and if all these might not contribute "some part of their substance as well as energy" to chemical change.[24] By the second translation of Boerhaave (1741), Shaw was expressing doubts as to the value of the "powers of attraction between the minima naturae" as a principle to solve the problems of chemistry, and the 1755 edition of his *Chemical Lectures* completes the transformation:

> . . . the more intelligent among the modern chemists do not understand by Principles those original Particles of Matter,

[23] Peter Shaw, tr., George Ernest Stahl, *Principles of Universal Chemistry, etc.* (London: J. Osborn and T. Longman, 1730).

[24] See F. W. Gibbs, "Peter Shaw and the Revival of Chemistry," *Annals of Science*, 7 (1951), 211-37, esp. 221-22.

of which all Bodies are by the mathematical and mechanical Philosophers supposed to consist. Those Particles remain undiscernible to the Sense . . . nor have their Figures and original Differences been determined by a just Induction. Leaving, therefore, to other Philosophers the sublimer disquisition of Primary Corpuscles, or Atoms, of which many Bodies and Worlds have been formed in the Fancy, genuine Chemistry contents itself with grosser Principles, which are evident to the Sense, and known to produce Effects in the Way of corporeal Instruments. And these grosser Principles are every Way sufficient to answer the Purposes of philosophical Chemistry.

. . . These sensible Principles, so far as we know them, are significantly expressed by the common Words, Water, Earth, Salt, Sulphur and Mercury: to which might be added the Air, if a Way were known to fix it, so as to render it more sensible, tangible, and corporeal.[25]

Boerhaave's practical empiricism must have had something to do with Shaw's reversal, while to credit to Stahl alone a revolution in thinking which is revealed only over a span of some 25 years is admittedly excessive. Nonetheless, it must be recognized that the introduction of Stahl's chemistry into England provided a signal for explicit anti-mechanism, not only in Shaw but in other chemists. Nor is this unexpected, for anti-mechanism was a major part of Stahl's chemical philosophy. The work of Stahl and the Stahlians has been insufficiently examined, partly because Stahl's own writings are prolix and obscure, but mostly because the history of eighteenth-century chemistry has been written to emphasize Lavoisier's oxidation theory of combustion. While there is no doubt that the opposing theory of phlogiston played a central role in Stahlian chemistry, in a very important sense that role was merely an example of a more general approach to the whole of chemistry. This approach attracted attention independently of phlogiston and, indeed, is embodied in the work of Lavoisier as well as in that of nearly every chemist of the second half of the eighteenth century. A survey of *Philosophical Transactions* articles on chemistry between 1731 and 1789 shows, for example, that phlogiston

[25] Peter Shaw, *Chemical Lectures, Publicly Read at London, in the Years 1731, and 1732; And at Scarborough, in 1733; For the Improvement of Arts, Trades, and Natural Philosophy* (London: T. and T. Longman, I. Shuckburgh, and A. Millar, 1755), 2nd edn. corr., pp. 146-47. I have not seen the first edition of 1734 to find whether this is a new expression of the later edition or not.

becomes a significant part of British chemical explanation only gradually and, to any marked degree, after mid-century, while a British encyclopaedia reference to phlogiston does not appear before the first *Britannica* of 1768.[26] Well before phlogiston seriously enters the scheme of chemical explanation, other parts of Stahl's chemical philosophy had already made their impact.

The most important of these were the related views of the place of mechanism in chemistry and the nature of chemical substances. The Stahlians did not so much reject the mechanical philosophy as deny its value for chemistry. The methods of mechanism, they felt, elaborated only the surfaces of things. Abstract explanations of phenomena from the figures and motions of particles were practically useless in explaining particular, experimental reactions. True, corpuscles must have shapes, but how does the necessarily rhetorical discussion of this fact assist the chemist in the laboratory? To say that a salt is a combination of water and one of two kinds of earth gives a real and true idea of what one means by salt; one knows what water and earth are, and can obtain a salt from any body whatever containing water and earth. But to describe a salt as consisting of pointed particles, longer than broad, angular, etc., gives no information useful in identifying a salt. The properties of chemical reactions are not explained by the form of particular indivisible particles, especially as we can never hope to perceive these even with the best microscopes. Mechanical hypotheses are only "amusing mathematical speculations" which do not lead to an understanding of laboratory experience. Moreover, motion, though a powerful instrument for chemical reaction, is insufficient to account for each special reaction. The chemist can appeal to the mechanical philosophy only to explain that which is common to all the phenomena he observes; to account for the diversity of reactions, he reserves the right to add different explications.[27] It is just this distinction between the abstract speculations of mechanical philosophers and the empirical requirements of the laboratory that Shaw repeats in his works published after his translation of Stahl. As this is a typical observation of the materialist-empiricist in Britain after about 1740, one cannot ascribe its application to chemistry entirely to Stahl, who acts to reinforce a prejudice, not

[26] Ambrose Godfrey Hanckewitz refers to phlogiston in his paper on phosphorus in *Philosophical Transactions*, 38 (1733), 58-70, but the first really serious use is that of John Huxham in his observations on antimony in *Philosophical Transactions*, 48 (1754), 832-69, after which phlogistic explanations are fairly common.

[27] Collected from Metzger, *Newton, Stahl, Boerhaave et la Doctrine Chimique*, pp. 101-04.

to create one; but the explicit connection of virulent anti-mechanism with the Stahlians is supported by the example of William Lewis. Lewis translated the *Chemical Works* of Stahl's disciple, Caspar Neuman, in 1759. In 1763 he wrote, in his *Commercium Philosophico-Technicum, or the Philosophical Commerce of Arts,* on the necessity of distinguishing between chemistry and mechanical philosophy. The latter, he said, considered bodies as entire aggregates, divisible into parts having the same properties as the whole, having knowable magnitudes and figures, and moving according to laws reducible to mathematical calculation. Chemistry was nearly the inverse; it considered bodies as composed of dissimilar species of matter, defined by particular qualities which may be separated and transferred. "These properties are not subject to any known mechanism, and seem to be governed by laws of another order."[28]

But if the mechanical permutations of ultimate particles are not to be used to explain chemical phenomena, what is to be substituted? In one sense, nothing at all. Stahl operationally defines chemistry as the resolution of compound bodies into their constituents and the reforming of those compounds by uniting the constituents. The chemist concerns himself not with the ultimate particles but with those aggregates which are sensible and combine in the laboratory. Here Stahl again departs from British practice by assuming that such aggregates form elements which retain their properties in and through the compounds they form. Among the many ambivalences toward Newtonianism revealed in George Cheyne's *Essay on Regimen* of 1740 was his declaration that aggregates of different particles formed elements which had specific operational or functional qualities which they never lose, however much the quality might be concealed in further combinations. Cheyne does not refer to Stahl for this opinion, and he may well have derived it elsewhere, for it owes much to Stahl's predecessors. Nevertheless, it was from Stahl and his disciples that the eighteenth century obtained a fairly clear notion

[28] Quoted by L. Trengrove, "Chemistry at the Royal Society of London in the Eighteenth Century—I," *Annals of Science,* 19 (1963), 191-92. William Lewis (1714-81) is a rare example of an English university graduate, nonmechanist, natural philosopher. Educated at Christ Church, Oxford, B.A. 1734, M.A. 1737, M.B. 1741, M.D. 1745, he practiced medicine in London, wrote on materia medica, and was consulting chemist to the Society of Arts, of which he was a founding member. F.R.S. 1745, author of works on practical chemistry and pharmacopoeia, he translated Hoffmann's *System of the Practise of Medicine* (published posthumously, 1783) as well as Neuman's chemical *Works*; described platinum for the *Philosophical Transactions*; and received a gold medal from the Society of Arts for his research essay on potashes.

of constant constituents in chemical operations, and that notion was based on the reification in some permanent material form, as principles or elements, of such qualities as acidity, alkalinity, combustibility (phlogiston), and metallicity (mercurial principle).

The Stahlian definition of these elements, as bodies so simple they cannot by any known method be decomposed, left much still to be desired. For one thing, the definition does not apply to that class of elements, fire, for example, or air, which enter chemical operations not as constituents but as instruments. These elements, like their equivalents, light, electricity, and magnetism, were substantialized physical hostages to chemistry and, because they confused the issue—what, for example, was the relation between instrumental fire and air and constituent phlogiston?—they were increasingly neglected by chemists of the Stahlian persuasion. Nevertheless, the definition, even when limited to constituent elements, still fails, for Stahl had no clear criteria for determining whether the products of chemical change were those of composition or decomposition. Not until the choice, by Lavoisier, of weight as the primary parameter of chemical change was that problem to be solved. In the meantime, the properties of bodies were presumed to reflect those of its constituent elements, and higher order compounds were assumed to form by the association of lesser compounds possessing the same elements. These two assumptions directed the theoretical activities of most chemists of the period. They permitted a kind of indirect analysis by which compounds could be arranged into classes, orders, genera, species, and varieties, with class and order designations based on presumed elementary composition; and from them developed the concept, which, far more than phlogiston, dominated the chemistry of the eighteenth century, the doctrine of affinities.

The notion that chemical composition and decomposition depended upon some inherent disposition of substances to unite in varying degrees had been employed prior to the seventeenth century; it is related to the physiological concept of the organizing principle, and may be traced back at least to Galen and the doctrine of humoral attraction by similitude. The idea had, however, been rejected by the kinematic corpuscularians until Newton gave it mechanistic sanction by relating chemical combination to inter-particulate forces, providing material for a list of attractive-force affinities in the last query of the *Opticks*. The first true table of affinities, however, was published by Étienne Geoffroy in 1718. Though Peter Shaw, among many others, early attempted to relate the table to Newtonian attractions, Geoffroy had not spoken of attraction. The word he used was "rapports," and the context of his work was elementary and Stahlian, not mechanistic and Newtonian.[29] During much of the eighteenth century there were persistent attempts to relate affinities, or, as they were sometimes called, elective attractions, to Newtonian forces of gravity, capillarity, or cohesion. Many of the most distinguished European natural philosopher-chemists—Buffon, Macquer, Bergman, Guyton de Morveau, and Berthollet, for example—were active in these efforts, but many more decried even the term "attraction" and retained the Stahlian aversion to mechanical assumptions. Most British chemists belonged to the latter group, agreeing with William Lewis that affinities were just those "laws of another order" than mechanical which distinguished chemistry from the mechanical philosophy. For these men, the use of affinity tables was merely as a descriptive device and as a means of refining the taxonomy of elements and chemical substances; chemistry remained stubbornly and irreducibly materialistic.

The importance of these Continental (and especially German) ideas to the chemistry of mid- and late eighteenth-century Britain, suggested in the writings of Shaw and Lewis, is particularly manifest in the teaching of the two most influential professors of chemistry of the period. Between them, William Cullen and Joseph Black dominated the teaching of chemistry at Glasgow and Edinburgh for nearly 50 years. Cullen's chemical doctrines are not easily discovered, for he published no general treatise on chemistry and, of his experimental papers read before the Philosophical Society of Edinburgh, only that on heat (or rather cold) was published.[30] As the paper "Of the Cold produced by evaporating Fluids, and some other means of producing Cold" contains an early description of adiabatic cooling, it has some historical interest, but, in the long run, Cullen's ideas on heat and cold were important only for their possible influence on Black's more famous discoveries and interpretations. For the more specifically chemical ideas of Cullen, we must rely on his biographer and the manuscript papers and notes left by his students.

29 E. Geoffroy, "Des Différents Rapports observés en chimie entre différentes substances," *Mémoires de l'Académie Royale des Sciences* (1718), 202-12. Geoffroy was one of the earliest French adherents of the chemistry of Stahl.

30 William Cullen, "Of the Cold produced by evaporating Fluids . . . ," *Essays and Observations, Physical and Literary, Read before a Society in Edinburgh (Edinburgh Essays)*, 2 (1756), 145-56. See Thomas S. Kuhn, "The Caloric Theory of Adiabatic Compression," *Isis*, 49 (1958), 132-40.

The tendencies of Cullen's teaching are suggested by his efforts to acquire the writings of Becher, Stahl, J. Bohn, and J. H. Pott for his lectures. He also used the work of Boerhaave (especially respecting the matter of heat), possessed a manuscript copy of the lectures of Boerhaave's successor, H. D. Gaubius, and inspired in Black an admiration for the empirical studies of Andreas Marggraf. But it was Stahl and the Stahlian Junker who made the greatest impression on him. Complaining to his students that most chemistry teachers were unsystematic and their systems incomplete, he said: ". . . I must take notice, however, that Dr. Stahl is one who has endeavoured to avoid these faults; he has taught chemistry with a more general view, and attempted to collect the chemical facts, to arrange them in a better order. . . . Perhaps we have the substance of Dr. Stahl's lessons in a book published by a disciple of his, Dr. Junker of Halle. . . . This is the fullest collection that I have met with, and I have made a good deal of use of it . . . [though] it is written in . . . a clumsy manner, is mixed . . . with much pedantic, trifling philosophy, and is . . . often inaccurate and superficial describing experiments. . . ."[31]

Given his enthusiasm for Stahlian chemistry, it is not surprising to find Cullen recommending the division of "the productions of chemistry" into "their proper classes, orders, genera, species, and varieties as is commonly done with respect to the objects of natural history." Species were to be determined by the properties or combination of properties peculiar to any particular matter and no other, while a variety of that species was produced when, retaining most of the species' properties, these were varied or added to, by a modification of the substance or the addition of other matter.[32] Typically, he also concerned himself with the doctrine of affinities, or of elective attractions, and at least as early as 1759 had developed a technique for representing elective attractions by diagrams of intersecting lines connecting the attracting substances in combinations of compounds and indicating, qualitatively, comparative strengths of attraction by letters (or numbers) placed between the open ends of the lines. Determination of selective decomposition and recomposition

[31] Thomson, *Life of Cullen*, I, 40-41. See also Leonard Dobbin, "A Cullen Chemical Manuscript of 1753," *Annals of Science*, 1 (1936), 138-56, esp. p. 143 for more praise of Stahl.

[32] Dobbin, "Cullen Chemical Manuscript." Cullen had the usual trouble of taxonomists with people possessing other systems of classification. See, for example, Donald Monro, "An account of some neutral Salts made with vegetable Acids, and with Salt of Amber, which shews that vegetable acids differ from one another . . . ," *Philosophical Transactions*, 57 (1767), 479-516, which cites Cullen's table of neutral salts to criticize it.

was suggested by summing and balancing the respective attractions.[33] One must emphasize that this reach toward an evaluative handling of elective attraction had nothing to do with Newtonian forces, for Cullen explicitly denies that elective attraction relates to the theory of interatomic attractive forces. "The qualities that chemistry considers are different from those which mechanics or the mathematical philosophy considers, both in the principles and the objects. To regard attraction as proportional to particle density is purely speculative, for we know nothing of the figure of the small parts of bodies."[34]

The only place where a Newtonian concept enters Cullen's teaching is with respect to the aether. In his lectures of 1762-63, Cullen declares that "Dr. Bryan Robinson's of Dublin, treatise on the aether of Sir Isaac Newton . . . is the only probable scheme of a chemical theory." Just what he can have meant by this statement is far from clear. He declares that by "observing the various states of the aether in bodies we shall understand the forms of solid and fluid, the various states of concretion, and consequently the various properties depending on those states," but admits that the hypothesis of the aetherial fluid presents some difficulties when considering the "doctrine of elective attraction." Yet he feels the difficulties vanish when the analogy of the electric fluid is considered.[35] Beyond showing that Cullen had applied to chemistry, as he did to physiology, the contemporary enthusiasm for Robinson's revival of aetherial notions and the presumed relation of aether and electricity, this passage gives little help in understanding how the aether was specifically related to chemistry.

An example of such a role for the aether can be found in a paper of Dr. John Bond, in the *Philosophical Transactions* for 1753, describing experiments on the replacing of copper by iron in solution. Bond obtained his doctorate at Edinburgh in 1751 and may have learned his chemistry there from Cullen's teacher, Andrew Plummer, or possibly, as Black later was to do, he studied first with Cullen at Glasgow. In any event, his general attitudes are those of Cullen. Transmutation is a "ridiculous doctrine, which destroys the essential qualities of bodies, which were im-

[33] See letter of Cullen to George Fordyce, October 1759, quoted by Thomson, *Life of Cullen*, I, 570-71.

[34] William P. D. Wightman, "William Cullen and the Teaching of Chemistry—II," *Annals of Science*, 12 (1956), 192-205, esp. p. 194.

[35] Quoted from manuscript lectures of Cullen, Manchester University Library, MSS CH.C.121.1, pp. 43-45, by Arnold Thackray, "The Newtonian Tradition and Eighteenth-Century Chemistry."

pressed by the Great Creator on all material substances, in order to distinguish them from each other, and therefore are intransmutable." The effect he describes is due to an active principle (the vitriolic) in the water and to the varying elective attractions of acid, copper, and iron. The activity of acids is best explained in a late publication of the "ingenious Dr. Robinson," his *Essay on the Operations of Medicines,* in which acids and light are proved to be the same thing, "for he infers from Sir Isaac Newton's philosophy, that whatever attracts or is much attracted, is light; therefore an acid is light."[36] That Cullen held this particular view is far from clear, though the parallel references to Robinson are suggestive. However, there is no evidence, from the work of Cullen's most famous student, Joseph Black, that the aether played an essential part in Cullen's chemical theory. In none of Black's published work does the aether enter the picture, except in the form of its lineal descendent, the fluid substance of heat. As Cullen retained his belief in aetherial explanations for physiology, he probably did so for chemistry, but clearly one could (as Black did) develop from Cullen's teaching a consistent approach to chemistry without calling on the aether.

Joseph Black's most significant original work in chemistry was done in connection with his thesis for the M.D. at Edinburgh. The background for the thesis was derived from his chemical studies at Glasgow and, when published in 1754 as *de Humore Acido a cibis orto, et Magnesia Alba,* the work was dedicated to Cullen. The experimental part of the thesis was then expanded and published as "Experiments upon Magnesia Alba, Quicklime and some other Alcaline Substances" in 1756.[37] The topic had been chosen in response to recent work on the lithontriptic properties of lime water. From the time that Stephen Hales had declared that urinary calculi contained the active principle of air and, therefore, a proper dissolvent for the stone could probably be found, Professors Robert Whytt and Charles Alston (1683-1760) of the University of Edinburgh had competed to demonstrate the efficacy of lime water for that purpose. Black prudently avoided entering an argument between two of his pro-

[36] John Bond, "A Letter . . . containing Experiments on the Copper Springs . . . ," *Philosophical Transactions,* 49 (1753-54), 181-90. There is no reference to Bond in the *D.N.B.,* Poggendorf, or the standard catalogues of graduates or matriculants of Glasgow or Edinburgh, but the *British Museum, Catalogue of Printed Books,* lists his Edinburgh doctoral dissertation, *de Incubo,* published in 1751.

[37] The dissertation has been translated by Crum Brown and published as "Joseph Black's Inaugural Dissertation. I, II," *Journal of Chemical Education,* 12 (1935), 225-28, 268-73. Joseph Black, "Experiments upon Magnesial Alba . . . ," *Edinburgh Essays,* 2 (1756), 157-225.

fessors, but their experiments had indicated that solutions of "alcaline earths" other than lime might be even better solvents. Black was disappointed in his choice of an alternative, for calcined magnesia alba did not form a "lime water"; that line of investigation was abandoned, and a routine discussion of the substance as an antacid was prefixed to his dissertation to satisfy its medical requirements. As a medical thesis, Black's work was trivial, but in the process of investigating the nature of alkaline substances, he developed chemical concepts which were highly significant.

Here again the original impulse can be found in the work of Hales, this time in his supposition that the "fire particles" in calcined lime, which were presumably the cause of its activity, were actually particles of air fixed in the substance of the lime and released when the lime was dissolved.[38] Hales' interpretation of the phenomena, consistent with his dynamic corpuscularity, was an explanation of the instrumental role of air in chemical processes. The air, temporarily fixed in an inelastic state in substances, was released by dissolution, recovered its property of elastic repulsion, and communicated this activity to other substances. Black arrived at an alternate explanation. By careful gravimetric experiments, he demonstrated that the activity resulted from the constituent presence in lime of precisely measured amounts of a unique species of air, which he named fixed air and whose properties he described. This work was effectively revolutionary on three counts: the constituent rather than instrumental role of air, the identification of a different species of air, and the systematic use of gravimetric techniques. For none of these has there been found an historically satisfying explanation either in Black's education in chemistry or in the authorities he used for this particular investigation. Neither Hoffmann, Geoffroy, Marggraf, Hales, nor Whytt understood the chemical role of air as a constituent one, nor did Cullen until after Black's experiments were completed. Boerhaave's confused merging of Hales' studies with his own, in the official version of his text, could have given rise to such conclusions, while Shaw's statement of 1734, that the chemists' elements might contribute substance to chemical change, suggests that these ideas were in the air, but Black was the first to demonstrate effectively their validity. Once the notion of a

[38] See Chap. IV. I have retained my assumption that Hales' statement is best interpreted for the modern reader by the substitution of "particles of air" for their Halesian near-equivalent "particles of fire." The wording with the "ambiguity" was, however, what was read by and influenced Whytt, Alston, and presumably Black.

constituent role of air is accepted, the concept of differing kinds of air may seem to follow—and perhaps it does, for after Black had described fixed air, other investigators soon followed with mephitic and inflammable airs. It is not obvious, however, why there might not earlier have been supposed various instrumental airs, while the concept of different constituent airs was not fully established until the work of Lavoisier was accepted.

As for the use of gravimetric techniques, now regarded as the most significant element of Black's work, neither he nor his contemporaries seem to have thought it remarkable. It has frequently been observed that the concept of conservation of substance, as determined by constant weight measurements, was not new in Black. As early as 1700, Geoffroy, reporting to Hans Sloane on some work of Wilhelm Homberg (1652-1715), casually refers to the use of the balance and assumes that variations of weight indicate the transfer of substance from one reactant to another.[39] Yet Black appears nearly alone in his critical application of weight and conservation of substance concepts until Lavoisier adopts weight as the primary chemical parameter for determining constituent changes in chemical reactions. It has been suggested that the argument between Whytt and Alston, on the nature of lime water, gave Black his procedural clue. Whytt had accepted the reading of Hales that quicklime owed its activity to the element of fire united to the earthy matter when heated and transferred to the lime water in solution. Alston, not believing in the substance of fire, showed that lime water acquired its virtue as a simple consequence of the solubility of lime in water. This demonstration of the substantial origin of the activity might have prompted Black to study the problem gravimetrically, but one must note that Black did believe (or was soon to do so) in a substance of fire and found no essential contradiction in the imponderability of that substance.[40] It would appear that Black's originality remains a problem for his biographers.

With the publication of his paper on magnesia alba and quicklime, Black ceased original investigations into what would, today, be called chemistry, turning his attention to studies of heat, doing industrial consulting, and lecturing on elementary chemistry, year after year, to steadily increasing classes. The nature of his teaching is, supposedly, embodied in the posthumous edition of his *Lectures on the Elements of Chemistry* (1803), the

[39] E. Geoffroy, "Part of a Letter . . . concerning the exact quantity of acid Salts, contained in acid Spirits," *Philosophical Transactions*, 22 (1700-01), 530-34.

[40] See Guerlac, "Joseph Black and Fixed Air," pp. 143-46.

complete reliability of which has previously been questioned. With the same reservations established for their use in discussing his theory of heat, the *Lectures,* supplemented by the earlier anonymous *Enquiry into the General Effects of Heat, with Observations on the Theories of Mixture* (1770), can, however, provide some clues as to his general chemical philosophy and its relation to theories of matter and its action.

As one might expect, Black's chemistry lectures seem to have developed naturally out of Cullen's course. Like Cullen (and Boerhaave), Black commenced his course with a historical discourse; like Cullen, he recommended Boerhaave's text for its section on heat—though his own work provided the primary materials for his lectures on the subject; and, like Cullen, Black's emphasis was on the classification of compounds and the concept of affinities. Black, however, though he thought Stahl's *Theoria Chemiae Dogmatica* a "work of great merit and ingenuity [*Lectures,* I, 232]," preferred the more sophisticated Stahlian chemistry of Pierre Joseph Macquer (1718-84). Macquer, probably the greatest of Stahl's French disciples, had written a text, *Elemens de Chymie Theorique et Practique* (Paris, 1749, 1751), which had the advantage of didactic simplicity and organization over Stahl's own work and offered a viable alternative to Boerhaave's text. It was this text of Macquer, translated into English as *Elements of the Theory and Practice of Chymistry* in 1758, which Black recommended to his students.[41]

Yet Black's lectures differ markedly from Macquer in a number of ways. The heroes of Macquer's text are Stahl and Geoffroy, and from their work he attempts to develop a systematic nomenclature based on a rational taxonomy and ultimately a mechanistic concept of affinities. Black was impressed by Stahl and Geoffroy, but his inclinations were more inductive and empirical. According to his editor, John Robison, Black could not "endure the title of a system to be given to any body of chemical doctrines yet published," and he persisted in calling his own discourses "Lectures on the Effects of Heat and Mixture" rather than "Lectures on Chemistry [*Lectures,* I, lxiv, 12]." Although he came to accept the oxidation combustion theory, he disliked Lavoisier's general approach for "its train of synthetic deduction," holding that chemistry was not yet a science and must rather content itself with building a set of facts on which true in-

[41] Pierre Joseph Macquer, *Elements of the Theory and Practice of Chymistry,* Andrew Reid, tr. (London: A. Millar and J. Nourse, 1764), 2nd edn.; first edition in English 1758, fifth and last, 1777.

ductions might be based [*Lectures*, I, 547]. In chemical taxonomy, Macquer had commenced with elements, described as those substances which the chemist finds incapable of being resolved into others [*Elements*, I, 1]. Black was disinclined to make even this basic assumption: "I do not pretend to determine what are the ultimate elements of bodies. I content myself with distinguishing and dividing the principal objects of chemistry into a number of classes, each of which comprehends substances that bear a remarkable resemblance or analogy with one another in their chemical properties, and differ from those comprehended in the other classes [*Lectures*, I, 334-45]." The different approaches did not have substantially different consequences. Macquer accepts, as a type of element, substances acting in an instrumental as well as constituent way, to produce change by mechanisms which operate in a manner "to which we are strangers, and concerning which nothing beyond conjecture can be advanced . . . these we neglect, resolving to keep wholly to facts . . . [*Elements*, I, 129]." In practice, Macquer devotes a substantial part of his text to the definition of various constituent substances by macroscopic physical characteristics or chemical properties—with what do the substances combine, how easily, and with what results? Neither he nor Black considers any microscopic characteristics (*e.g.*, particle size, shape, or motion) as relevant considerations. In the end, though Macquer might further subdivide substances by their elements of air, earth, fire, and water, both men essentially agree on their systems of classification into salts, earths, inflammable substances, metals, and waters.

Oddly enough, although William Cullen, by 1766, had added to his system of classification a new division of "aerial bodies," because "we are now sure that they enter into the composition of even solid bodies," and Macquer had an undifferentiated element of air, Black has no such class.[42] In spite of his major responsibility in achieving recognition of the constituent role of gases in chemistry, Black finds no necessity for constituting a class of airs, or gases, and even questions the propriety of doing so, as gases are but various kinds of matter combined with latent heat [*Lectures*, I, 345]. Nor does Black emphasize the use of weight in describing chemical operations, though, of course, his descriptions to his students of the magnesia alba, quicklime, etc., experiments gave them a classic example to follow. In fact, Black was willing to accept the possibility of particular kinds of

[42] William P. D. Wightman, "William Cullen and the teaching of Chemistry—II," p. 200.

matter not possessing the property of gravitation and not even to worry if the loss of the inflammable principle (phlogiston), increased rather than decreased total weight.[43]

There are few better examples than Black of the adage that the scientist who denies theory is merely unconscious of the theory he uses and will, therefore, be inconsistent in its application. In the development of his heat theory and in that of the constituent role of gases, Black's materialism had given him implicit guidance in the interpretation of experimental phenomena; in his approach to affinities he found no such guide, for he was unable to accept the materialist notion of mutual conformity of elementary constitutions. Macquer's treatment of affinities, in the *Elements,* is an empiricist-inductivist one, in which the disposition of some bodies to unite with others is examined in a series of seven propositions relating the presumed affinities of two or more substances to the varying combinations which ensue when they are mixed [*Elements,* I, 12-13]. This approach is that which Cullen had also adopted and which Black attempted to maintain, but it was an unstable position from which some change, attempting to explain the phenomena, was obviously required if chemistry was to be an analytical science rather than an empirical art. Such a change is revealed in Macquer's discussion of the problem which appears in his influential, quasi-anonymous *Dictionnaire de Chymie* (Paris, 1766). In the *Dictionnaire,* Macquer was prepared to indulge those conjectures on the possible relationship of affinities to Newtonian laws of attraction, now popular, after a period of reaction, with some of Europe's best philosopher-chemists.[44] Black recommended Macquer's *Dictionary* to the use of his students and neither he nor his British materialist contemporaries wanted to deny the influence and authority of Newton, but he had an aversion to mechanism as used in chemistry. The result was a discussion of affinities full of ambiguity and self-contradiction which, however, reveals the inductivist view of Newton attained by late eighteenth-century British natural philosophers to support their anti-mechanistic materialism.

Black's dilemma is perhaps most clearly revealed in the anony-

[43] See, for example, Douglas McKie, "On Thomas Cochrane's MS Notes of Black's Chemical Lectures, 1767/8," *Annals of Science,* 1 (1936), 101-10.

[44] [P. J. Macquer], *A Dictionary of Chemistry, containing the Theory and Practice of that Science* (London: T. Cadell and P. Elmsly, 1777), 2nd edn., tr. James Keir, 1st edn. 1771, Vol. I, Article: "Affinity," Vol. II, Article: "Gravity." Note that Macquer's *Dictionnaire* first appeared the year after Buffon argued for a gravitational theory of chemical affinities in his *Historie Naturelle.*

mous *Enquiry* of 1770, where Robison's neo-mechanistic editorial work, present in the *Lectures,* is missing. In the latter part of the *Enquiry,* Black presents, as a praiseworthy development, Newton's extension, by way of the doctrine of gravitational attraction, of Bacon's and Boyle's attempt to explain chemical mixtures mechanically. Indeed, Black declares that extension to be "the only theory which is almost universally accepted at this day" and criticizes the French for substituting "affinity" for attraction, as this intimates a similarity between uniting bodies [96]. Black goes on, however, to praise the objections of the Germans, who, "Enemies to theory, and assiduous only in accumulating facts and experiments . . . were disgusted at the extravagant lengths to which this theory has been pushed by some of our own countrymen, particularly by Dr. Freind and his Chemical Lectures [97]." And he continues by denying that chemical attractions are at all related to those of gravity, magnetism, or electricity. The "attractions" of chemistry act only between very small particles of matter and to distances and with forces inconceivable and imperceptible to the senses [98-99].

The *Lectures* merely develop the same confusion in greater detail. Black adopts the concept of particulate attraction, and even polarity, to explain the various forms of crystals [I, 45-47]. With considerable reluctance, he tentatively accepts William Cleghorn's description of the substance of heat as particulate, mutually repulsive, and attractive to other matter. But he disliked this approach as being that of a mechanician rather than a chemist and compared it to Gowin Knight's essay on magnetism as an example of mechanistic interpretations of phenomena which initially appear to explain and then are found continually to involve inconsistencies [I, 516-17]. Again he introduces the topic of elective attractions with praise of Newton, and the statement that "Sir Isaac's theory . . . is now . . . well established wherever the science of chemistry has made any progress, and . . . we shall find it most extensively useful to us in assisting us to understand a great variety of chemical phenomena . . . [I, 267]." It is in this context that he discusses the general aspects of elective attractions, but as he develops the concept, it again becomes clear that Newtonian attractions, such as cohesion, gravitation, etc., are different from those of chemistry and act only to balance, counteract, and confuse the operations of chemical combination [I, 272-78]. He then declares that chemical science is obstructed by speculations on the causes of affinity "and particularly by the attempts of ingenious men to explain the chemical operations by attractions and repulsions. . . . I may venture to say that no man ever got a clear and really explicatory notion of chemical combination by the help of attractions [1, 282-83]." So consistent was Black's basic objection to the mechanistic implications of elective attraction that he abandoned the use of affinity diagrams, learned from Cullen, because these suggested the notion of bodies interconnected by levers movable about the point of intersection, and levers were mechanical not chemical [1, 544-45]. Black recommends that chemists forego speculation about ultimate internal action and direct their attentions to external phenomena. Even in the mechanical philosophy, attraction was merely a metaphorical term; in chemistry, attraction and affinity should mean no more than a certain faculty of combining. "Let chemical affinity be received as a first principle, which we cannot explain any more than Newton could explain gravitation, and let us defer accounting for the laws of affinity, till we have established such a body of doctrine as he has established concerning the laws of gravitation [1, 283-84]."

For practical purposes eighteenth-century British chemistry ended, as did the eighteenth century itself, with a French revolution. The "chemical revolution" of Antoine Laurent Lavoisier (1743-94) was achieved over a number of years, but its manifesto is most completely expressed in his *Traité Élémentaire de Chimie,* first published in Paris in the year of the revolution, 1789. Translated into English by Robert Kerr, as *Elements of Chemistry,* in 1790, the fifth and last contemporary edition of that translation in 1802 coincided with the last futile efforts of Joseph Priestley to stem the tide of the new chemistry.[45] The transformation of a premature mechanical science into a taxonomic, materialistic one was complete.

The complete nature of Lavoisier's "chemical revolution" and its development has yet to be sufficiently described, but some pertinent elements of its origin and influence seem clear.[46] Professor Henry Guerlac has shown how little aware Lavoisier was of near-contemporary British work in chemistry as he began his own, and his ignorance of the earlier speculations of the dy-

[45] Antoine Lavoisier, *Elements of Chemistry,* tr. Robert Kerr, with a new Introduction by Douglas McKie (New York: Dover Publications, facsimile reprint of the 1790 Edinburgh edition, 1965).

[46] While we wait for the completion of Henry Guerlac's analytical studies of Lavoisier, we must content ourselves with the tantalizing and useful preliminary studies now essentially collected in his *Lavoisier—The Crucial Year* (Ithaca: Cornell University Press, 1961).

namic corpuscularians was surely even more complete. His entrée into chemistry was made by way of geology and mineralogy, to which the work of German chemists was most applicable, and through the lectures of Guillaume Rouelle (1703-70), with the Baron d'Holbach a major avenue for the introduction of Stahl's ideas into France. In spite of the well-advertised differences between Lavoisian and Stahlian chemistry, they share a great many common attitudes. Lavoisier's view of the function of chemical experiments, "to decompose natural bodies so as separately to examine the different substances which enter into their composition [176]," is characteristic of Stahl, as is his definition of the chemical element as the last point which analysis is capable of reaching [xxiv]. Lavoisier's aversion to the discussion of ultimate, indivisible atoms (previously quoted p. 191) might have been paraphrased from any of several Stahlian chemists, while his division of chemical combinations into "mixts," composed of two simple elements, and secondary, tertiary, etc., compounds formed by the union of "mixts" [150], is admittedly taken directly from Stahl. Even his explicit avoidance of a discussion of affinities can be seen to have Stahlian, or at least anti-mechanist, implications. The ingenuous admission of his disinclination to venture where his colleague, Guyton de Morveau, was working and the excuse that affinities, or elective attractions, being the chemical equivalent of transcendental geometry, had no part in elementary chemistry must be set against the fact that he uses the concept of affinities throughout the *Elements* and admits that this is, of all branches of chemistry, "the best calculated of any part . . . for being reduced into a completely systematic body [xxi-xxii]." Lavoisier did do some work on measuring numerical affinity coefficients, presumably to supply some of the data he felt was missing to a foundation of affinity theory. He subsequently (c. 1792) prepared segments of a "Cours de chimie expérimentale rangée suivant l'ordre naturel des idées," in which he refers to the explanations by Bergman and Guyton of affinities, but these fragments give no indication of his own views on their cause.[47] In the *Elements*, however, he adopts an argument by similitude, holding, for example, that combustible substances "ought . . . to attract or tend to combine with each other" because they, "in general, have a great affinity for oxygen [109]." It is hard to believe that the person making this simplistic (and incorrect) empirical observation was in sympathy with the increasingly

[47] See Maurice Daumas, "Les Conceptions de Lavoisier sur les Affinités Chimique et la Constitution de la Matière," *Thales*, 6 (1949-50), 69-80.

mechanistic connotations of the affinity speculations which he refused to discuss in the *Elements*. There is no indication that Lavoisier, any more than Stahl, thought of chemistry as operationally mechanical.

To insist on the anti-mechanist character of Lavoisier's chemistry may seem paradoxical in view of the heavily mechanistic nature of the first twenty-five pages of the *Elements*. These pages, however, relate to a privileged substance in Lavoisier's list of elements, the element of heat or caloric, and it is precisely in the unique position he gives this substance that he affirms the anti-mechanism of his chemistry and, at the same time, makes his first significant advance from the position of Stahl. For Stahl, fire and air were chemical instruments, while water and the earths were chemical constituents. Rouelle had adjusted to post-Stahlian developments by including all elements as compositional as well as instrumental factors—and, incidentally, had thoroughly confused the distinction between the matter of fire and that of phlogiston. Lavoisier extracted all of the instrumental aspects of chemical elements and placed them in the single physical substance of caloric. This was the vehicle of repulsion, the cause of fluidity and elasticity, the producer of light and of dimensional change. It is probable that his use of caloric as a means of heat transfer, capacity, and change of state was instrumental in winning Black's acceptance of Lavoisian notions of combustibility. But it is essential to note that caloric (along with the possible element of light) was an aberrant member of the table of elements and played no constituent part in chemical change. As the British materialists had used the aether to rid themselves of the embarrassment of forces, so Lavoisier used caloric, in a purely physical role, to rid himself of the embarrassment of chemical instruments or operators.[48]

The isolation of caloric from the elements of Lavoisier's chemistry is clearly indicated by its lack of weight, for it was in weight, its measurement and changes, that he found the essential character of chemical analysis. The only place where quantitative

[48] I am initially indebted to the work of Hélène Metzger, and particularly her "Newton: La Théorie de l'Emission de la Lumière et la Doctrine Chimique au XVIIIème Siècle," *Archeion*, 11 (1929), 13-25, for the interpretation of Lavoisier's elements of heat and light as uniquely physical. Joseph Agassi, "Towards an Historiography of Science," *History and Theory*, Beiheft 2 (1963), has added the suggestion that caloric was Lavoisier's material response to Boscovich's mechanization of the particles of matter, as a means for preventing ultimate contact of particles. This proposal makes Lavoisier substantially more interesting as a theoretician, but should be supported by some evidence that he took Boscovich that seriously.

argument enters his chemistry is in the quantification of substance, and this invariably was to be achieved by weight. "We may lay it down as an incontestable axiom that, in all the operations of art and nature, nothing is created; an equal quantity of matter exists both before and after the experiments; the quality and quantity of the elements remain precisely the same. . . . Upon this principle, the whole art of performing chemical experiments depends [130-31]." ". . . the usefulness and accuracy of chemistry depends entirely upon the determination of the weights of the ingredients and products both before and after experiments . . . [297]." The "material proof" of the accuracy of analysis was the exact equivalence of "the whole weights of the products taken together, after the process is finished," to the weight of the original substances submitted to analysis [393]. The identity of conservation of weight with that of substance was so unnecessary of proof, that Lavoisier frequently assumes weight conservation in his experiments: *e.g.*, "As no gravitating matter could have escaped through the glass, we have a right to conclude, that the weight of the substance resulting from the combustion . . . must equal that of the phosphorus and oxygen employed . . . [56]." Lavoisier was the first person explicitly and repeatedly to insist upon conservation of substance-conservation of weight considerations in chemistry and it was this insistence (a primary materialist criterion) which changed his definition of the chemical element, otherwise trivially familiar to his contemporaries, from banality to an operational concept of use in laboratory procedure.

Finally, we come to that part of Lavoisier's chemical revolution which he, himself, regarded as the major contribution of the *Elements*, the new nomenclature and taxonomy. This emphasis was more than an accident developing from the expansion of the *Méthode de Nomenclature Chimique* into a text. Lavoisier's repeated citations from Condillac, in the preface to the *Elements*, show that it meant more to him than a simple distinguishing of substances from one another, assigning them names, and putting them into classes. For him, as for Condillac, language was an analytical method; the "distinctions are not merely metaphysical, but established by nature [xxv]," and once established, the new language could be expected to throw "great light upon all the operations of art and nature [81]." The names of the elements were to be framed "in such a manner as to express the most general and most characteristic quality of the substance [xxv]." Once these were distinguished by appropriate terms, "the names of all compounds derive readily and necessarily, from the first denomination [53]." From the name of a compound alone, one should be able, instantly, to determine what substances entered into the combination, in what proportion, and to what degree of saturation [xxx].

It was a magnificent (if not entirely attainable) objective, and one for which subsequent generations of chemists have cause to thank the originators. It also evaded much of what previously had been thought the task of a chemist and made explicit a host of previously ill-defined problems. This is perhaps best illustrated by the example of oxygen and oxidation, for though this most strongly indicates the errors in his system, it was the pivotal example in Lavoisier's *Elements* and the one on which the revolutionary character of his work is most generally based. It has frequently been observed that phlogiston was the inverse of Lavoisier's oxygen, but this observation, by inverting the actual order of events, conceals some essential evolutionary characteristics of the theory of oxidation. When phlogiston was deprived, through the use of caloric, of its physically instrumental properties, it became a constituent substance very like oxygen, and Lavoisier once explicitly observed that he should not mind being viewed as a phlogistonist who considered vital air (*i.e.*, oxygen plus caloric) to be phlogiston.[49] But oxygen, as the name implies, is much more than the agent of combustion. It is the property of combustible substances to become acids by their combination with this element which, in fact, constitutes acidity and is therefore to be named oxygen, from the Greek for acid-former [61, 51]. Acid now becomes a generic term in which each member is distinguished from all others by the name of the substance, base or radical, which has combined with oxygen to make the acid [66]. This being now a matter of definition, it necessarily follows, though "we have not yet been able to compose or decompose [the] . . . acid of sea-salt, we cannot have the smallest doubt that it, like all other acids, is composed of oxygen with an acidifiable base [71]." And it further follows that "vegetable and animal oxyds and acids . . . differ . . . According to the number of simple acidifiable elements of which their radicals are composed . . . According to the proportions in which these are combined together: And . . . According to their different degrees of oxygenation . . . [192]."

[49] In the notes translated with the second edition of Richard Kirwan's *Essay on Phlogiston* (London, 1789), pp. 20-21. I am indebted to the observation of Joseph Agassi, "Historiography of Science," note 91, for this reference.

In spite of Condillac and Linnaeus, in spite of Lavoisier, taxonomy, it seems, is not a simple arrangement of facts, nor is it an end in itself. Lavoisier's new nomenclature and taxonomy was an attempted rationalization of empiricism. His elements were the substantialization of the chemists' qualities with no attempt to explain their origin. Compounds possessing certain qualities were assumed to contain appropriate elements, and in a proportion reflecting the intensity of the quality. Compounds similar in chemical behavior were similarly constituted; compounds different in behavior had, practically by definition, different compositions. This was, for its time, a substantial advance over previous terminologies, but it was not empirical. It was, in fact, precisely the materialism which had previously replaced mechanization in electricity, magnetism, heat, and physiology.[50] Seen in this context, Lavoisier's chemical revolution is surely one of the least revolutionary of any to effect scientific change, for it was the purest expression of that tendency away from mechanism and toward materialism which characterized the natural philosophy of the second half of the eighteenth century. It was just his aversion to mechanism and fondness for empirical materialism which enabled Joseph Black to become the first prominent British chemist to adopt Lavoisier's new chemistry. Of his British peers, Cavendish and Priestley, for example, were neo-mechanists, deriving their impulse to investigation from Newton and from Hales; they never adopted Lavoisian chemistry. Their work was a continuation of the tradition of the scientific revolution of the seventeenth century, which had introduced the mechanical philosophy. Lavoisier's revolution was, in effect, a final ratification of Stahl's opposition to mechanization and its acceptance in Britain deflected, for nearly half a century, any significant efforts toward the development there of a physical chemistry.

[50] For the emphasis on the taxonomic character of Lavoisier's chemistry, I am initially indebted to Charles C. Gillispie, *Edge of Objectivity* (Princeton: Princeton University Press, 1960), pp. 202-50, but particularly to his unpublished paper, "Devoid of Atoms," delivered at the Conference on the History of Eighteenth-Century Chemistry (Paris, 1959). For the interpretation of this as materialistic, Dr. Gillispie, however, cannot be blamed.

PART III

NEO-MECHANISM

1760-1815

CHAPTER TEN

Forces, Fluid Dynamics, and Fields

MATERIALIST EXPLANATIONS FOR PHENOMENA were to retain their superiority over mechanist ones throughout the second half of the eighteenth century. Nor is this particularly surprising. Quality-bearing substances had always been easier to conceive than quality-causing mechanisms and, with the application of conservation considerations, such substances had become easier to quantify as well. No doubt the questions the materialists answered were not those that the dynamic corpuscularians had asked, and, in thus evading the problems of reductionism, the materialists clearly were betraying the hopes of a century of anti-Aristotleian natural philosophers. But for most of the subjects of primary concern to the eighteenth century, the mechanists had scarcely advanced beyond questions, while the materialists achieved resolution of problems the dynamic corpuscularians had scarcely been able to define.

Nonetheless, during the course of the second half of the century there appeared in Britain a group of natural philosophers who were not satisfied. Although they had, perforce, to come to terms with the successes of materialism, these successes seemed to them essentially incomplete, not wisdom in moderation but complacence in ignorance. Content with neither the substitution of taxonomy for analysis nor the quantification of quality by conservation of substance, these men returned to the mechanist aspirations of earlier generations. Naturally their work was distinguished by its originality, but, as it could not be comprehended in the rubric of customary materialism, much of it was ignored by contemporary readers and has been misunderstood by later expositors. It is one of the chief advantages of matter theory as an instrument of historical analysis, that it brings to the work of such men as John Michell, Henry Cavendish, William Herschel, Joseph Priestley, and James Hutton, a new measure of coherence and finds, in the interpretation of them as neo-mechanists in a period of dominant materialism, an explanation for the reception accorded their work.

That some measure of mechanistic speculation was continued in Britain into the second half of the eighteenth century has already been illustrated in the work of Gowin Knight. Knight's *Attempt to demonstrate . . . all the Phaenomena in Nature*

(1754) had, however, aetherial overtones, and its infinite complexities tended rather to repel than attract supporters. Of far greater significance was the work of the Croatian, Roger Joseph Boscovich (1711-1787), whose *Theoria Philosophiae Naturalis, redacta ad unicam legem virium in natura existentium* (Vienna, 1758; Venice, 1763) was nearer in spirit than Knight's *Attempt* to the traditions of Newtonian mechanism, while its concepts were easier to visualize. Mathematician, poet, astronomer, diplomat, engineer, Jesuit priest, and metaphysician, Boscovich wrote more than a hundred books and papers, but those relating to matter theory essentially began with his *de Viribus vivis* of 1745 and culminated in the *Theoria*, published in its second, definitive, edition under Boscovich's supervision, in 1763.[1]

Although the *Theoria* contains speculations on the nature of space and time, the relation of body and spirit, and the connection between mind and God, its primary function is to detail Boscovich's version of the dynamic corpuscular faith that all the phenomena in nature can be explained in terms of homogeneous, particulate, primary matter, its motions, combinations, forms, and forces [185]. To Boscovich, however, these particles are indivisible and non-extended geometrical points, possessing inertia as individuals, and, as pairs, a tendency to approach one another at some distances and recede from one another at others [20-21]. Rather than supposing these tendencies to result from a multitude of different forces—gravitational, cohesive, etc.—he declares that there exists, for each point with respect to any other, a single, continuous, action-at-a-distance curve of attractive and repulsive accelerations, a radially symmetric, single valued function of distance, asymptotic toward repulsive infinity at the center, crossing the zero-acceleration axis many times before sensible distances are attained, after which the curve approximates, within any measurable degree of accuracy, the attractive, inverse-square, gravitational curve of Newton and, like it, is asymptotic to zero acceleration with distances at least as great as those of the planetary and cometary systems [53-54].

Much has recently been made of the kinematic as opposed to dynamic character of Boscovich's theory and to the fact that

[1] The *Theoria* was translated into English by J. M. Child and published in a massive parallel-text edition, in folio, by the Open Court Publishing Company in 1922. Child's English text has since been extracted from its inconvenient form and republished as *A Theory of Natural Philosophy* (Cambridge: M.I.T. Press, 1966), which is the version I have used. For a biographical sketch of Boscovich and a selection of some recent enthusiasms respecting him, see Lancelot Law Whyte, ed., *Roger Joseph Boscovich, S.J., F.R.S., 1711-1787* (London: George Allen and Unwin, 1961).

"Boscovich's atomism is not a materialist theory, but a geometrical one, concerning itself with structural patterns, their changes, and rates of change."[2] On the whole, the first assertion seems an unnecessary and anachronistic refinement of his intentions, while the second is nearly as true for all the corpuscularians. Boscovich's "elements of primary matter" do achieve dimension solely through their infinite repulsive action at close proximity, and it is hard to assign them mass. Boscovich, however, assigns them inertia and declares that, "The mass of a body is the total quantity of matter pertaining to that body; and in my theory this is precisely the same thing as the number of points that go to form the body." When he writes that "the idea of mass is not strictly definite and distinct, but . . . quite vague, arbitrary and confused," he is primarily concerned with the difficulty of defining what parts of matter do actually pertain to its constitution and are therefore to be included in assessing its mass [139-40]. No more than Newton will Boscovich speculate on the origin of inertia or of his "accelerative propensities"; they may depend upon "some arbitrary law of the Supreme Architect, or on the nature of points itself, or on some attribute of them. . . . I do not seek to know . . . [and] see no hope of finding the answer [21]." His "determination" of bodies to approach one another is almost a paraphrase of the "endeavour of bodies to approach each other" which Newton had earlier defined as attractive force in the *Principia* [Bk. I, Sec. XI, Scholium]. It seems reasonable, therefore, to say that Boscovich's curve is a curve of attractive and repulsive forces. This is the way he used it and it is the way he was understood in the eighteenth and nineteenth centuries.

Boscovich's theory, then, is a dynamic, if not corpuscular, extension of dynamic corpuscularity and can be meaningfully compared to the theories of his predecessors.[3] At once obvious

[2] See "Boscovich's Atomism," in Whyte, *Boscovich*, pp. 102-26, esp. p. 108.

[3] Such a comparison raises the issue of a possible dependence on these predecessors, and, to my mind, this is not impossible. Boscovich's explanation that the theory grew naturally out of a combination of Newton (especially Query 31), Leibniz, and the Law of Continuity appears to me a typical, scientific *post-facto*, neat, rationalization of the complicated process of discovery. His British citations are limited to the works of Newton and Maclaurin (though he mentions Franklin and Hales), but he does not refer to all the writers, even European, that he had read. The internal evidence of his reading of Boerhaave, for example, is strong, and he probably knew Hales only in that way, but Boerhaave's name is not mentioned. As he could not read English, the work of Rowning would be closed to him, but Newton, both Keills, Freind, Hales, Desaguliers, and Maclaurin would all have been available to him either in Latin or in French translations. Much work remains to be done on Boscovich's reading at the Collegium Romanum, where he studied and taught.

similarities to the theories of such British writers as John Rowning and Jean Théophile Desaguliers appear. They share a conviction in the ultimate homogeneity of all matter; their primary particles are surrounded with alternating spheres of attractive and repulsive forces—with, for Rowning as for Boscovich, the innermost sphere repulsive and the outermost that of gravitation. For Boscovich, as for his predecessors, the diversity of matter results from the varying combinations of ultimate particles into those of a second order, possessing less force, and these into still higher order particles, of decreasing forces, until particles of a sensible magnitude are achieved [144]. And, of course, these men share the conviction that all phenomena can be explained in terms of these particles, their sizes, shapes, combinations, and the distribution of forces resulting from these combinations. Conceptually, there is more equivalence than difference in their approaches, for all the disappearance, in Boscovich, of the fundamental, extended particle into a geometrical point.

In the general realization of his conception, however, Boscovich has a clear superiority. By reducing the various distance-dependent forces of Newton, Keill, Freind, Rowning, etc., to a single force-curve represented graphically, Boscovich identifies a variety of parameters which give his theory a functional flexibility unrecognized in earlier speculations. The interface between attractive and repulsive spheres had, for example, previously gone unnoticed, but Boscovich emphasizes that the places where his force-curve crosses the zero axis are limit-points of no force, and, moreover, are alternately limit-points of cohesion, where particles can rest in stable equilibrium, and of noncohesion, where the equilibrium is unstable. Use of the concept of cohesive limit-points adds plausibility to the conception of chemical elements having substantial permanence, to regularities of compound formation, crystallization, etc. [73]. No notice had previously been taken of the various thicknesses of attracting and repelling spherical shells, but in Boscovich's theory, the areas under the arcs drawn between consecutive limit-points represent energies (defined by Boscovich as velocity-squared). By denying that the curve need be regular, he obtains potentially different "energy levels" between different limit-points, and as the curve can approach the axis at different rates, explanations of elasticity, vibrations, varying conditions of transformation of state, radiation, etc., are also made possible [73, 76-78]. No wonder Boscovich's theory exerted a continued fascination on physical model-builders long after the earlier speculations of the dynamic corpuscularians were forgotten.

In a way, however, the very fertility of the theory becomes finally frustrating. The variety of available parameters provides a way in which the general mechanism of any phenomena might be conceived, and this could (and did) inspire the investigations of British scientists as late as Lord Kelvin and J. J. Thomson. But the infinite possibilities of the curve make it, in any specific and detailed application, infinitely useless as a way of realizing the Newtonian dictum: from the motions find the forces, from the forces derive the phenomena. Boscovich states that he proves geometrically, from the combinations, what phenomena, or what species of bodies, ought to arise, rather than creating special combinations for particular properties [24], but this is true only in the most general way. Even in considering the simplest curve, relating two points, he admits that "There are indeed certain things that relate to the law of forces of which we are altogether ignorant, such as the number and distances of the intersections of the curve with the axis, the shape of the intervening arcs, and other things of that sort; these indeed far surpass human understanding . . . [49]." Yet these are precisely the things that need to be known, and if values cannot be assigned them for a two-particle force-curve, how is one to evaluate and define the character of that for combinations of points into solids? No more than the dynamic corpuscularian could finally assign value to the sizes, shapes, motions, and forces of his particles, could Boscovich begin physically to evaluate the basic parameters of his force-curve. When one examines this part of his theory, it takes on an aspect of unreality. It is frequently not at all clear, when Boscovich discusses the character of his curve, whether he is treating something which has physical properties or only metaphysical and mathematical ones. He justifies the multiplicity of limit-points, physically, with general examples of vaporization, effervescences, fermentations, and soft substances [43], but is far more detailed and concerned with justifications drawn from the mathematical nature of curves, the greater probability that any curve will cross an axis and that many times, than that it would be parallel or cross but once [51-52]. In a long appendix on the "Analytical Solution of the Problem to determine the nature of the Law of Forces [206-12]," he is content to write a general equation, summing two power-series with term coefficients representing functions of limit-point values. He does not, because finally he cannot, explicitly describe how one might assign any physical values to any of these coefficients, even the most simple.

This specific sterility is even more apparent when Boscovich turns to explicit problems of natural philosophy, where he evidently had little personal or detailed knowledge. Most of the direct applications of his theory relate to phenomena scarcely more clearly defined than in Newton's queries to the *Opticks.* Much as Descartes and Boyle before him, he seems intent on adapting his theory to the nature of any generally accepted explanation of his day. A force-curve arc in the form of a rectangular hyperbola will explain both the elasticity of air, according to Newton's inverse-power-of-the-distance repulsion, and its capability of being fixed, as shown by Hales; details are not given. The lack of compressibility of water indicates particles near limit-points on any of an infinite variety of classes of curves crossing the axis nearly perpendicularly [131]. He concedes he cannot explain the growth of animals and plants—this being due to a "life-principle" insinuating itself into and passing along the fine tubes of the fibers [63]; but his theory will comprehend the existence of very tenuous aethereal matter and, indeed, "suggests . . . the idea of these dispositions of matter, such as are most of all capable of explaining the difficult and compound phenomena of Nature [140, 182]." Fire "is some kind of fermentation [*i.e.*, vibrating agitation], which is acquired, either more especially, or even solely by some sulphureous substances with which the matter forming light ferments very vigorously, if it is concentrated in sufficiently great amounts [165]." The cause of heat consists "of a vigorous internal motion of the particles of fire. . . . Cold may be produced by a lack of this substance, or by a lack of motion in it. Also there may be particles which produce cold by their own action. . . [179-80]." His explanations of the chemical operations of solution, precipitation, liquefaction, etc., are less detailed than Freind's and lack any attempt to become quantitative. The discussion of electricity is simply an assertion that his theory permits the derivation of Franklinian principles, and the suggestion that the fluids of fire and electricity differ only "in the fact that one occurs in conjunction with actual fermentation . . . while the other is suitable to setting up of fermentation. . . [181-82]." The futility of his explanations is best revealed, however, in the discussion relating to the phenomena of magnetic force, beginning with the statement that these can be reduced "to a mere attraction of certain substances for one another." He continues with the standard corpuscular declaration that the attractive force must be due to structure, adds that attractive and repulsive poles are consistent with his theory, and ends: "A somewhat greater difficulty arises from the huge distances to which this kind of force extends. But even this can take place through some intermediate kind of exhalation, which owing to its extreme tenuity has hitherto escaped the notice of observers, and such as by means of intermediate forces of its own connects also remote masses; if perchance this phenomenon cannot be derived from merely a different combination of points having forces represented by that same curve of mine [182-83]."

In almost every critical problem of eighteenth-century natural philosophy, where substance had replaced force as an explanatory device, Boscovich's explanations accept substance and assert only that these substances can, in some undisclosed way, be reduced to the combinations of geometrical points and their summed force-curves. Now it is not really necessary, for the policy success of Boscovich's theory of matter in relating diverse phenomena and stimulating experiment, that Boscovich himself most successfully apply it to those ends and for specific problems. Nevertheless, his failure to do so explains the casual extrapolations from "true" Boscovicheanism employed by his successors who did apply the theory, and, more significantly, it explains the continuance of an indigenous dynamic corpuscularity in people who knew and even generally approved of Boscovich's theory.

One of the earliest and most influential of these was John Michell.[4] Michell is perhaps best known as the "Father of Seismology," for his work, *Conjectures concerning the cause and Observations upon the Phaenomena of Earthquakes,* first printed in the *Philosophical Transactions* of 1760 and separately published the same year. Containing many of the elements of

[4] John Michell (1724-93), educated at Queen's College, Cambridge; B.A. 1748 (fourth wrangler), M.A. 1752, B.D. 1761. He became a fellow of Queen's in 1749 and was tutor and lecturer there in arithmetic, geometry, Greek, and Hebrew from 1751 to 1763. Named Woodwardian Professor of Geology in 1762, he vacated both professorship (without having lectured) and fellowship in 1764. He held various rectorships to 1767, when he settled as rector of Thornhill, near Leeds, where he died. F.R.S. 1760, he wrote on geology, optics, astronomical observations, and mathematical instruments. The standard, now deficient, biography is Sir Archibald Geikie's *Memoir of John Michell* (Cambridge: Cambridge University Press, 1918), to which should be added Clyde L. Hardin's "The Scientific Work of the Reverend John Michell," *Annals of Science,* 22 (1966), 27-47; and esp. Russell McCormmach's "John Michell and Henry Cavendish—Weighing the Stars," *British Journal of the History of Science,* in press, which I read in manuscript through the courtesy of the author.

present earthquake theory—faulting of strata, wave propagation of shock through the earth's crust, and use of observations of place and time to locate the epicenter—although differently arranged and inter-mixed with contemporary vulcanism, the work has since greatly impressed such geologists as Lyell, von Zittel, and Geikie. The major part of Michell's work, however, related to questions of natural philosophy where he revealed himself as one of the very few late eighteenth-century British scientists with sufficient command of both mathematics and experimental techniques to advance the application of dynamic corpuscularity.

Michell was aware of Boscovich's theory at least as early as 1760, when Boscovich, then touring England and complaining that few Englishmen had yet read the *Theoria,* recorded meeting Michell at Cambridge and described their discussions of artificial magnetization and Michell's "Theory of molecular magnets."[5] According to Priestley, whose first knowledge of Boscovich was almost surely acquired by way of Michell, Michell had, however, already reached a theory of matter, that dissolved particles into inertial centers of Newtonian attractive and repulsive forces.[6] That he used such a theory with rare ingenuity is apparent in his later papers; some support to his prior independence in achieving it can be found in his *Treatise of Artificial Magnets,* first published in 1750.[7]

Like most of Michell's published work, the *Treatise* is tantalizing in its suggestions of undescribed experiments and concealed ideas. Intended as a description of a practical method of making artificial magnets (the so-called "double-touch" procedure similar to that of Gowin Knight and later of John Canton), there is little theoretical discussion. Michell declares, however, that the ability to make iron permanently magnetic is related to its hardness, that the degree of magnetical power achieved is limited by the "Inability of the Materials . . . to retain any more," and that a magnet is divided into sections, paralleling the axis, each of whose action endeavors to counteract the rest, "as superposed magnets will do [12-13]." More significantly: "Each pole attracts or repels exactly equally, at equal distances, in every direction"—which is "utterly inconsistent" with the hypothesis

[5] Whyte, *Boscovich,* pp. 65, 68, 130.

[6] Priestley, *History of Optics,* p. 392.

[7] J[ohn] Michell, *A Treatise of Artificial Magnets; In which is shewn An easy and expeditious Method of making them, Superior to the best Natural Ones; and of changing or converting their Roles* (Cambridge: W. and J. Mount, and T. Page, J. and P. Knapton, C. Buthurst, W. Thurlbourn, T. Merrill, J. Fletcher, and J. Hildyard, 1751), 2nd edn. corr. and improved.

of a subtle magnetic fluid—and this attraction and repulsion equally decrease "as the Squares of the distances from the respective Poles increase"—as experiments he has performed make probable, but with insufficient evidence to confirm with sufficient exactness [17-19]. How one wishes Michell had published that bulky theory of magnetism which he had here deferred as "not of interest to Artificers and Seamen [2]."

For insight into the theory of matter which enabled Michell to deny hypotheses of subtle fluids, we must turn from the standard problems of the eighteenth-century repertory to the neglected subject of light. Optical theory was not a popular area for speculation in eighteenth-century Britain. For mechanists and materialists alike, Newtonian light theory had become a national treasure, rather to be preserved and defended against Continental criticism than to be further developed.[8] Except for Dollond's experimental discovery of achromatic lens-combinations, only in the work of Thomas Melvill and John Michell is there significant progress from the work of Newton. The extension, in both cases, is a mechanistic one and, in both cases, is done against a knowledge, at least, of Boscovichean matter theory.

Melvill's work can, unfortunately, be briefly described, for he published only two papers before his death at 27 and the first of these is repeated in the second.[9] His "Observations on Light and Colours" of 1756 is the first motions-forces-phenomena study of optics to appear since Newton's own work. Melvill accepts the existence of the aether, on the grounds of heat radiation, but argues against a wave-theory of light and uses light particles and their attraction and repulsion, at a distance, of other bodies as an instrument of scientific investigation. His earlier paper had argued that if this attractive power operated in analogy with gravitation, then differently sized light particles would receive equal accelerations toward refracting bodies and, therefore,

[8] For the general attitude toward optics, see Peter Anton Pav, "Eighteenth-Century Optics: the Age of Unenlightenment," unpub. Ph.D. diss., Indiana University, 1964. Benjamin Franklin's agreement with Euler in attacking particulate light and supporting a wave theory is a rare exception to British approval of Newtonian theory. See Cohen, *Franklin and Newton*, pp. 318-23, for a description of Franklin's somewhat ill-digested views and the typical contemporary response to them.

[9] Thomas Melvill, "Discourse concerning the Cause of the different Refrangibility of the Rays of Light," *Philosophical Transactions*, 48 (1753), 261-70, and "Observations on Light and Colours," *Edinburgh Essays*, 2 (1756), 2-90. Melvill (1726-53) was a student of Divinity at the University of Glasgow, 1748-49, and is listed as M.A. on the title page of the "Observations." His philosophical interests seem to have developed in conversation with Alexander Wilson, later first professor of astronomy at Glasgow, and his work showed great promise before his early death in Geneva.

be equally refrangible. He proposed, instead, that the different colors traveled with different velocities and suggested that observations of the emergent satellites of Jupiter could provide a test for the theory. That James Short's observations failed to confirm the hypothesis is less significant than the fact that a testable theory, based on deductions from motions and forces, had been produced, and the failure of that test did not discourage Melvill in the creation of other, similarly based hypotheses.

The "Observations" contains what must be one of the earliest British references to Boscovich, where Melvill argues that the concept of indivisible points, "endued with an insuperable repulsive power, reaching to a finite distance," as described in the dissertations *De viribus vivis* (1745) and *De lumine* (1748), does not solve the problem of light being able to penetrate bodies, for these points would be "as subject to interfere, as solid particles of finite magnitude [18]." Criticizing the views on heat of Boerhaave, Rutherforth, and Nollet, Melvill provides an estimate of the rates of particle vibration corresponding to heat, because light particles, suffering multiple reflections in the opaque bodies which they heat, must travel the distances between particles of the body (supposed not greater than one 12,500th of an inch) with such velocity as to cause them to vibrate at least 1.25×10^{14} times a second [19]. As the lustre of water drops on leaves demonstrates that the water is suspended from the surfaces by a repulsive power, from the shapes of the surfaces of water drops (or better, of the water surrounding a floating repelling body), the law of the repulsive force might be determined. He also recommends, for the same purpose, an investigation of the spontaneous motions of light bodies on the surfaces of fluids, in varying combinations of positive and negative meniscuses at container wall and adjacent to the floating body. "From the motions, find the forces." And, almost incidentally, he notes the characteristic colors of burning substances (particularly the bright yellow of sodium) and observes the distinct terminations between colors, an immediate transition, not a gradual one, in the prismatic spectrum.

With Melvill's death, only Michell remained to employ similar mechanistic notions in the design of experimental optical inquiry. Michell's views of the relation between light and matter were first revealed, at second-hand, in the *History of Optics* of Joseph Priestley, published in 1772.[10]

[10] Joseph Priestley, *The History and Present State of Discoveries relating to Vision, Light and Colours* (London: Joseph Johnson, 1772). See also Robert E.

Priestley's dependence on Michell in the preparation of the *History of Optics* is revealed by his repeated acknowledgments in the work itself and in his correspondence. Michell assisted in reading proof, gave advice on interpretations of phenomena, and supplied Priestley with original discussions of experiments and of theoretical significance. Although Priestley was so impressed with Boscovich's theory that he devotes pages to describing it, "on account both of the *novelty* and *importance* of it [393]," he gives Michell's explanation of the "fits of easy reflection and transmission," not that of Boscovich. Michell, Priestley tells us, had, early in life and without knowledge of Boscovich, developed a "scheme of the immateriality of matter" after reading Baxter *on the immateriality of the soul.* Baxter had conceived matter, as it were, as bricks cemented by immaterial mortar, but to be consistent these "bricks" must, in turn, be lesser bricks likewise cemented, and so on *ad infinitum.* Michell observed that as the bricks themselves were impossible to perceive, every effect ascribed to them might better be thought the result of the immaterial, spiritual, and penetrable mortar [391-92].[11] An example of such an effect can be seen in the production of colors by thin plates. Suppose every particle of the thin medium to exert, at alternate intervals, attraction and repulsion relative to the particles of light, these intervals being of different magnitude for light particles producing different colors. Now if the thickness of the transparent medium in which these particles of matter are uniformly distributed, is such that the attracting and repelling intervals for extreme particles coincide for a particular color (the intermediate particles mutually destroying one another's effects), then that color will be reflected and others transmitted. And as the thicknesses where the intervals coincide, and those where they counteract, will vary for different colors, thin plates will produce regular bands of color, from which the producing thicknesses and consequently interval relationships might be determined [309-11].

Schofield, ed., *A Scientific Autobiography of Joseph Priestley,* Nos. 39 and 40, for the association of Michell and Priestley in preparation of the *History of Optics.*

[11] This description of the origin of Michell's ideas is too circumstantial entirely to ignore, but I find it hard to see how one goes obviously from Baxter's mortar to particles surrounded by alternate intervals of attraction and repulsion. Lack of additional information makes further explanation impossible; but Priestley's use of illustrations for the *History of Optics* derived from Rowning, his reading of Rowning while a student at Daventry (see Schofield, *Scientific Autobiography,* p. 6), and the use of Rowning as one of the texts at Cambridge when Michell studied there, all suggest a reasonable path of development for both Priestley's and Michell's adoption of Boscovichean ideas.

To answer Euler's (and Franklin's) criticism that the sun would become depleted in emission of material particles of light, Priestley describes Michell's experimental determination of the momentum of light particles. Light, concentrated by a concave mirror, is directed to strike a copper plate suspended at the end of a counterpoised wire, centrally pivoted on an agate cap and needle. From the size of the mirror and plate, weight of the apparatus, speed of light, and speed of rotation of the plate under impact from the light, the quantity of matter striking the plate was determined to be no more than one twelve hundred millionth part of a grain, and by easy extrapolation, it was shown that the sun had emitted about 670 pounds of matter in the 6,000 years of its existence—not enough to reduce its semi-diameter more than 10 feet in that time, assuming it to be formed of matter no more dense than water [387-90].

The theological implications of the six thousand years of the universe may well have been Priestley's, but the assumption that particles of light might be treated as any other particle of matter in the solution of a physical problem was characteristically Michell's. The best, and most ingenious application of this assumption begins also in the *History of Optics*. There Michell presents, in an essay, "Of the Force by which light is emitted from luminous bodies, and of that which, being lodged at the surfaces of bodies, is the cause of what is generally called their impenetrability," an elaborate argument by which the short-range attractive or repulsive forces between light and matter might be estimated. He computes the velocity at which a body, freely falling from infinity, would reach the surface of the sun and compares this to the velocity of light. Assuming the distance over which an emitting force might work to be less than a hundredth of an inch, he compares the emitting force to the gravitational at the surface of the sun and at the surface of the earth. Noting that the velocity of light could be little diminished by the gravitational attraction of the sun, he suggests that the forces exerted on light particles to reflect or refract them must be larger than the force of the sun's gravitation and greater than the force of gravity on the earth by about 1.9×10^{19} to one.

The next stage of Michell's application of optical dynamics transferred the problem to sidereal astronomy and depended on work done for an earlier astronomical paper. In his "Inquiry into the probable Parallax, and Magnitude of the fixed Stars," published in 1767, Michell had suggested methods of determining stellar magnitudes and distances from the quantity and intensity of starlight.[12] First assuming, for computational purposes, that the fixed stars are, at a medium, equal in magnitude and natural brightness to the sun, Michell shows that the parallax of the sun, were it removed approximately two hundred thousand times further away until its brightness equaled that of Saturn, would be less than two seconds. This then gives an order-of-magnitude distance for that class of stars having the brightness of Saturn and explains the failure to observe their parallax. Having thus demonstrated that direct determination of stellar distances is unlikely, he then proposes an indirect method and here illustrates the dynamic fertility of his physical and mathematical imagination. In the first significant application of probability arguments to a physical problem, Michell shows that the chances are many million millions to one that the stars are grouped, not by chance, but according to some general law—which Michell assumes to be that of gravitation—and that the sun is a member of one such systematic grouping. It is, therefore, reasonable to suppose that stars of the first and second magnitude, not obviously part of other groups such as nebulae, belong to the same group as the sun and that size and distance comparisons based on the sun's have physical justification. As the quantity of light emitted by a star is a function of its mass and, masses being equal, its brightness is a function of its surface (or density), a concerted effort to compare brightness and quantity of starlight for a series of stars would, at least, permit a comparative classification of stellar sizes and distances.

The most notable part of this paper (apart from its use of probabilities) was, however, based on Michell's extension of gravity to the stars. This had previously been assumed, though never proved, and Boscovich had even suggested that his force-curve, at such distances, might reach toward another, infinitely repulsive, asymptote. Michell applies gravitational law, without question, to the case of stellar doublets which, from probability considerations, were almost certainly physically related to one another gravitationally and must therefore be expected to rotate with respect to one another, to maintain dynamic stability. From the periods of revolution and apparent elongations, the relation between density and apparent diameters of the central stars might be computed and actual masses determined when actual

[12] John Michell, "An Inquiry into the probable Parallax, and Magnitude of the fixed Stars, from the Quantity of Light which they afford us, and the particular Circumstances of their Situation," *Philosophical Transactions*, 57 (1767), 234-64.

distances become known, all to be compared with total light and brightness.

Now none of the values Michell uses in his estimates—apparent diameters, periods, brightnesses, total light received, etc.—were then measured, and some of them were unmeasurable. The supposition of gravitationally rotating binaries was entirely hypothetical, and the whole concept of examining the forces, densities, masses, and distances of the stars was unique to British sidereal astronomy. The enterprise was, however, in the best tradition of Newtonian dynamics, and its inspiration can be seen in almost every endeavor of the greatest of "British" sidereal, observational astronomers of the eighteenth century, the transplanted Hanoverian, William Herschel. But before relinquishing the field entirely to Herschel, Michell made one last attempt to "weigh" the stars. For his paper, "On the Means of discovering the Distance, Magnitude, etc., of the Fixed Stars, in consequence of the Diminution of the Velocity of their Light," published in 1784, Michell combines the techniques first explored in the paper of 1767 with the optical mechanics introduced through Priestley in 1772.[13] Noting that Herschel's recent catalogue of stellar binaries had confirmed the predictions of his 1767 paper, Michell again demonstrates how the motions (period and elongation) of those pairs that show rotation permit the determination of density-apparent diameter relations for the central star. The quantity of matter in the star and its real magnitude could then be determined if only the velocity of free-fall at its surface were known. This was clearly unmeasurable, but, as he had shown in Priestley's *History of Optics*, another and indirect way of achieving the same determinations was possible. The mass of the sun was too small sensibly to reduce the velocity of its emitted light, but that of a star, of equal density but with a diameter 500 times that of the sun, should be sufficient to pull back all the starlight emitted, by its own gravity, and intermediate sized stars would reduce the velocities of their light in proportion to their masses. While an exact measure of the phenomenon was not to be expected, there was a method by which a comparison of starlight velocities might at least be made, confirming the phenomenon and giving some indication of stellar magnitudes. Accepting Newton's theory that refraction resulted from

[13] John Michell, "On the Means of discovering the Distance, Magnitude, etc., of the Fixed Stars, in consequence of the Diminution of the Velocity of their Light, in case such a Diminution should be found to take place in any of them, and such other Data should be procured from Observations, as would be farther necessary for that Purpose," *Philosophical Transactions*, 74 (1784), 35-57.

a short-range force alteration of the velocity of light, the amount of refraction would vary with incident velocity. If Michell were correct, the angular separation of the stars in a rotating binary, as seen through a prism, would vary with suitably changed prism positions, for the central star of a rotating pair would be considerably the more massive and its light correspondingly the slower.

Michell's paper of 1784 was sent in the form of a letter to Henry Cavendish, at whose instigation Nevil Maskelyne and William Herschel each attempted, and failed, to find the variation predicted by Michell. And there, for the most part, the matter was dropped. Neither Michell nor Cavendish believed that failure to find the effect was a disproof of the theory; they argued instead that it showed no stars to be of sufficient magnitude measurably to diminish the velocity of light, while Michell also supposed it possible that light (and perhaps the electrical fluid) was not as affected by gravity as other substances. Cavendish later attempted to find a "bending of ray of light which passes near the surface of another body by the attraction of that body," but with the failure of Michell's scheme, the audacious attempt to combine the two most exact sciences of his day—optics and planetary theory—within a Newtonian central-forces frame appeared to come to an end.[14] Appearances, however, are notoriously deceptive and the reappearance of many of Michell's subjects of investigation in the career of Sir William Herschel suggests a need for analysis of his work in the same theoretical frame.

That Herschel had any theoretical frame for his researches may come as a surprise to those familiar only with the legendary Herschel, the professional musician, self-taught in the sciences, whose patience in pure observational astronomy was rewarded in the discovery of Uranus and embodied in a series of charts presumably devoid of theoretical implications. Although the exact and patient neo-Baconian observer is the familiar caricature of the eighteenth-century British scientist, it might be an acceptable, if exaggerated, portrait of Herschel were it not for the incongruities which appear as soon as his career is given a more than superficial examination.[15] There is, for example, an almost

[14] My discussion of Michell's 1784 paper is heavily derived from Russell McCormmach's study, "John Michell and Henry Cavendish—Weighing the Stars."

[15] Sir William Herschel (1738-1822), born in Hanover and trained in music by his bandmaster father. He visited England with the band of the Hanoverian Guard in 1755 and returned permanently in 1756 or 1757 to take up various musical posts until 1766, when he settled at Bath as organist. Self-educated in the sciences, he read Locke, Maclaurin's fluxions, Smith's optics, and Ferguson's astronomy for

one-to-one correspondence between the problems for sidereal astronomers suggested in Michell's work and Herschel's "observational" career; those studies of binary and other combinations of stars, the systematic classification of stars by their brightness, the investigation of nebulae, the determination of the proper motion of the sun and its assignment to a particular system of stars—all were related to crucial questions first raised by Michell in 1767.

It is, of course, possible to adopt a program of experimental research without accepting the theoretical suppositions that initially suggested it, but that is almost as unlikely as experiment without any theoretical presupposition, and Herschel can be shown to have indulged speculations throughout his scientific career. During the early years, while his "important discoveries" were first being communicated to the Royal Society, he was also reading to the Bath Philosophical Society what have variously been described as a "torrent of indifferent philosophical and scientific papers" and his "scientific wild oats."[16] During his life, and since, there have been persons who praised Herschel the theorist, but the majority have agreed with the contemporary criticism by the *Edinburgh Review*, which spoke of his "loose and often unphilosophical reflections, which give no very favourable idea of his scientific powers, however great his merit may be as an observer."[17]

Herschel, of course, would never have accepted the idea that observations could be separated from theory. In 1801, he argued, as he had earlier done in his paper "On the utility of speculative enquiries," read to the Bath Society in 1780, that "Observations are made with no other view than to draw such conclusions from them as may instruct us in the nature of the things we see." In 1811, he named explicitly, as the "ultimate object of all my ob-

relaxation. He began to construct his own telescopes in 1773 and eventually was to "manufacture" them for sale. In 1780 he was introduced into the Bath Philosophical Society by Sir William Watson (the younger) and in 1781 became F.R.S. and Copley medalist for his discovery of the planet Uranus; he ultimately was to write more than 69 extensive astronomical and physical memoirs for the Royal Society and Royal Astronomical Society. In 1782 he became court astronomer to George III and held that position until his death.

[16] Michael A. Hoskin, *William Herschel and the Construction of the Heavens* (London: Oldbourne Press, 1963), p. 24; and *William Herschel: Pioneer of Sidereal Astronomy* (New York: Sheed and Ward, 1959), p. 16.

[17] Review of Herschel's "Observations on the two lately discovered Celestial Bodies," *Edinburgh Review*, 1 (1802-1803), 430. Hoskin is one of the more recent scholars to praise Herschel as a daring theorist as well as an outstanding observer; see his *Herschel: Pioneer*, pp. 27-28.

servations," knowledge of the "construction of the heavens."[18] Even without that statement, the purpose of his work would have been apparent, for the words "construction of the heavens" are repeated again and again in titles of his papers from as early as 1789. It is, moreover, clear from those papers that he has in mind the structuring of sidereal systems under the influence of attracting and repelling forces of some kind, which operate not only on the stars themselves, but on "nebulous matter" and "luminous substance" in the creation of those stars [*Papers*, II, 210-11, 478-80, 481-83, 484]. Like Michael Faraday, Herschel restrained the *publication* of his most speculative ideas until his later years and after having had a nervous breakdown, and, like Faraday, it can be shown that the roots of those ideas go back to his earliest thinking as a scientist.

Those "wild oats" which Herschel sowed before the Bath Philosophical Society were, after all, not the ingenuous raptures of an impressionable adolescent. Herschel was in his forties when he read those papers at Bath, and they represented ideas he had been formulating for years and to which, by inference, he was repeatedly to return until he gave their consequences explicit expression in his papers after 1811. A reading of the 31 papers delivered at Bath reveals that five of them refer in some detail to those mechanical theories of matter described by Boscovich, Michell, and Priestley—three of the papers being almost entirely devoted to that subject.[19] There were, moreover, at least two other papers relating to such central forces which Herschel wrote, but did not publish.[20] It would be impossible, on reading these papers, to suppose Herschel peculiarly gifted for philosophical speculations on natural phenomena and the nature of matter. In-

[18] See William Herschel, in J.L.E. Dreyer, *The Scientific Papers of Sir William Herschel* (London: Royal Society and Royal Astronomical Society, 1912), II, 149, 249.

[19] "Observations on Dr. Priestley's Optical Desideratum 'What becomes of light' " (his second paper) read Feb. 4, 1780; "On the central powers of the particles of matter," read Feb. 18, 1780; and "On Central Powers," read Jan. 1781. These papers are printed, for the first time, in Dreyer's introduction to volume one of Herschel's *Papers*, but I have been privileged to see them in the original transcripts by the Bath Society's secretary, in the personal possession of Wesley C. Williams, Curator of rare books and manuscripts in science, Case Western Reserve University.

[20] One of these, "On Central Powers," was written sometime shortly after June 1789 and was published by Dreyer, *Herschel Papers*, I, cxi-cxiv, with a note added by Herschel late in life that Cavendish had thought it very similar to the theory of Boscovich; the other, "Remarks on Dr. Priestley's Disquisition on Matter and Spirit," written about 1779 and now preserved in manuscript in the Library of the University of Texas, Austin, contains a section very similar to the fragment, "Interference of Light," published by Dreyer, *Herschel Papers*, I, cxiv-cxvii.

deed, it is apparent that his information during the 1780s was based only on a reading of Priestley's *History of Optics,* and his understanding even of that sketchy version of Boscovich's and Michell's theories was imperfect. Nevertheless, his criticisms are reasonable (though occasionally based on misunderstanding) and, moreover, show just that concern with forces at great distances to be expected of the astronomer. His major criticisms of these theories are a denial that geometrical points or "immaterial matter" can possess physical properties and an objection to the single force-curve (mistakenly for its lack of central symmetry and continuity, but also) for its variation, without sufficient reason, from attraction to repulsion and vice versa. Herschel prefers actual, divisible, particles of substance endowed with a multiplicity (he suggests at least nine, including cohesion, gravity, deflection, inflection, and refraction of light, "elective attraction," and a possible weak repulsive power at great stellar distances) of coexistent powers of attraction or repulsion, all spherically symmetric functions of distance and each extending to infinity. Apparent "spheres" are, therefore, the consequence, at various distances, of the summation of all the powers, positive and negative, at those distances.

Although most of our direct evidence relates to the years between 1780 and 1789, Herschel's interest in central powers cannot be confined to that period. His commonplace book cites the reading of Priestley's *History of Optics,* with specific reference to its discussions of Boscovich, Michell, the momentum of light particles, and the hypotheses of attraction and repulsion, from about 1775 and shows that he pursued those ideas by a reading of Desaguliers, 'sGravesande, Gowin Knight, and Bryan Higgins' *Essay concerning Light* (1776).[21] And the concern with Boscovichean-like central powers was retained at least as late as 1810, for Herschel possessed a copy of the first edition of Boscovich's *Theoria* acquired from the library of Henry Cavendish who died in that year.[22]

To trace the specific influence of central-force concepts on Herschel's observational design and interpretation would require the detailed knowledge of an historian of late eighteenth- and

21 Herschel's Commonplace Notebook is preserved in the collections of the Linda Hall Library, Kansas City, Missouri. The critical references appear on pp. 83-84, between references to reading in the *Philosophical Transactions* for 1774 and for 1777, and are followed, pp. 84-85, with renewed discussion of Newton's belief in forces acting on light particles and Priestley's explanation of fits of easy transmission and reflection.

22 See Whyte, ed. *Boscovich*, p. 125n3.

early nineteenth-century astronomy, but some indication that they had such an influence can be derived from Herschel's paper, "A Catalogue of a Second Thousand of new Nebulae and Clusters of Stars; with a few introductory Remarks on the Construction of the Heavens," published in the *Philosophical Transactions* for 1789. There Herschel writes:

> If we find that there is not only a general form, which . . . is a sufficient manifestation of a centripetal force, what shall we say when the accumulated condensation, which every where follows a direction towards a center, is even visible to the very eye? Were we not already acquainted with attraction, this gradual condensation would point out a central power, by the remarkable disposition of the stars tending towards a center. In consequence of this visible accumulation, whether it may be owing to attraction only, or whether other powers may assist in the formation, we ought not to hesitate to ascribe the effect to such as are *central*. . . .
>
> I am fully aware of the consequences I shall draw upon myself in but mentioning other powers that might contribute to the formation of clusters. . . . The idea of other central powers being concerned in the construction of the sidereal heavens, is not one that had only lately occurred to me. Long ago I have entertained a certain theory of diversified central powers of attractions and repulsions; an exposition of which I have even delivered in the years 1780, and 1781, to the *Philosophical Society* then existing at Bath. . . .[23]

Like others of his contemporaries, Herschel was to find that such suggestions met with disbelief or lack of understanding. His later papers are less bold in proclaiming the nature of the forces responsible for the evolution and construction of the heavens. Nonetheless, the evidence seems clear that Herschel had found in a neo-mechanistic theory of matter, developing out of Newtonian dynamic corpuscularity, a concept which had stimulated and directed his observations, a compass which had led him into previously unexplored regions of space.

In his lack of easy competence in mathematics, there was a fundamental limitation to William Herschel's application of central-force mechanisms to natural philosophy, and some justification exists for the failure of his contemporaries to recognize the

[23] *Herschel Papers*, I, 333-34. The paper on "Central Powers" published by Dreyer and mentioned in note 20, is clearly a consequence of these remarks.

roots of his speculations. No such justification can be found for similar misapprehensions of the work of Henry Cavendish.[24] Recognized in his own day as uniquely proficient in "the knowledge of mathematics, chemistry, and experimental philosophy," Cavendish's few published papers were read for their experimental disclosures, but the theoretical implications of his work bewildered his contemporaries.[25] The posthumous revelation of his unpublished researches confused his readers still further, and their failure to understand him was embellished by anecdotes about his personal idiosyncrasies until this most creative scientist of the eighteenth century was described as without imagination, the unity of his work was lost in apparent diversity, and his desire for completeness and for quantitative verification was transmuted into whim and a compulsion for weight, number, and measure.[26]

In fact, Cavendish's analytical imagination was unequaled in Britain in the years between Newton and Maxwell, but he lacked that ingenuity which invents new problems. His researches, therefore, tended to be elaborations of the ideas of others, which he defined with a precision and developed to an extent beyond the conception of the originators. It is possible to find, in the work of Hales, Black, Franklin and Watson, Michell, and Priestley, the immediate origin of the projects he pursued so successfully, but the ultimate inspiration of Cavendish's work was Newton. Only when Cavendish is seen as a literal Newtonian, working in the dynamical method of the *Principia* and on the problems

[24] Henry Cavendish (1731-1810), educated at Hackney Seminary and Peterhouse College, Cambridge, 1749-53, when he left without taking a degree. Thereafter he lived in or near London, essentially alone after the death of his father, Lord Charles, increasingly reserved and devoting all of his time to science. Nothing is known of his early scientific training, though his earliest scientific studies may have been guided by his father, a gifted experimentalist. F.R.S. 1760, he published no books and fewer than twenty articles during his nearly 50 years of research. He earned a permanently high reputation in the history of science for the few published papers on electricity, pneumatic chemistry, and the density of the earth, while his unpublished papers later revealed him to be equally proficient in mathematics, mechanics, heat, astronomy, and geology.

[25] The estimate of Cavendish's abilities is that of John Playfair, cited in George Wilson's *Life of the Honble. Henry Cavendish* (London: for the Cavendish Society, 1851), p. 166. Playfair did not know Michell; otherwise his estimate was sound, and it was generally accepted.

[26] This characterization is summarized from that by George Wilson, *Cavendish*, pp. 20, 185-86, 444. Wilson devotes some 25 pages (164-190) to recounting stories in support of these allegations. His biography is not only out-of-date, it was written primarily to sustain Cavendish's claim to the discovery of the composition of water by a man who admired but neither liked nor understood his subject. A. J. Berry's *Henry Cavendish: His Life and Scientific Work* (London: Hutchinson, 1960) is based on Wilson and is little better.

posed by the queries of the *Opticks,* does he cease to be the enigma that both impressed and bemused his contemporaries.[27]

Commensurate with his fundamental dependence upon the *Principia,* Cavendish's earliest original research, of which we have record, was on dynamics. In "Remarks relating to the theory of Motion," prepared sometime before 1765, he concerns himself with analysis of the motions of large assemblages of small particles, defined only in terms of their positions, mobility, inertia, and the forces of attraction and repulsion between them.[28] As the motions of such particles are not directly observable, their short-range forces are not immediately deducible, and an indirect approach is required. To that end, Cavendish creates the concept of mechanical momentum (representing mass times square of velocity). He demonstrates that if the forces, attractive or repulsive, are symmetric functions of distance only, there is a conservation of mechanical momentum plus such "additional momenta" as is stored in elastic compression, raised elevation, etc., whenever the system of particles is isolated and no "force" is lost by friction, imperfect elasticity, or the impinging of unelastic bodies.

Mechanical momentum was not, for Cavendish, a concept of energy as an independent entity; it was a derivative quantity by which one might study Newtonian forces. Nonetheless, in one of the more obvious applications of the concept, Cavendish achieved (what no previous Newtonian had been able to do) a mechanical analysis of heat. In the second of five corollaries which follow from the derivation of mechanical momentum, he examines Newton's declaration that heat is the vibration of the particles of bodies. Conservation of mechanical momentum is applied to an understanding of the thermal interaction of two bodies—the heat lost by one was the loss of mechanical momentum of the particles of that body, while the heat gained by the other was the gain in mechanical momentum by those of the other, and neither heat nor total mechanical momentum was

[27] The view of Cavendish is almost entirely that of Russell McCormmach, who is preparing a new and much needed biography. I have depended on an early version of his "Henry Cavendish: Rational Natural Philosopher," *Isis,* in press, and his "Electrical Researches of Henry Cavendish," unpub. Ph.D. thesis, Case Institute of Technology, 1967.

[28] This paper was not published until early in this century, when it was included in a collection of published and some (but by no means all) of the unpublished papers, *The Scientific Papers of the Honourable Henry Cavendish F.R.S.*, I, *Electrical Researches,* ed. James Clerk Maxwell, rev. Sir Joseph Larmor; II, *Chemical and Dynamical,* ed. Sir Edward Thorpe (Cambridge: at the University Press, 1921). "Theory of Motion" is in *Papers,* II, 415-30.

gained or lost in the combination. The concluding section of this paper reveals, however, an acute consciousness of the inadequacy of his understanding of the dynamics of heat phenomena. There exist a variety of instances (*e.g.*, fermentation, dissolutions, and combustions) in which heat is lost or gained by systems of bodies in a way which could not be interpreted as variation in "additional momenta" or communication with other bodies. Cavendish suggests that the explanation might lie in different kinds of interparticle forces, not symmetric or strictly distance-dependent, but he is not prepared to abandon Newtonian forces and abandons the paper instead, to pursue the problem along other lines.

Turning from theory to experiment, his next paper (also unpublished [*Papers*, II, 327-47]), "Experiments on heat," uncovered still more problems for a dynamical theory of heat. Investigation of heat exchanges between bodies revealed a dependence upon the nature of the substance (i.e., he independently discovered the concept of specific heat), where he had supposed only the mass of the particles (and thus the density of the substance) would be involved. He also discovered the phenomenon of "latent heat" (though he objected to the use of that term, when it was subsequently introduced, as implying the concept of heat as a substance [*Papers*, II, 151]). This he attempts to explain as a result of different specific heats for different states of a body which he sees as the effects of mechanical change. He observes that "all matter generates cold in changing from a non-elastic state to an elastic state and generates heat in the reverse process" [*Papers*, II, 347], and exemplifies his observation in the release of factitious airs from a body as well as the change of water from fluid to vapor. These are equivalent mechanically, and cold is generated in each case.

Another attempt from the motions to find the forces between particles is described in a paper on the "Theory of Boiling," which apparently also dates from this early period [*Papers*, II, 354-62]. Arguing, as Rowning and Desaguliers had argued before him, that water particles must repel one another at minute distances, but attract at some greater distance, he supposes these distances to vary differently with heat. Vaporization was then a function of the strengths and distances of the attracting forces. The heat at which particles boil off should provide a measure of their binding forces, while those particles in cohesive, capillary attraction to the sides of the vessel should require more heat for vaporization than the particles attracting only one another.

In the end, Cavendish published none of these papers. Each

was, in a sense, incomplete, for a mechanical theory of heat should depend only on the mechanical properties of bodies, and Cavendish was unable to explain mechanically the substantial qualities of bodies which produced the heat anomalies of specific heat, fermentations, dissolutions, and combustions. His next major research, conceptually as well as chronologically, was therefore an investigation of the specific differences between uniquely identifiable collections of particles in minimum interaction with one another; he turned to an experimental investigation of factitious airs. This work was embodied in "Three papers, containing Experiments on factitious Airs," and, as they contained information previously unknown about such "airs," he permitted their publication.[29] Although Cavendish acknowledges the prior studies of Joseph Black, the concepts which inform his work are those of Stephen Hales, and the phenomena he is primarily concerned to investigate are just those which potentially involve changes in "additional momenta"—the release of airs from inelastic confinement in substances and the consequent changes in temperature, the fermentations, dissolutions, and combustions—which had given him trouble in his theory of heat. Because he worked by weight and measure, this first paper published by Cavendish shows him in the guise of a chemist, identifying the gases he produced by their specific gravities, solubilities, inflammabilities, etc., and this is what has impressed the students of his work, but to Cavendish these are physical parameters as well, and their immediate relation to his studies on heat demonstrates a congruence of research interests, not a diversity.

There is a similar congruence in his next major researches, unpublished as well as published. "Chemical" studies had multiplied rather than reduced the problems of dynamic explanations; then, just after the publication of his work on gases, Joseph Priestley's *History of Electricity* (1767) appeared with its description of the prototype of a Halesian elastic fluid—the fluid of electricity. And not only were these particles of a fluid which repelled one another and were capable of being fixed in the pores of bodies, there was the compelling (to Cavendish, though apparently to him alone) discovery that the elastic forces between these particles were long-range, inverse-square, forces similar to Newtonian gravitation, not the short-range, immeasurable forces which had previously given him so much trouble [*History*, 372]. Over his entire career, Cavendish was to discover only one sub-

[29] In the *Philosophical Transactions* for 1766; reprinted in *Papers*, II, 77-101.

stance which he could thoroughly reduce to mathematics, the permanently elastic fluid of the matter of electricity.[30]

Almost at once Cavendish began his investigations (theoretical and experimental) on the dynamics of the electric fluid. This was to be his most extensive research, covering not only a published theoretical paper, "An Attempt to explain some of the Principal Phenomena of Electricity, by means of an Elastic Fluid," which appeared in the *Philosophical Transactions* for 1771 and an experimental paper, "An account of some attempts to imitate the effects of the Torpedo by Electricity," in the *Philosophical Transactions* for 1776 [*Papers*, I, 33-81, 194-210], but also two preliminary drafts for the papers of 1771 and sufficient subsequent studies, in so formal a state, as to indicate that he proposed to write a book upon the subject. A detailed analysis of material so abundant and so fully developed is impossible here, but some of the characteristic and significant features can be noted. Here, for example, the influence of the *Principia* is most obvious. Cavendish did not quote that work extensively, but he developed his own "book" in an almost explicit parallel to the form of that model. From the particles of matter (albeit a special form of matter) and the forces between them, he set about to deduce the phenomena of electricity, beginning with general propositions, without reference to any single form of the law of forces, and proceeding through theoretical application and observational justification to a precisely defined law of electrical forces.

The immense advantage given his work by this mathematical-dynamical analysis is immediately evident, and the contrast between Franklin's conceptualization of electric fluid as substance and Cavendish's as a carrier of dynamic forces is manifest. No more than Franklin did Cavendish ever consider the electric fluid to be convertible into ordinary matter, though Cavendish at once (and independently of Æpinus) develops the parallel electrical forces—of attraction between particles of ordinary matter and those of electricity and mutual repulsion of each for their own particles. In Cavendish's dynamical theory of electricity, one can see a natural effort to reconcile the material fluid theories, so popular and so successful in the mid-eighteenth century, to the more Newtonian dynamic corpuscularity of an earlier age. Here, there is no discussion of the aether, and no denial of action-at-a-distance, but the first consistent development of a fluid dynamics

30 This is the conclusion of Russell McCormmach, "Electrical Researches," whose work I here summarize. Vol. I of the *Papers* is entirely devoted to electricity.

for the imponderables. From the implications of the dynamics of an elastic fluid, there is developed a mathematical treatment of electrical phenomena requiring more than simple variation in quantity. Initially deriving his treatment from density suggestions published by William Watson, Cavendish soon describes the quantity in terms involving the compressibility of the fluid. In this treatment, Cavendish admits, explicitly in published as well as unpublished statements, a connection with his studies of gases. The state of electric fluid in bodies is related to the capacities of those bodies, the variation in capacity is compared to the change of pressure of a gas in a tube, where the total quantity of gas remains constant while the pressure is changed (perhaps in part only of the tube) by heating, and motion of the fluid is according to pressure gradients.

The introduction of compressibility as well as quantity of fluid makes it possible to handle the problems now described in terms of electric potential and shows the work on electricity to be a culmination of Cavendish's studies on dynamics, heat, and factitious airs. To subsequent readers of the work, Cavendish appeared to have achieved remarkable things, but the major part of this work remained unpublished. And again the explanation appears to be that, for Cavendish, the research was incompletely realized. His published paper of 1771 had suggested, what his unpublished papers prove, that the form of the force law was that of the inverse-square. But the application of the theory showed to Cavendish that electricity was not purely mechanical in any clearly definable way. He could not explain mechanically why different substances should vary in their abilities to transmit electrical fluid; his theoretical predictions of the electrical capacities of glass plates and jars did not agree with measured values—again the nature of substance was involved. Although there is little evidence that he accepted the notions of Boscovich, in Cavendish's attempts to explain what Faraday was later to call specific inductive capacity, he proposed a theory of stratified layers of substance, where the electrical particles might be movable or fixed according to some such rule of intervals as suggested by Michell's theory of thin films and color. By such an argument from analogy, he achieves *ad hoc* agreement with experiment, but the generality of his dynamic design is gone.

Nor was there any indication that Cavendish's efforts would be appreciated by his contemporaries. The theoretical paper of 1771 was essentially ignored, while the experimental paper of 1776 excited comment for the dramatic ingenuity by which man could imitate what a fish did naturally, and for its estimated experimental values of electrical conductivity. Cavendish continued his personal studies of electricity for a time, but his next published work returned to phenomena involving gases. Known primarily as his papers on the composition of water and of nitric acid, Cavendish's "Experiments on Air," published in the *Philosophical Transactions* for 1784 and 1785 [*Papers*, II, 161-81, 187-94], were initiated by a renewed study of the change of airs from an elastic to an inelastic state accompanied by the generation of heat. Characteristically, the immediate impulse to the investigation was provided by the work of others. In his *Experiments and Observations on Various Branches of Natural Philosophy*, published in 1781, Joseph Priestley had described experiments, performed with John Warltire, on the explosive mixture of common and inflammable air to produce dew and heat.[31] Warltire had proposed the experiments as a means of determining the weight of the substance of heat, but Cavendish had never believed heat to be a substance and regarded the experiments as a combination of electricity, the transition from elasticity to inelasticity, and the production of heat and light—in other words, the investigation drew together the basic themes of a quarter of a century of Cavendish's research. Once again, as Cavendish repeated and extended these researches, the internal momentum of experimental investigation combined with the meticulous character of his research resulted in discoveries of basic importance to chemistry. These discoveries were, for Lavoisier, fundamental in achieving the "chemical revolution," but Cavendish never accepted that revolution. The reasons are far from entirely clear, but some clues exist in the mode in which Cavendish reported his "discovery" of the composition of water. He did not decide that water was composed of dephlogisticated and inflammable air, for he thought a reasonable alternative suggestion was that dephlogisticated air was water minus phlogiston and inflammable air water plus phlogiston. Perhaps this alternative was proposed to supply a solution to the strange phenomena of two elastic airs becoming fixed in one another. Finally, one may note that Cavendish declined to adopt any specific explanation for the production of dephlogisticated air from red precipitate of mercury until he was "much better acquainted with the manner in which different substances are united together in compound bodies. . . ."[32]

[31] Priestley, *Experiments and Observations on Natural Philosophy* (London: J. Johnson, 1781), II, 395*-398*.

[32] Cavendish, *Papers*, II, 175n; see pp. 179-81 and 325-26 of the same volume for some of his explicit objections to Lavoisier's system of chemistry.

In chemistry, as in heat and electricity, Cavendish remained a mechanist, seeking answers to questions of forces.

His final major research was, in a way, a collaboration with his deceased friend John Michell. Using a torsion-balance designed by Michell for the purpose, Cavendish performed the famous experiments, reported in his "Experiments to determine the Density of the Earth," and again achieved fame for the precision of an experiment.[33] All of his life Cavendish had attempted to obey the Newtonian injunction: from the motions, find the forces, from the forces derive the phenomena; in an apparently diverse range of subjects he had devised experiments and interpreted the results in order to reduce to mechanics the separating areas of chemistry, heat, light, and astronomy. Inevitably he had failed, but it is surely a tragedy that the magnificence of his failure should be concealed in separate, and empirical, interpretations of the experimental chemist and electrician.

The discovery of this relation between mechanistic theory and experiment in the work of Henry Cavendish may help to illuminate a similar relationship in that of Joseph Priestley.[34] For Priestley, like Cavendish, is described as having "no ruling purpose behind his work," and, like Cavendish, he spent most of his productive scientific career making experiments conformable to mechanistic design.[35] But Priestley was not a mathematician; though he knew something of the *Principia,* he was not inspired by its mathematico-deductive formalisms. If Priestley was a mechanist, it must have been from a very different design from that of Cavendish.

The general nature of such a design is discernible in the preface to Priestley's first published scientific work, the *History of Elec-*

[33] Henry Cavendish, "Experiments to determine the Density of the Earth," *Philosophical Transactions* for 1798; *Papers*, II, 249-86.

[34] Joseph Priestley (1733-1804), educated for the dissenting ministry at Daventry Academy, 1752-55, became a minister, teacher, leader of the Unitarian movement in Britain, and author of tracts and treatises on theology, politics, history, education, psychology, and metaphysics, as well as on science, an activity of his leisure. F.R.S. 1776, Copley medalist 1773, he wrote and experimented on electricity and optics in addition to the studies on gases for which he is most famous. A prominent liberal political leader, he was mobbed from Birmingham in 1791 and went into exile in the United States in 1794. The latest biography is that of F. W. Gibbs, *Joseph Priestley, Revolutions of the Eighteenth Century* (Garden City: Doubleday and Co., 1967); the most detailed picture of his scientific work is that contained in Robert E. Schofield, ed., *A Scientific Autobiography of Joseph Priestley.*

[35] The quoted description of Priestley's work is that of A. N. Meldrum, *The Eighteenth Century Revolution in Science—First Phase* (Calcutta: Longmans, Green, 1930), p. 60, but is typical of Priestley criticism. The most recent example is the continuously snide treatment by Partington, *History of Chemistry*, III, 237-97.

tricity of 1767: "Hitherto philosophy has been chiefly conversant about the more sensible properties of bodies; electricity, together with chymistry, and the doctrine of light and colours, seems to be giving us an inlet into their internal structure, on which all their sensible properties depend."[36] It can scarcely be coincidence that Priestley's research career, thereafter, took him from electrical studies to a *History and Present State of Discoveries relating to Vision, Light and Colours* and thence to a course of research into the nature of gases. Clearly the nature of matter, its action and internal structure, is an appropriate path to an understanding of Priestley's science.

The specific nature of that path is, however, anything but clearly marked. In all that stream of apparently ingenuous scientific writing which issued from Priestley's pen, to demonstrate his conviction that scientists ought "immediately communicate to others whatever occurs to them in their inquiries," there are only two explicit references to the nature of matter and its action.[37] While writing the *History of Optics,* Priestley was impressed with Boscovichean notions and described them in some detail, but Boscovich's name is never again mentioned in Priestley's scientific writings. Nor, indeed, is the nature of matter again discussed until 1794 and the appearance of his *Heads of Lectures on a Course of Experimental Philosophy.*[38] In this synopsis of his lectures to the students of New College, Hackney, Priestley is revealed as a pragmatist, adopting the successes of materialist explanations while trying to retain some measure of mechanism. The changes in properties of bodies, Priestley says, may result from substantial (i.e. material) causes, such as "the addition of what are properly called substances, or things that are the objects of our senses," as acids added to alkalies produce neutral salts; or by "a change in texture in the substance itself, or the addition of something that is not the object of our senses," as steel becomes magnetized by the touch of a magnet [4-5]. The term "substance," however, is "merely a convenience in speech," as the object in which properties inhere, and certain shared properties define substance in general. These properties are extension, and therefore infinite divisibility, and powers of attraction and repulsion. Apparent impenetrability and *vis inertia* are to be ascribed to prior causes,

36 Joseph Priestley, *History of Electricity* (reprint 1966), pp. xiv-xv. These sentiments are repeated in the same work, II, 78-80, where chemistry, electricity, and optics are linked as keys to the latent and less obvious properties of bodies.

37 Joseph Priestley, *Experiments and Observations on different Kinds of Air* (London: J. Johnson, 1775), 2nd edn., pp. vi-vii.

38 Priestley, *Heads of Lectures, etc.* (London: J. Johnson, 1794).

probably related to those powers, whose balance keeps the component parts of all substances together. Gravitation, cohesion, and chemical affinities are examples of attractive powers, the causes of which are unknown [9-12].

These "properties of all matter" are not inconsistent with Boscovichean mechanism, but they do not necessarily imply it, and their enumeration falls far short of the detail necessary to implement mechanistic explanations of phenomena. Nor do these, in fact, follow, and the text primarily relates to problems of substantial change, to definitions of the chemical nature of material substances, to electricity as a fluid, to light as particles of matter. Priestley does not here commit himself on that pivotal question, whether "the cause of heat be properly a substance, or some particular affection of the particles that compose the substance that is heated [135]." On the basis of these lectures, it would be impossible to declare that Priestley was a mechanist. Indeed, so far as any of his original scientific work is concerned, one would never know that he had deeply committed himself to a Boscovichean description of matter in that one area of greatest interest to him, his theology.

In 1774 Priestley began a series of philosophical, metaphysical, and theological publications relating to the doctrine of free will, the operations of the mind, and the relation between body and soul. Few of these seem to have been read by historians of science, yet they all involve scientific principles, and two of them, *Disquisitions relating to Matter and Spirit,* first published in 1777 with a second edition in 1782, and *A Free Discussion of the Doctrines of Materialism and Philosophical Necessity* of 1778, directly concern the nature of matter.[39] In these two works, Priestley denies the traditional dichotomy between matter and spirit; spirit or soul being nothing more than bodily organization of matter, dying as the body dies and achieving reorganization in a resurrection of the body at that second coming of Christ explicitly promised man in God's scriptural revelation. Now Priestley, himself, calls this position a materialist one, but what is the nature of this "materialism"? He denies that matter can be characterized by any of those properties: solidity, impenetrability, sluggishness, or inertness, generally used to distinguish it from spirit. By the accepted (Newtonian) rules of philosophizing, Priestley argues, all of these properties can be ex-

[39] Joseph Priestley, *Disquisitions. . . .* (London: J. Johnson, 1777), and *Free Discussion . . . in a Correspondence between Dr. Price and Dr. Priestley* (London: J. Johnson and T. Cadell, 1778).

plained in terms of Boscovichean point-atoms with their surrounding alternate spheres of attractive and repulsive forces. He even goes beyond Boscovich to insist that inertia itself can be explained mechanistically rather than assumed *a priori,* and matter becomes nothing but the geometry of space and the interaction of forces.

By resolving the sensible properties of matter into space and those forces of attraction and repulsion which represented, to Bentley or Rowning, for example, the continuing power of God in the universe, Priestley serves as a link between the Cambridge neo-Platonists of the seventeenth century and the nineteenth-century Platonism of German nature philosophy. And as the focus of Priestley's mechanism was his theology, so its origin, mediately or immediately, was this pre-Boscovichean religious metaphysics. In the only even quasi-scientific response to the *Disquisitions* published in Britain, Richard Price, like Priestley both natural philosopher and dissenting minister, objected to the definition of matter as extension and force and repeatedly raised the fundamental materialists' question: what thing is it that is extended and possessed of powers? Priestley's reply, never understood by Price, was that substance has no meaning save as a collection of properties.[40] Though this has overtones of the positivist attitude which Sir Philip Hartog thought might be ascribed to Priestley, the statement has identifiable antecedents in that period, nearly three-quarters of a century earlier, from which Priestley drew his initial mechanistic affinity for Boscovich. During his schooling at Daventry Academy, while he was reading Newton and Rowning for pleasure, he was studying, as a school text, the pneumatology of Philip Doddridge, original founder of the Academy.[41] Available then only in manuscript, the work was subsequently published as *A Course of Lectures on the Principal Subjects in Pneumatology, Ethics, and Divinity,* and in it are to be found numerous tutorial references to the authors from whom sections on such subjects as: "The dependence of the mind on the body," "motion not essential to matter," and "matter not self-existent" are drawn.[42] These authors include Baxter (from

[40] *Free Discussion,* pp. 17, 20, 45-47. Neither Boscovich's objections, on theological grounds, nor those of Herschel, on science but characterized by jejune sophisms and misunderstanding, were published.

[41] Sir Philip Hartog, "Newer Views of Priestley and Lavoisier," *Annals of Science,* 5 (1941), 1-56; for Priestley's studying of Doddridge, see *Memoirs of Dr. Joseph Priestley* (London: J. Johnson, 1806), pp. 178-84.

[42] Philip Doddridge, *A Course of Lectures . . . Pneumatology . . . with References to the most considerable Authors on each Subject* (London: J. Buckland, J. Rivington, R. Baldwin, *et al.,* 1763).

whom John Michell claimed to have derived his mechanism); the Cambridge neo-Platonists, More and Cudworth; Newton and Locke; such early Boyle lecturers as Bentley, Clarke, and Derham; and Cheyne and 'sGravesande. Doddridge's *Lectures* also declares: "We can have no conception of any substance distinct from all the properties of the being in which they inhere, for this would imply that being itself inheres, and so on to infinity [2]." Doddridge might never have subscribed to Priestley's spiritual mechanization of body and soul, but his compendium of religious metaphysics introduced Priestley to philosophers for whom such concepts would have represented a logical, though perhaps extreme, expression of the relation between natural philosophy and natural theology.

Feeling as he did about religion—he always insisted that he was first and foremost a minister and once maintained the "superior dignity and importance . . . [of] *theological* studies to any other whatever"—Priestley would naturally carry into his science some part, at least, of the theory of matter so important to his theology.[43] His original scientific writings were, however, usually cast in a Baconian mode, in which judgment as to cause was to be suspended and "real knowledge" attained by going from "bare facts" to more, analogous, facts.[44] There is no place in such a work for an explicit declaration of basic theory. Nor can theory always be inferred from his assumptions, for sheer virtuosity in the multiplication of experiments lends to much of Priestley's research a kind of independence of any basic conceptual scheme. It cannot, therefore, be assumed that matter theory alone provides an explanation by which all of his experiments can be coordinated, but many apparently diverse trains of experiment, many apparently anomalous expressions and attitudes can be rationalized by the hypothesis that Priestley designed his experiments to investigate the nature of matter and preferentially interpreted his results to achieve a mechanical explanation of phenomena.

Although a substantial part of Priestley's *History of Electricity* reflects only his conviction that good history is retrospective journalism, much of it conveys a suggestion of mechanism—his statement of interest in matter theory and his deduction of the inverse-square law, already cited, a few of his interpretations of the work of others, but particularly the nature of his own re-

[43] Joseph Priestley, *Experiments and Observations on various Branches of Natural Philosophy* (London: J. Johnson, 1786), III, x.

[44] Joseph Priestley, *Experiments and Observations*, pp. x-xi.

searches, experiments which were diversified and continued into subsequent editions of the *History* and papers for the *Philosophical Transactions*.[45] His work on electricity was guided by the counsels of Benjamin Franklin, John Canton, and William Watson, but even without the influence of these three of its major architects, Priestley would have had to take the fluid theory of electricity seriously. Nor could the judgment of the historian be ignored by the natural philosopher, and, in spite of occasional references to the "supposed electric fluid [*e.g.* II, 7]," Priestley's own work was built about that concept. He did not, however, have to accept the aetherial overtones frequently given the fluid theory. He adopts the arguments against "electric atmospheres," except as the "spheres of action of any electrified body," and refers approvingly to spheres of electrical attraction and repulsion. He also accepts Æpinus' modification of Franklin's theory [I, 293-94, 303-06; II, 7, Index: "Æpinus"]; although, because of his own deduction of the inverse-square law, and probably also his distrust of mathematics, he dismisses most of Æpinus' theoretical work:

> . . . the result of many of his reasonings and mathematical calculations cannot be depended upon; because he supposes the repulsion of elasticity of the electric fluid to be in proportion to its condensation, which is not true, unless the particles repel one another in the simple reciprocal ratio of their distances, as Sir Isaac Newton had demonstrated in the second book of his *Principia* [II, 35].

The mathematical disability, which contributed to Priestley's undervaluing of Æpinus, also restricted the range of his experimental inquiry, for he could not derive the consequences of possible states of electric fluid and subject the derivations to experimental verification. He had, however, a remarkable experimental imagination, which he used to design instances in which the variation of such mechanical parameters as configuration and forces of elasticity might produce variation in the motion and action of electric fluid. He tried an experiment to see if "condensing air in a glass tube lessens its power of electrifying," he asked if electrifying a plate of glass might not compress it, changing its density; if refractive indices, which vary with density, might not, therefore, vary with the amount of charge on a glass plate; if

[45] The recent reprint, Robert E. Schofield, ed., Joseph Priestley, *History of Electricity*, 3rd edn. of 1775 (New York, 1966), with its appendices, contains all Priestley's significant electrical publications.

the electrical capacity of glass was a function of its density.[46] His most completely realized train of electrical research was a study of conductivities. From his first pseudo-discovery, of the conductivity of carbon dioxide, to his last major electrical paper, on that of charcoal, his experiments show a conscious effort to vary the conditions of bodies and examine the comparative ease of motion of the electric fluid through them. He compared the conductivities of different substances, of various states (solid, liquid, and vaporous) of the same substance and under various conditions of temperature, hardness, density, and dimension [II, 192-214, 368-71, appendix I].

His final conclusion appears to have related the matter of electricity to the principle of inflammability (phlogiston) and conductivity to the completeness of union between phlogiston and its base. On the surface, this appears to be a materialist explanation, but that interpretation is open to doubt. In an earlier series of beautifully conceived experiments, Priestley had attempted to exhibit changes of momentum of the substance of electricity as it moves in a discharge. His report of these experiments illustrates the danger of relying on Priestley to be open and exhaustive in published work. His *Philosophical Transactions* paper (of 1769) merely records the failure to detect any momentum change; an unpublished comment on the same experiments adds, "You will think I talk strangely, but I own, that these experiments, together with those of the insulated hot glass, mentioned p. 716 of my book, and others too long to recite here make me inclined to think, there is no electric fluid at all, and that electrification is only some [new] modification of matter of which any body consisted before that operation."[47] Only Priestley, during the eighteenth century, appears to have made this ultimate mechanist assumption that electricity was merely a form of ordinary matter and he, perhaps, for only a short time. But the evidence here is clearly incomplete. One cannot but wonder, for example, if some of those other experiments "too long to recite here" might have included the beginnings of those on the "lateral explosion" in which Priestley was to conclude that there were circumstances in which the effect of electric discharge was produced without apparent transfer of the substance of electricity (possibly by the simultaneous influx and efflux of fluid, so as to leave no charge) [II, 336-42, appendix 2].

Priestley and Cavendish each performed electrical experiments

[46] See *Scientific Autobiography*, No. 20, and *History*, II, 71.

[47] *History*, II, 344-45; *Scientific Autobiography*, No. 16.

and asked questions markedly unlike those of any of their contemporaries. When Cavendish performed his experiments and asked his questions, we know that he did so in pursuance of a mathematical-mechanical view of electricity. It seems likely that Priestley's similar questions and experiments had a similar motivation, but, unable to develop the deductive part of the task, he remained true to his concept of Baconianism and experimented from "bare fact" to analogous facts, for someone else, better endowed than he, to develop into real knowledge. In electricity Priestley could illustrate a variety of motions under a range of conditions and constraints; he could not, from these, find the forces.

Nor, of course, could he do so in other branches of natural philosophy. This deficiency is particularly demonstrated in his next major work of science, the *History of Optics*, of all subjects next to mechanics, then the most mathematized, but which Priestley hoped, nonetheless, to make "perfectly intelligible to those who have little or no knowledge of Mathematics."[48] Naturally the *Optics* was Priestley's weakest and least successful work. He lacked the discrimination to appreciate and emphasize the most significant work of others, and the subject afforded him the least inducement to original experimentation. Of the experiments relating to light, which he performed or proposed be performed, few, if any, were not referable either to his earlier studies on electricity or to his research on gases. But it was for its relationships with electricity and chemistry that Priestley was initially interested in optics, for these relationships and the insight they might give to the internal structure of matter.

His work on light illustrates Priestley's suspicion of aetherial hypotheses and his preference for mechanist explanations:

> . . . philosophers are now almost universally agreed, with Sir Isaac Newton, that these effects are produced by certain powers of attraction and repulsion, extending beyond the surfaces of bodies. But, not being content with resting the matter here . . . Newton . . . proposes a conjecture concerning the physical cause of this attraction and repulsion; but his hypothesis [of the aether] seems to labour under as many difficulties as the hypothesis of the mechanical production of light without attraction or repulsion [378].

His interest in examining changes of motion of light under various constraints is illustrated in both electrical and pneumatic ex-

[48] Priestley, *History of Optics*, p. viii.

periments: those relating indices of refraction to degree of electrification, already cited; those comparing proportions of prismatic colors of electric sparks with those of flame, possibly to determine "what that heterogeneous matter is, that is roused into action by the rapid passage of the electric fluid"; those "to ascertain whether the reflection, refraction, or inflection of light be affected by electrification, or charging"; and those on the refractive powers of different kinds of airs.[49]

By the time the *History of Optics* reached publication, Priestley was reading to the Royal Society his first paper on the properties of gases. With the considerable success of his experimental studies on air, he abandoned the notion of writing a history of all experimental philosophy, to concentrate on reporting his own work rather than interpreting that of others. The *Philosophical Transactions* papers and the six volumes of *Experiments and Observations* are still cast in Priestley's favorite Baconian mode, but each of the volumes contains also a preface which extends the discussion of natural philosophy beyond the "facts" reported in the text. These, plus the range of experiments discussed, the extensive scale in which they are reported, and the span of years over which the work was done—amounting to more than thirty years from the time he first began seriously "to take up some of Dr. Hales' inquiries concerning air" to his last experiments before his death—all these provide a major opportunity for analytic investigation of the rationality behind Priestley's experimentation.[50]

In fact, the wealth of material is such that only extracts can be taken from it. Now the internal integrity of scientific inquiry frequently provided its own justification for his pneumatic experiments, and, insofar as Priestley's researches related to chem-

[49] The experiment on prismatic colors is proposed in the first edition, *History of Electricity* (London, 1767), p. 726, and by the third edition of 1775, II, 366, the reason for the experiment has been omitted. The experiments on reflection, etc., are mentioned in a letter of January 1768, *Scientific Autobiography*, No. 20. Those on refractive powers of gases are described in Priestley's *Experiments and Observations on different Kinds of Air* (London: J. Johnson, 1777), III, xx, 365.

[50] The reference to Hales is from a letter, Priestley to Theophilus Lindsey, of February 21, 1770, printed in J. T. Rutt, *Life and Correspondence of Joseph Priestley* (London: R. Hunter, 1831-32), I, 112-14. The bulk of Priestley's experiments on gases eventually appeared in one volume or another of his *Experiments and Observations on different Kinds of Air* or the *Experiments and Observations relating to various Branches of Natural Philosophy*, all published by J. Johnson, of London, between 1774 and 1786. As Priestley himself frequently referred to these as six volumes of a single series, I will do the same. I have used: *Experiments and Observations*, I, 2nd edn. 1775; II, 2nd edn. 1776; III, 1777; IV, 1779; V, 1781; and VI, 1786.

istry, they inevitably called for those materialist explanations which, in chemistry even more than in electricity, were enjoying so signal a success. Nonetheless, one might reasonably inquire how many of Priestley's pneumatic studies actually involved him in what he might call chemistry. He never claimed to be a chemist; the titles of his pneumatic studies reflect their Baconian emphasis on experiment and observation, and both the vocabulary he used and the parameters he measured reveal their Halesian, mechanistic origin.[51] Like Hales, Priestley heated substances and thus "released" air which they had "contained" in an unelastic form; he examined processes which "diminish" airs, changing them from an elastic to an unelastic state or disposing them to "deposit" other airs which existed in them. Typically the parameter he measured was change in volume, for it is this which is involved in variation of distance-dependent elastic forces. Priestley appears never seriously to have understood the importance of Lavoisier's arguments based on chemical weight. But these arguments imply conservation of weight and, as Boscovich suggests and as de Marignac and Mendeleev were later to assert, the weights of groups of elementary particles need not equal the sum of the weights of the individual particles themselves.[52] Priestley never mentions this possibility, but, as none of his publications includes theoretical arguments, his silence is not evidence that he was incapable of understanding such reasoning, which could derive directly from the relation of weight to configurations of point-atoms and their positions on the Boscovich force-curve.

Clearly Priestley was well aware of the mechanistic implications of pneumatic studies. In 1775, he described them as ". . . not now a business of *air* only . . . but . . . of much greater magnitude and extent, so as to diffuse light upon the most *general principles* of natural knowledge. . . . we may perhaps discover principles of more extensive influence than even that of gravity itself . . . [*E and O*, II, vii-viii]." In 1777, he explained: "The reason of my great expectations from this mode of experimenting is simply this, that, by exhibiting substances *in the form of air*, we have an opportunity of examining them in a less compounded state, and are advanced one step nearer to their primitive elements." Even when he concedes that a general theory of kinds of air will involve a knowledge of the "several ingredients" which enter into

51 For further discussion of this point see Robert E. Schofield, "Joseph Priestley, Natural Philosopher," *Ambix*, 14 (1967), 1-15.

52 See W. V. Farrar, "Nineteenth-Century Speculations on the Complexity of the Chemical Elements," *British Journal of the History of Science*, 2 (1965), 297-323.

their composition, he insists also that such a theory would require knowledge of the cause of their combination in the form of air, and the reason of their resolution into substances of other forms and properties [*E and O*, III, ix, x].

A large number of Priestley's pneumatic experiments were clearly designed to find information other than chemical. He examined the intensity of sound as transmitted in different kinds of air, to see if this was a function of any property of the gas other than density [*E and O*, v, 295-99]. He concerned himself with problems of gaseous diffusion, trying, for example, to determine if heat, cold, condensation, or rarefaction might facilitate the mixture of various airs.[53] He published accounts of experiments on the comparative heat expansions of various kinds of air [*E and O*, I, 253; III, 345-48; v, 359] and on the thermal conductivity of gases [*E and O*, v, 375-78]. Some three-quarters of a century later, James Clerk Maxwell was to investigate the same phenomena (sound transmission, diffusion, heat expansion, and conductivity) in gases, mathematically as well as experimentally, when he assumed a corpuscular theory of gases, with forces between particles, for his kinetic theory of heat.[54]

That Priestley also entertained the notion of a mechanical theory of heat, in spite of his later equivocation in the *Heads of Lectures*, is demonstrated by his statement that "probably . . . the heated state of bodies may consist of a subtle vibratory motion of their parts [*E and O*, I, 181]." Acting on this belief, he supposes that heat is generated in chemical separations, as in ignition, by "the action and reaction, which necessarily attends the separation of the constituent principles, exciting probably a vibratory motion in them [*E and O*, I, 160-61]." In 1780, he notes an observation that "*air* parts with its heat in condensation" and in 1799 reports on some experiments to measure the latent heat in airs by dissolving them in water and measuring changes in solution temperature.[55] When Henry Cavendish earlier made the same observation and experimented on the release of heat in the "fixing" of gases in solids or liquids, he did so as a meas-

[53] His earliest reference to diffusion is in a letter of 1768, *Scientific Autobiography*, No. 20; the last is published in "Experiments relating to Change of Place in different Kinds of Air, through several interposing Substances," *Transactions of the American Philosophical Society*, 5 (1802), 14-20.

[54] See, for example, Alex. C. Burr, "Notes on the History of Experimental Determinations of the Thermal Conductivity of Gases," *Isis*, 21 (1934), 169-86.

[55] See Priestley to Benjamin Vaughan, 26 March 1780, *Scientific Autobiography*, No. 87; and Joseph Priestley, "On the Transmission of Liquid in Vapor, #6. Of the proportion of latent heat in some kinds of Air," *Transactions of the American Philosophical Society*, 5 (1802), 10-11.

ure of the "mechanical momentum" stored in the elasticity of the gases. Priestley had no concept of mechanical momentum, but, as Desaguliers had earlier shown, the idea of latency in elastic repulsive bonds does not require mathematics, and Priestley surely possessed the requisite mechanical sophistication.

In 1777 Priestley noted his failure to confirm the conversion of water into earth, but explains: "I went upon the idea, that the change in consistence in water was brought about by extending the bounds of the repulsion of its particles, and at the same time preventing their actual receding from each other, till the spheres of attraction within those of repulsion should reach them. The hypothesis may still be not much amiss, though I did not properly act upon it [*E and O*, IV, 408]." In 1782, he commenced a series of experiments which he viewed as the conversion of water into permanent air, because, as he said in a letter to Josiah Wedgwood, "I had also a general idea that if the parts of any body were rarified beyond the sphere of attraction they will be in a sphere of repulsion to each other."[56] In the face of evidence which convinced most of their younger contemporaries, neither Cavendish nor Priestley could accept Lavoisian chemistry, and each gave essentially the same reason. As early as 1774, Priestley had expressed his opinion that the difference between an air which was inflammable and one which extinguished flame might lie in the mode of combination of constituent principles as well as in the proportional quantity of each [*E and O*, I, 261]. His conviction and primary interest in such modes of combination never flagged. In 1802 he wrote:

> Indeed a knowledge of the elements which enter into the composition of natural substances, is but a small part of what it is desirable to investigate with respect to them, the principle, and the mode of their combination: as how it is that they become hard or soft, elastic or non-elastic, solid or fluid, etc., etc., etc., is quite another subject of which we have, as yet, very little knowledge, or rather none at all.[57]

The root of Priestley's disagreement with Lavoisier was not the existence of phlogiston, but rather the difference between a ma-

[56] *Scientific Autobiography*, No. 100.

[57] Joseph Priestley, "Miscellaneous Observations relating to the Doctrine of Air," *New York Medical Repository*, 5 (1802), 264-67. This statement is a variation of another of 1793, *Experiments on the Generation of Air from Water* (London: J. Johnson, 1793), pp. 38-39, in which he declares that constitutional differences of substances depend as much on mode of arrangement as on the elements of which they consist.

terialist and a mechanist. To achieve successful explanations of chemical phenomena, Lavoisier abandoned physical reductionism for the jig-saw puzzle problems of permutation and combination of substances. To Priestley the solution of such problems was comparatively unimportant. He ended his scientific career, as he had begun it, with a declaration of faith in the fundamental significance of determining the ultimate constitution of matter in its mechanistic modes and operations.

Perhaps it is because James Hutton's most famous work was done in geology, perennially a study in natural history rather than of natural philosophy, that Hutton has so seldom been included among the speculative scientists of late eighteenth-century Britain.[58] Even geologists and historians of geology have treated his theory of the earth in isolation and, unable then to trace the progress of Hutton's development of his principles, have fallen back on that refuge of the Baconian true-believer, the suggestion that those principles grew out of the observations Hutton is supposed to have made in England and Scotland while pursuing agricultural interests.[59] Hutton's contemporaries would not have agreed, and while few went so far as Richard Kirwan, to suggest that he had manufactured observations and falsified data, most of them saw, as W. D. Conybeare was later to see, evidence in Hutton's work of preconceived hypotheses.[60]

Now John Playfair, Hutton's friend and biographer, tells us that Hutton did not confine his attention to a theory of the earth, "but had directed it to the formation of a general system, both

[58] James Hutton (1726-97), educated at the University of Edinburgh, including its medical school, 1740-47; in Paris, 1748-49; and at Leyden, where he obtained the M.D., 1749. On returning to Scotland, he applied himself to a study of agriculture, said to have directed his attention to mineralogy and geology, as his partnership in a sal-ammoniac manufactory redirected his interests in chemistry. In 1768 he settled in Edinburgh, where he joined his friends Joseph Black, Adam Ferguson, John Robison, and others as an original member of the Royal Society at Edinburgh. Most famous for his work in geology, he also published essays on farming, meteorology, medicine, chemistry, and metaphysics. My section on Hutton is derived directly from the work of Dr. Patsy A. Gerstner, who has condensed her unpublished M.A. research essay on Hutton into a paper, "James Hutton's Theory of the Earth and his Theory of Matter," *Isis*, 59 (1968), 26-31.

[59] See, for example, the comment of W. T. Fitton, "Review of Elements of Geology by Charles Lyell," *Monthly Review*, 69 (1839), 443, and that of Sir Archibald Geikie, *The Founders of Geology* (New York: Dover Press reprint of the 2nd edn. of 1905, 1962), p. 299.

[60] For the contemporary reception of Hutton's work, including Kirwan's criticism, see Charles C. Gillispie, *Genesis and Geology* (New York: Harper Torchbook, 1951), p. 55. Conybeare's statement, nearly unique among opinions of later geologists, is cited by Wm. H. Hobbs, "James Hutton, the Pioneer of Modern Geology," *Science*, 64 (1926), 265.

of physics and metaphysics."[61] Characteristically, that system is not described in the most widely read of Hutton's works, the *Theory of the Earth.* In his studies of geology, Hutton concerned himself "not with causes but effects," and, like Priestley, he wrote other works as vehicles for his metaphysics—especially the scarcely known *Dissertations on Different Subjects in Natural Philosophy* and *An Investigation of the Principles of Knowledge.*[62] These three works: the *Theory, Dissertations,* and *Investigation,* taken together, join metaphysics, natural philosophy, and geological observations into a whole which rationalizes Hutton's *Theory of the Earth,* but they cannot be said to make that "whole" a clear one, for nothing Hutton wrote was clear. Thomas Thomson, writing of Hutton's ideas, declares that they are set out in a "manner so peculiar, that it is scarcely more difficult to procure the secrets of science from Nature herself, than to dig them from the writings of this philosopher."[63] No doubt much of the problem lies in the tortured obscurity and prolixity of Hutton's literary style, but some of the difficulty is traceable to his use of mechanist distinctions and arguments which might have been familiar to the Keills or Stephen Hales, but were lost on his contemporaries and which vanished in the "translation" of Hutton, prepared for nineteenth-century readers by John Playfair, the *Illustrations of the Huttonian Theory of the Earth* of 1802.

Hutton's departure from Mosaic geology, for example, which earned him the title of "atheist," was not so much "uniformitarian" in an evolutionary sense as it was a conviction that Divine order in the universe must be evidenced by continual action. To assure life on earth, land must be destroyed for the soil from which plants grow to feed animals, but if land is destroyed to

[61] John Playfair, "Biography of James Hutton," *Transactions of the Royal Society of Edinburgh,* 5 (1805), 74.

[62] Hutton's geological theory was first read to the Royal Society of Edinburgh and published in abstract in 1785. The complete paper (from which, p. 213, the reference to cause and effect was taken) appeared as, "Theory of the Earth; or an Investigation of the Laws observeable in the Composition, Dissolution and Restoration of Land upon the Globe," *Transactions of the Royal Society of Edinburgh,* 1 (1788), 209-304. The paper was then expanded into a book, two volumes of which were published as *Theory of the Earth* (Edinburgh: Cadell, Junior and Davies, London; and William Creech, 1795), while a third volume remained unknown and unpublished until 1899. The other works (probably those Hutton had completed in manuscript before 1781—see Playfair, "Biography of Hutton," p. 74n) are: James Hutton, *Dissertations, etc.* (Edinburgh: A. Strahan, and T. Cadell, London, 1792), and *Investigation of the Principles of Knowledge and the Progress of Reason, from Sense to Science and Philosophy* (Edinburgh: A. Strahan, and T. Cadell, London, 1794).

[63] [Thomas Thomson], Article: "Chemistry," *Supplement,* 3rd edn. of the *Encyclopaedia Britannica,* 1801.

assure life, it must be restored for the same reason. Any explanation of geological phenomena must, therefore, assume a mechanical, cyclical process in which damage to the earth is repaired by the same powers by which it was formed ["Theory," 215-16]. He finds those powers in the fundamental nature of the substances from which all matter is formed, but unlike the corpuscularians, he does not go to ultimate particles. Such explanations, Hutton feels, permit only the statement that large bodies are made of smaller ones. Instead he makes a neo-mechanist distinction between body, characterized by weight, extension, motion, and resistance; and matter, which is the cause of these characteristics and which he identifies with force [*Dissert.*, 315]. Matter is indefinitely extended or unextended according as it is treated metaphysically or physically; the greatest quantity of matter may be compressed in the smallest space or the smallest quantity may occupy the greatest.[64] Two kinds of matter exist: gravitating matter, which is equivalent to the forces of attraction and union; it is necessarily associated with body, and its effects (whether gravitational, cohesive, or concretive) are determined by the various distances of the parts of body [*Dissert.*, 388-403, 661-62]. The second kind of matter Hutton calls solar substance. It represents the power of repulsion and separation, can be translated among bodies, and takes the form of light, specific or latent heat, or electricity, depending upon whether it is contained in bodies, to act there in opposition to gravitating matter, or is not. Light is solar substance in free motion, electricity that occupying body surfaces, heat that which acts only in internal connection with body [*Dissert.*, 491-503].

"Of these two different kinds of matter, natural bodies are . . . composed; and it is from various modifications of these two general principles . . . that [they derive] all their qualities . . . [*Dissert.*, 623]." Given only gravitating matter, tending always to union and rest, there would be no motion in the universe; given only matter of heat, tending always to separation, only random motion would ensue [*Dissert.*, 262-63, 434]. And from this conclusion, so reminiscent of Stephen Hales, the application of these principles to the cyclical processes of geology follows. At one time the material of the earth tends toward the bottom of the sea, toward union and rest. Then the equally forceful power of heat

[64] This statement, which might have been quoted from John Keill, is paraphrased from Hutton, *Dissertations*, p. 315. Hutton's evident familiarity with the writings of early dynamic corpuscularians may have been reached through his reading of John Harris' *Lexicon Technicum*, his earliest source of chemical information.

takes control, the land is raised above the sea, until a height is attained such that the power tending to union resumes control.

And as the theory of matter explains the quality of Hutton's cyclical geological processes, so it explains his characteristic emphasis on heat as the principal agent active in the earth. For it is essentially through the interaction of heat, in its different forms, with gravitating matter, that the various states of body (solid, liquid, or vaporous) are produced. The effects of external forces are limited because the expansive power increases rapidly as the volume of body diminishes. When the distances are such that the power of concretion is added to those of gravitation and cohesion, body becomes hard [*Dissert.*, 661]. When gravitation is in control, but concretion is balanced by latent heat, fluidity ensues, and, as the power of concretion is further weakened through the expansive power of specific heat, latent heat will prevail over concretion, until it balances both cohesion and gravitation, and a vaporous state will follow [*Dissert.*, 636-37]. Then, when the distances become too great for the effective action of repulsive powers, the forces of attraction resume their activities, and dissolution is replaced by union.

It is no wonder that Sir Archibald Geikie could not find, in the *Theory of the Earth*, the mechanism by which the expansive power of heat gave rise to elevation of the earth's crust. Nor is it surprising that Hutton's contemporaries found this heat, not maintained by any of the causes by which heat was derived and acting intermittently, was a "substance with which we have no acquaintance."[65] Probably a reading of the *Dissertations* and the *Investigation* would not have brought these critics understanding. Hutton's terminology was confused, but his conceptualizations would, to the materialists, have been even more so. Matter which was not substance but force might just have been conceivable to the corpuscularians earlier in the eighteenth century; it was possible to such neo-mechanists as Priestley in the second half of the century; it was to be familiar only to Kantians, nature philosophers, and field theorists in the nineteenth century.

[65] Archibald Geikie, *Geology at the Beginning of the Nineteenth Century* (London: Geological Society, 1907), p. 10; J. A. deLuc, "First Letter to Dr. James Hutton . . . on his Theory of the Earth," *Monthly Review*, 2 (new series, 1790), 210; and [Anon.], Review of Playfair's *Illustrations*, *Edinburgh Review*, 1 (1802), 202-203.

CHAPTER ELEVEN

Interregnum, 1789-1815

DURING THE NINETEENTH CENTURY mechanism again reached an ascendency, but with explanations quite unlike those which had characterized the dynamic corpuscularity of eighteenth-century mechanical philosophy. Whether in electrodynamics, thermodynamics, or chemical dynamics, the emphasis was on forces and energy, their manifestations, modifications, and interactions; next to nothing was said (or believed) of a material substratum in which such forces might inhere. The "classical" physical sciences of the nineteenth century retained the dynamics, but not the corpuscularity of the eighteenth and, in their development, were dependent upon frames of thought characteristic of their own age. Nonetheless, in their inception, these dynamical theories represent a culmination of eighteenth-century mechanism. Bringing to their research concepts learned in the previous century, certain transitional nineteenth-century British natural philosophers exhibit, particularly in their early work, the continuation of an indigenous neo-mechanist tradition which was to stimulate and direct the experiments of a new era of British science. In these final pages, the intellectual baggage carried by a few of these scientists into the nineteenth century will be described. The tracing of further development of their ideas, under additional influences and with new accretions of information, is properly to be left to historians of nineteenth-century scientific thought.

One of the most important of these transmitters of eighteenth-century thought was John Robison, whose lectures at Edinburgh began that strain of Scottish Boscovicheanism which continued through the work of John Leslie and Thomas Thomson into that of Lord Kelvin and James Clerk Maxwell.[1] Robison's lectures were never completely published, but there is ample evidence of his commitment to mechanism in the work of his students and in his miscellaneous writings.

[1] John Robison (1739-1805), educated at the University of Glasgow, where he graduated in arts in 1756. He succeeded Joseph Black as professor of chemistry in 1766, resigned in 1770 to take a post in Russia, but returned to Scotland in 1773 as professor of natural philosophy at the University of Edinburgh, where he remained until his death. Primarily known as a teacher, he edited Black's chemical lectures, wrote for the *Encyclopaedia Britannica,* and published one volume of an incomplete text, *Elements of Mechanical Philosophy* (1804), which covers only his lectures on mechanics.

The beginning of Robison's interest in Boscovich may be dated with some precision. In 1778, one of his and Black's students, William Keir, published a dissertation, *de Attractione Chemica,* which discusses forces of chemical attraction without referring to Boscovich.[2] Now Keir's definition of chemical attraction is very like that of Joseph Black. As attraction, in general, is the power by which bodies tend to move together, so chemical attraction is that power which obtains between unlike substances, over distances too small to be apparent to the senses. It acts in opposition to the forces of gravity, cohesion, capillarity, and elasticity; is the only species of force to which there is no corresponding repulsion; and, though great in magnitude, has not yet been measured. From this point, Keir (but not Black) proceeds to an explication of chemical process by way of attraction. His most interesting discussion comes in the development of the proposition that ". . . the less there is of any one component part in proportion to the rest, the more strongly will it be attracted by the rest, for the force exerted by each particle of one component will then be divided among fewer particles of the other." Saturation is a state of equilibrium between chemical attraction (varying in inverse proportion to the quantity of any one component) and resisting mechanical forces.

In view of Black's declaration that understanding of chemical combination was not aided by the use of force concepts, it is reasonable to ascribe Keir's use of these arguments to the influence of Robison's contrary convictions.[3] Hence Keir's failure to refer to Boscovich (surely an obvious authority for speculations on forces) is an indication that Robison was not yet acquainted with Boscovich's ideas in 1778.[4] In 1779, another Edinburgh student, William Cleghorn, published his dissertation, *de Igne.* In the introduction to his work, Cleghorn writes, "I regret that I did not make a thorough study of Boscovich's Theoria philosophiae naturalis before I completed my studies. He has not really, it seems to me, come to a just conclusion about fire; but there are several effects of fire which, in my opinion, confirm his general theory and, in turn, are conveniently explained by

[2] A. M. Duncan, "William Keir's *de Attractione Chemica* (1778) and the Concepts of Chemical Saturation, Attraction, and Repulsion," *Annals of Science,* 23 (1967), 149-73.

[3] See Chap. 9.

[4] The only reference Keir makes to an authority uncommon to students of Black is to Bryan Higgins, whose *Philosophical Essay concerning Light* was an influence, also, on Herschel's adoption of force hypotheses.

it."[5] Cleghorn's description of the nature of heat—a fluid *sui generis,* conserved in total amount from the Creation, with repulsive force between its particles and attractive forces to the particles of almost all other bodies [43]—was called, by Black, the most probable of any that he knew. Black also, however, called the approach that of a mechanician and compared it with Gowin Knight's, as one which appeared to explain phenomena but continually involved inconsistencies. Like Black, Cleghorn argues against a motion theory of heat, but, unlike Black, he continues with a heat-fluid mechanics. As the force laws of heat repulsion and attraction were not known (and as Cleghorn was not a mathematician), his fluid mechanics could not become a quantitative, analytical theory like Cavendish's electrical-fluid mechanics, but its qualitative development reveals a concern for the interaction of forces.

Adopting from Macquer (and possibly William Keir) the notion of saturable forces of attraction—" . . . if any new [attraction] is introduced, every former [attraction] is diminished [25]"—Cleghorn explains how all the phenomena of heat arise from changes in repulsive or attractive forces, due to changes of state in bodies. For example, as fluidity or vaporization occurs when the attraction between the particles of fire and other bodies overcomes the attraction of cohesion, so fusion or vaporization produces cold. For the attraction of cohesion being destroyed, the force attracting fire will increase, drawing fire particles from the surroundings. Similarly, congelation produces heat, by releasing fire particles as an increase in the force of cohesion decreases the force attracting fire [25, 27, 29]. Fermentation and putrefaction consist in a change in the arrangement of particles, introducing new attractions which contribute to decreasing the attraction of fire [37]. And heat is produced in chemical mixtures (combinations) by release of fire particles due to the attraction of the particles of combining bodies [39]. For authorities to support his views, Cleghorn cites Black (with whose opinions he must sometimes reluctantly disagree), Irvine, and Crawford. More significantly, he quotes from the last Query of Newton's *Opticks;* he paraphrases that observation of Hales that without fire (repulsive force) nature would become a "desolate and inert mass," while too much fire would disperse everything into vapor and nature into chaos [15]; he refers approvingly to

[5] Douglas McKie and Niels H. deV. Heathcote, "William Cleghorn's *De Igne* (1779)," *Annals of Science,* 14 (1958, published 1960), 1-82, contains the Latin text, and English translation, and a commentary. The quote is taken from p. 11.

spheres of attraction and repulsion as a means to the explanation of evaporation [47].[6] And, almost gratuitously, he cites Boscovich as showing that bodies, put into a state of agitation, may increase their power of attracting fire, which thus explains heating by friction [57].

Insofar as Cleghorn's theory received any public attention, it did so through Robison's edition of Black's chemical lectures (1803), and it was in that edition that Robison first publicly sided with Keir and Cleghorn against Black's view of the nature and utility of chemical attraction.

> We cannot form a conception of any thing opposing a mechanical force, but another mechanical force. Hence we must infer that the corpuscular powers or qualities which ultimately produce the chemical phenomena, are not different from the forces which produce the fall of heavy bodies, and the communication of motion by impulse. . . . The conjecture of Sir Isaac Newton, expressed in one of his reflections at the end of his treatise on Optics, seems fully verified; and it becomes highly probable that all bodies consist of atoms endowed with one and the same power of attraction and repulsion, varying according to a determined law of the distances. The ingenious essay on this subject by the Abbe Boscovich merits therefore the careful perusal of every philosophical Chemist.[7]

Should any such "philosophical chemist" have become curious about this essay by Boscovich, he need only have read the detailed description, provided by Robison, in the *Supplement* to the third edition of the *Encyclopaedia Britannica* (1801). There Robison is explicit in his praise of Boscovich's system and in his association of it with Newtonianism. Boscovich is "one of the most eminent mathematicians and philosophers of the present age." Although Robison does not like the notion of forces without "any substratum, any thing in which they are inherent as qualities," he thinks the system, as a whole, "well adapted for explaining the phenomena of nature," even were it nothing but a mere hypothesis while, in fact, it is founded on a chain of strict reasoning from evident principles. "It is more than probable,

[6] In their editorial notes McKie and Heathcote observe that the quotation from Newton's *Opticks* is taken from Query 23 of the Latin edition of 1706, rather than from 31 of the 1717 English edition or its 1719 Latin translation. Cleghorn also cites Rowning, but the editors believe the reference is second-hand, derived from a work by W. Hales.

[7] Joseph Black, *Lectures on the Elements of Chemistry*, ed. John Robison, I, 516-21.

that had Newton lived to be acquainted with the Boscovichean theory, he would have paid to it a very great regard. This we may conjecture from what he says in his last question of optics. . . ." And the article ends with several lengthy quotations from Query 31.[8]

These views are repeated and extended in the collection of Robison's articles, published posthumously by his friend and former student, David Brewster. Not only does the *System of Mechanical Philosophy* have a long section devoted to a careful explanation of Boscovich's theory and its uses, throughout the work there are other, favorable, references to Boscovich and to the application of force hypotheses as opposed to aetherial ones. Boscovich's theory is "undoubtedly one of the most curious productions of the last century, filled with original and ingenious notions of natural things. . . . If we shall ever acquire the knowledge of a true theory, it will resemble Mr. Boscovich's in many of its chief features."[9] Newton "very early hinted his suspicions that all the characteristic phenomena of tangible matter were produced by forces which were exerted by the particles at small and insensible distances [III, 642-43]." With the exception of Euler's unsuccessful attempt, no person trying to explain phenomena by means of elastic and vibrating fluids or aethers has described the nature of the undulations or shown how such vibrations are connected with, or even influence, the phenomena. "We think it our duty to remonstrate against this slovenly way of writing . . . It has been chiefly on this faithless foundation that the blind vanity of men has raised that degrading system of opinions called MATERIALISM . . . [III, 728-29]." But Robison is not prepared to be purely dynamical. "Surely . . . the active powers ascribed to matter are conceived as the endowments, or the attendants of something that is different from a mere point in space [I, 299]."

Given Robison's evident commitment to mechanical explanations of phenomena and his considerable facility in mathematics (his *Supplement* article provides a better explanation than Boscovich's of the analytical expression for the force-curve), it is not surprising that he was one of the few persons who understood,

8 [John Robison], Article: "Boscovich," *Supplement, Encyclopaedia Britannica,* third edition (Edinburgh: for Thomson Bonar, 1801), I, 96-110, esp. 104, 106n, 109-10.

9 John Robison, *A System of Mechanical Philosophy,* ed. David Brewster (Edinburgh: for John Murray, London, 1822), I, 267. The major section on Boscovich extends from pages 267 to 302, and incorrectly appears, from the headtitles, to continue on to p. 367. See also II, 269, 524.

and appreciated, what Æpinus and Cavendish had attempted in their fluid-dynamical theory of electricity. In the same volume of the 1801 Britannica *Supplement* in which he had so highly praised Boscovich, Robison declares his admiration for Æpinus and Cavendish and claims to have taught their electrical theory to his students. This article is also reprinted, in extended form, in the *System* of 1822. "It is to the Hon. Mr. Cavendish that we are indebted for the satisfactory, the complete (and we may call it *the popular*) explanation of all the phenomena. Forming to himself the same notion of the mechanical properties of the electrical fluid with Mr. Æpinus, he examined, with the patience, and much the address of a Newton, the action of such a fluid on the fluid around it, and the sensible effects on the bodies in which it resided. . . ."[10]

By the time of his death in 1805, Robison had transferred into early nineteenth-century scientific literature detailed information on the work of Cavendish and Boscovich, he had emphasized the Newtonian character of particulate-force considerations, giving some indication of how these might be applied, and he had decried aetherial hypotheses as materialistic. Through the agency of David Brewster, one of the most respected of mid-nineteenth-century British scientists, these views were continued, as potential authority, into the period of popular British dynamical metaphysics, after 1820. And, through the agency of another student, John Leslie, such views were applied to continuing scientific research and ultimately were maintained in the lecture rooms of Edinburgh.[11]

[10] Robison, *System*, IV, "Electricity," 1-204, esp. 110. Robison is less than kind to Priestley in this article, for Priestley, in 1801, represented to the Tory Robison all the worst aspects of Revolutionary fervor. Robison, who claims to have discovered the law of electrical forces, does not mention that Priestley had also done so, nor explain that Priestley's failure to appreciate Æpinus was based, in part, on the latter's incorrect inference of such a law. Robison's article is, however, the only extended treatment by a British scientist of the work of Æpinus and the only one of that of Cavendish until Maxwell published his edition of Cavendish's electrical papers in 1879.

[11] Sir John Leslie (1766-1832), educated at St. Andrews, 1780-85, and at Edinburgh, to which he migrated as a student of divinity. Under the influence of Black and Robison he changed his interests to science and graduated in arts in 1788. After private tutoring and lecturing, he became professor of mathematics at the University of Edinburgh in 1805 and, in 1819, succeeded John Playfair as professor of natural philosophy. Rumford medalist for his researches on heat, he wrote also on electricity, meteorology, instrumentation, and mathematics. In 1820 he became corresponding member of the Institut Royale de France. The most complete study of Leslie's work is that of Richard G. Olson, "Sir John Leslie: 1766-1832. A Study in the Pursuit of the Exact Sciences in the Scottish Enlightenment," unpub. Ph.D. diss., Harvard University, 1967.

Leslie's first paper, "Observations on Electrical Theories," can be quickly dismissed. Written in 1791, it was not published until 1824, well after the critical period for influencing early nineteenth-century science.[12] Moreover, the paper should not have been published even then, as it reveals all the worst aspects of Leslie's scientific work and few of the best. Primarily concerned with the action of electrified bodies on ambient air, it characteristically generalizes from this restricted base, without regard for phenomena left unexplained by a new theory. It shows no awareness of the instruction, presumably given him by Robison, on the electrical work of Æpinus and Cavendish, and the value of the whole is not enhanced by Leslie's ostentatious failure to change any of his opinions during the thirty-three years between the paper's preparation and its publication. Some suggestion of a fundamental mechanistic approach may be found in Leslie's denial of electrical fluid and acceptance of action-at-a-distance, but this is minimized by his insistence that electricity is communicated only by the intervention of some medium, usually air. The definition of electricity as a state or condition, of which every species of matter is susceptible [19-20, 38], the hints of accelerating and retarding forces between particles, and the denial of absolute contact of bodies, are indications of mechanism, but this is shown more clearly in better work, published earlier.

His next two papers, "On Heat and Climate," written in 1792 but unpublished until 1819, and "On Capillary Action," published in 1802, are more satisfying examples of neo-mechanism, although the belated publication of the former limited its influence on early nineteenth-century science.[13] Both papers are Boscovichean—the former avowedly so, with explicit reference to the "ingenious and profound theory of the late Abbé Boscovich ["Heat," 10]." Each maintains the essential sameness of all matter, the non-existence of the aether, and the use of attractive and repulsive forces, alternating with distance, between the particles, as the ultimate explanation of all phenomena. "Could we ascertain the gradations at near distances, we might determine structure, affinities, and mutual operations of bodies with the same certainty as we compute the revolutions of the planets ["Capil.," 194]." In the paper on heat, Leslie declares the elementary particles to be mathematical points—"the external world has a

[12] John Leslie, "Observations on Electrical Theories," *Edinburgh Philosophical Journal,* 11 (1824), 1-39.

[13] John Leslie, "On Heat and Climate," *Annals of Philosophy,* 14 (1819), 5-27; "On Capillary Action," *Philosophical Magazine,* 14 (1802), 193-205.

real existence, which yet consists of mere forces and *loci* [10]"—while in that on capillary action, he is "indifferent" as to whether they be regarded as points, atoms, particles, or molecules, as their dimensions, if any, do not enter into considerations respecting cause [193].

The application of these principles to capillarity is fairly clear. Indeed, as we have seen, capillary action had been a standard example for the dynamic corpuscularians until James Jurin's account of that phenomenon appeared; an account which, Leslie claims, will not bear the slightest examination. To Leslie, capillarity not only affords an intuitive example of the reality of attraction it also is an intermediate link between mechanical and chemical phenomena and provides an opportunity experimentally to establish instances of the approximation and transition of the motions caused by corpuscular forces [194]. It is on these grounds that Leslie proceeds to a determination of the perpendicular force of attraction between the surface of glass and the film of fluid which adheres to it. He also notes the differences in specific attractions of various kinds and concentrations of liquids and recommends a capillary hydrometer to test the strength of solutions; he derives a relation between fluid density, specific attraction, and the height of the capillary column, and, on the assumption that capillary absorption of a fluid in porous bodies approximates chemical combination, he even observes the rise in temperature to be expected of such a reaction.

Leslie's application of mechanical principles to the phenomena of heat is less straightforward, especially as he denies that heat consists in the intestine motions of the particles of bodies and insists that it (like electricity) cannot be communicated save through some medium [5, 6]. His theory of heat is put together, with typical neo-mechanistic eclecticism, of parts of Boscovich, sections of Hutton, and bits, perhaps, of Priestley and Michell. Heat is a substance, probably the same as that of light in a state of combination with bodies [6]. He will not decide whether light particles are primaeval points or simple combinations of them [10], but these particles react with bodies, are reflected, refracted, or absorbed by them. The fluctuations of reflection and transmission are explained (as Michell and Priestley explained them) by the alternations, with distance, of the attractions and repulsions between light particles and those of transparent matter. The absorption of light gives to bodies their contained heat. The heat of the earth, for example, is derived entirely from absorbed rays from the sun, as Mr. Hutton declares, and without the repulsion of these particles, at very small distances, the universe would be collected into a single point [10, 19]. All heating processes are to be explained by the differences between the rates at which the repulsive forces between particles of heat and the forces of attraction to other particles vary with distance. As attraction increases inversely with distance more slowly than repulsion, the mechanical compression of parts of a body will drive particles of contained heat into other parts. On recoil, the depleted sections will abstract heat particles, by attraction, from contiguous matter, especially the ambient air. The result is an increase in amount of heat substance in a body and an explanation of heating by percussion or friction [13]. The relative proportions of forces of attraction by various bodies for heat should be called their specific attraction for heat (capacity being an inadequate if not misleading term) and this attraction is a function of the amount the body can dilate. Hence air should have a wide variety of specific attractions for heat, making it a primary medium for heat transmission and explaining the changes in body temperature produced by contact with air [14, 17-18].

Leslie's paper on heat was published too late significantly to influence early nineteenth-century British science. The leading ideas in that paper, however, had already appeared, in an extended form, in Leslie's most famous work, that for which he received the Rumford Medal of the Royal Society, his *Experimental Inquiry into the Nature and Propagation of Heat,* published in 1804.[14] For his *Inquiry,* Leslie ". . . seldom ventured into the region of conjecture, but . . . patiently sought to determine the facts with accuracy, and laboured to deduce the consequences by a close train of argument . . . [xiii]." Now there is no doubt that the experiments Leslie performed were useful and impressive and the instruments he designed for those experiments were valuable acquisitions to the armory of experimental science. It is the more remarkable, therefore, that the theory which "resulted" from his investigations is essentially the same as that expressed in the paper on heat written twelve years earlier (though not published until fifteen years later).

The transmission of heat is "an actual flow or impulsion of some corporeal substance . . . quite distinct from the subtlety and extreme tenuity usually ascribed to aether and other imaginary fluids . . . [27]." "Such was the sway of metaphysical prejudice, that even Newton, forgetting his usual caution, suffered himself

[14] John Leslie, *An Experimental Inquiry, etc.* (London: J. Mawman, and Bell and Bradfute, Edinburgh, 1804).

to be borne along. In an evil hour he threw out those hasty conjectures concerning aether, which have since proved so alluring to superficial thinkers, and which have in a very sensible degree impeded the progress of genuine science [136]." In fact, the properties of bodies result merely from the different arrangements and configurations of their integrant parts. Attraction extends indefinitely through space and constitutes an original and absolute principle of nature. So also do the other dispositions of the particles of matter, those attractive and repulsive actions which alternate repeatedly with distance within very narrow limits [117-20]. This is the system of Boscovich, "the happiest and most luminous extension of the Newtonian system [515]," which Leslie discusses in detail, with a description of the "curve of primordial action" and its positions of stable and unstable equilibrium. But though Leslie accepts the Boscovichean concept of action-at-a-distance, and even avoids using the term "force," speaking instead of the particles exerting "a certain varying energy" upon one another, he discards the principle of continuity and that of pure geometricity. "Nature presents always individual objects, and proceeds by finite steps or differences." Boscovich's curve is a smoothed approximation to the serrated line, "whose gradations correspond to the breadth of the ultimate corpuscles, or the successive limits of action [515]."

The "shapeless hypothesis" that heat consists in certain intestine vibrations will not bear examination. Does the whole of the heated mass vibrate, or are there partial oscillations? On either supposition, "what precise idea can we annex to the degree of *heat*? Is it determined by the magnitude, the frequency, or the force of the pulses [140]?" Such a theory can explain neither the transmission of heat (which would require a sympathy of harmonic vibrations) nor the conservation of heat, for motion ceases and vibrations relax and die away [141]. Heat is an elastic fluid, extremely subtle and active, whose particles mutually repel while they attract those of other bodies; it is only light in a state of combination [149-50, 162]. Its specific attraction for other bodies "no doubt results from the collective energies of the primordial corpuscles, as modified by their peculiar internal arrangement," but as we are entirely ignorant of the nature and effects of elementary structure, it can be ascertained only by actual experiment [166]. A body evolves heat in being condensed, for its specific attraction is thereby diminished; any change in constitution or internal arrangement of a substance changes its attraction for heat [167-68]. The resistance which heat encounters in its passage through the interior of bodies "originates wholly from certain reiterated subsultory motions or expansions, impressed on the connected particles of the recipient [191-92]." The diffusion of heat through the atmosphere must be referred to the oscillations, or vibratory impressions, excited in the atmosphere [215].

Like other Boscovicheans, Leslie has here used that hypothesis primarily as a qualitative model suggestive of mechanisms, not as a frame whose parameters might be evaluated for a quantitative theory of phenomena. Nonetheless, his *Inquiry* is an experimental justification of mechanism, for he was led by mechanistic principles to open up the field of what became radiant energy investigations. The instruments he designed and the experiments he performed to determine the emissive, absorptive, and reflective powers of various substances might ultimately be separated from the theory which had suggested them, but no one, at the time, was left in doubt that there was such a theory or that Leslie felt it had been confirmed. And the passage of time did not reduce Leslie's faith in a mechanistic approach to phenomena. His *Elements of Natural Philosophy,* first published in 1823, does not mention Boscovich by name, but it is permeated with Boscovichean terminology. In a work limited to "somatology" (i.e., study of the essential properties of all bodies) and mechanics —including that of fluids—all problems are approached as assemblages of atoms, or physical points, connected together by a system of mutually attractive and repulsive "energies" modified only by the mutual distances of the particles and the positions of stable and unstable equilibrium along a primordial line of physical action.[15]

How far Leslie's course at Edinburgh, reflected in the *Elements,* may generally have influenced the new generation of British natural philosophers is unknowable, but clearly that, or his original researches, or both, had some specific influence on the course of nineteenth-century British science. One of Leslie's students and his successor in the chair of natural philosophy at the University of Edinburgh, James David Forbes, carried on his research in radiant energy and discovered the polarization of radiant heat. Another student, Thomas Graham, later professor of chemistry at University College, London, is said to have held the "bold theory" that all so-called elements might be only forms of one primordial substance. Graham won the Keith prize

[15] John Leslie, *Elements of Natural Philosophy* (Edinburgh: Oliver and Boyd, and Geo. B. Whittaker, London, 1829), 2nd edn., pp. 37, 220, 289, and passim.

of the Royal Society of Edinburgh for his discovery of the laws of diffusion of gases—the motions of those particles of the ambient elastic medium which had so fascinated Leslie. And those laws were employed by James Clerk Maxwell in his work on the kinetic theory of gases, which assumed forces of attraction and repulsion between particles. Maxwell, in addition, knew Leslie's work on capillary action and described it as "original and powerful . . . even now very little out of date."[16]

Lest it be assumed that the only transmitters of eighteenth-century mechanism to the nineteenth century were Scots and Boscovichean, we should consider some examples of the varied work of Thomas Young, who was neither.[17] Young is most widely known for his three papers, read before the Royal Society in 1801, 1802, and 1803, which introduced the concept of optical interference and revived the wave theory of light. Young's theory is essentially neo-mechanist, in that it requires a particular substance, the "luminiferous aether," possessing mechanical properties. He did not at first, however, specifically describe those properties, and the vituperous criticism his theory received drove him into near silence on the subject until after the work of Fresnel was published in 1816. Not until 1817, when he suggested transverse vibrations, in the article "Chromatics" he was writing for a *Supplement* to the fourth edition of the *Encyclopaedia Britannica,* did Young seriously consider the possible constitution of elastic media capable of transmitting the vibrations of light. From there the tendencies of aetherial explanations ran counter to the corpuscular, action-at-a-distance, attractive and repulsive force, principles of eighteenth-century mechanism; only much later in the century, as the mechanical properties of the aetherial con-

[16] See, for example, J. Clerk Maxwell, "On the Dynamical Evidence of the Molecular Constitution of Bodies," *Journal of the Chemical Society,* 13 (series 2, 1875), 493-508; and W. B. Hardy, "Historical Notes upon Surface Energy and Forces of Short Range," *Nature,* 109 (1922), 375-78.

[17] Thomas Young (1773-1829), precocious child of a wealthy Quaker family, was educated privately until he commenced the study of medicine in London, 1792, under Baillie, Cruikshank, and Hunter. He entered the University of Edinburgh in 1794, in 1795 went to Göttingen, M.D. 1796; entered Emmanuel College, Cambridge, 1797, M.B. 1803, M.D. 1808. F.R.S. 1794, Foreign Secretary of the Royal Society, 1802; one of eight foreign members of the Académie Royale des Sciences, 1822. Professor of natural philosophy, Royal Institution, 1801-1803. Most famous for his work in optics, Young was a polymath, with interests in languages (including the ancient Egyptian), medicine, and most subjects of physics. The standard biography, long out of date, is that of George Peacock, *Life of Thomas Young, M.D. F.R.S. etc.* (London: John Murray, 1855), which has been modernized but not substantially improved in Alex. Wood's *Thomas Young, Natural Philosopher 1773-1829* (Cambridge: Cambridge University Press, 1954), completed by Frank Oldham.

tinuum were lost in mathematical equations, was the luminiferous aether truly assimilated into nineteenth-century dynamical theory.[18]

It is, therefore, in his other work, most completely represented by the long-unappreciated *Course of Lectures on Natural Philosophy* (1807), that Young's general acceptance of mechanistic principles can best be seen.[19] Based on the lectures he delivered as professor of natural philosophy at the Royal Institution in 1801 and 1802, the mechanistic ideas expressed in the *Course* must obviously be derived from a study of eighteenth-century science, but their immediate origin is uncertain. Young's knowledge of scientific literature was so inclusive as to provide no precise information on particular influences. One of the most valuable features of the *Course of Lectures* is a classified, critical bibliography of works in natural philosophy, extending to four hundred quarto pages and including more than twenty thousand references. Here are to be found articles and books by Scholastics and empiricists, kinematic and dynamic corpuscularians, and materialists. Why, from such a plethora, did Young select the mechanist emphasis for his lectures? It cannot have been the influence of Boscovich's hypothesis, for Young says of that, ". . . the grand scheme of the universe must surely . . . preserve a more dignified simplicity of plan and principles, than is compatible with these complicated suppositions [I, 615]." Possibly he acquired a predilection for the mechanical approach during his studies in Germany, where he read some of the works of Kant, Goethe, Schiller, and Herder—all of them significant in the development of the dynamical mystique of *naturphilosophie*. But Young was not favorably impressed by Kant's *Critique of Pure Reason*, and his reading of the other writers appears to have been limited to their literary works.[20] Moreover, the nature of Young's mechanism has a simplicity not attained by German metaphysics. The evidence available at the moment is insufficient to support any confident assertion on the origins of Young's particular scientific preconceptions.

In any event, whatever may have been his reasons, Young chose to present, in his *Course of Lectures*, the essential features of a mechanistic theory of matter. All matter was probably originally of one kind and owes its different appearances only to

[18] For Young's work on optics see Peacock, *Life of Young*, pp. 138-87, 369-402.

[19] Thomas Young, *A Course of Lectures on Natural Philosophy and the Mechanical Arts* (London: J. Johnson, 1807), 2 vols.

[20] See Peacock, *Life of Young*, pp. 94-95, 113.

the form and arrangement of its parts. The properties of bodies are by no means as simply reduced to general laws as the more mathematical doctrines of space and motion, but the essential properties may be enumerated as: extension and divisibility, repulsion or impenetrability, inertia, and gravitation; while most of the accidental properties are consequent upon cohesion. Particles may be described as spheres of repulsion extended without penetrability, which, like particle contact, is impossible due to repulsive forces. Attempted demonstrations of infinite divisibility are applicable to space only. It is not impossible that a division of atoms or single corpuscles might be reached, beyond which their constitution, on which their existence as matter depends, would be destroyed, but it is unlikely that matter could thus be annihilated in the common course of nature [1, 606-08].

In his application of these principles, Young is less consistent. Although it is probable that the minute actions of the particles of matter are regulated on the same mechanical principles as the motions of the heavenly bodies, nothing "in the unprejudiced study of physical philosophy . . . can induce us to doubt the existence of immaterial substances," which are the causes of attractions of various kinds [1, 605, 610]. Among such immaterial substances, Young lists electrical fluid, magnetism, universal aether, and "higher still perhaps are the causes of gravitation, and the immediate agents in attractions of all kinds which exhibit some phaenomena apparently still more remote from all that is compatible with material bodies . . . [1, 611]." He denies, however, the existence of a substance of heat, a caloric "liable to no material variations except those of quantity and distribution," pervading the substance of all bodies according to their capacities for heat and their actual temperatures, and commonly supposed to be the general principle of repulsion. The production of heat by friction affords an unanswerable confutation to this doctrine, and, if heat be not a substance, it must be a quality, and this quality can only be motion. Newton believed that heat consists in the minute vibratory motion of the particles of bodies, communicated through an apparent vacuum by the undulations of an elastic medium, and recent arguments in favor of the undulatory nature of light support this doctrine, it being only necessary to suppose the vibrations and undulations constituting heat to be larger and stronger than those of light [1, 653-54]. Strangely enough, though Young was the first person consistently to employ the term "energy" to mass times the square of velocity

(Cavendish's mechanical momentum) and show that it was proportional to work, defined as force exerted through a distance [I, 78-79], he failed to apply these concepts to a study of heat.

In fact, Young's most successful uses of mechanistic principles tended to be sophisticated versions of just those most successfully attempted by early eighteenth-century dynamic corpuscularians —in elasticity, capillarity, and body-fluid physiology. This, and his tendency to use geometrical rather than analytical mathematical techniques, suggests the limitations, but also perhaps the influences, of his youthful self-education in the sciences. In his lecture on passive strength and friction, Young develops the concept of modulus of elasticity, with specific reference to the cohesive and repulsive forces between particles of matter [I, 135]. These elasticity investigations were extended into a study of the strength of materials and the design of structures for articles for the *Encyclopaedia Britannica* and reports for the Admiralty, and his work is memorialized in the term "Young's Modulus." His work on capillary attraction is made ambiguous by a characteristic obscurity in phrasing and mathematical derivation, but it explicitly uses the concept of attractive forces between particles of fluid and container. A later article, on cohesion and its physical foundation, written for the *Encyclopaedia Britannica* in 1816, included speculations on the corpuscular forces in the interior of matter which Lord Rayleigh was to admire more than half a century later.[21] Even in his physiology, Young betrays some of this early inspiration. A memoir on hydraulics and another on the functions of the heart and arteries, published in the *Philosophical Transactions* for 1808 and 1809, are connected by arguments which would have seemed familiar to the iatro-hydrodynamicists of a century earlier, though both experimental techniques and methods of analysis permitted a modernization of work which had otherwise ceased with Stephen Hales.[22]

No doubt Young's *Course of Lectures* was a tour de force, as were the articles which he subsequently wrote for the *Encyclopaedia Britannica.* It was said, years later, that no researcher could safely neglect to consult Young's writings, which contained the seeds of many significant discoveries. Nonetheless, the *Lectures* were not successful, and their significance became apparent only later when their basic concepts had otherwise again become so familiar as to overcome the obscurity of their presentation. It is not in Young that the most significant English transitional in-

[21] W. B. Hardy, "Surface Energy and Forces of Short Range," p. 378.

[22] See Peacock, *Life of Young*, pp. 417-21.

fluence can be found, but in Young's successful rival as a popular lecturer at the Royal Institution, Humphry Davy.[23]

Davy is a crucial figure in the case for the continuation of eighteenth-century mechanism into nineteenth-century dynamical thought. Not only was he personally involved in the early stages of the dynamic theories of heat and chemistry (and through his protégé, Michael Faraday, in that of electricity, as well), he was the beau ideal of Romanticism and, by his lectures, his publications, researches, and, ultimately, his administrative decisions, he exercised an influence second to none on the sciences of early nineteenth-century Britain.[24] If a strain of native, eighteenth-century scientific ideas can be found in Davy's early writing and in the speculations he brought to his laboratory for confirmation, then the argument for relating eighteenth-century mechanism to nineteenth-century dynamics is immeasurably advanced regardless of the influences, in whatever accent, license, or degree, detected in his later work.[25]

Davy's earliest original scientific work was described in an "Essay on Heat, Light, and the Combinations of Light, with a New theory of Respiration," written during 1798.[26] Examination of the experiments described in this "Essay" indicates that they do not prove what Davy thought they proved—even for those relating heat and motion. But this, if anything, enhances the value of those experiments and Davy's conclusions from them as

[23] Sir Humphry Davy (1778-1829), educated in grammar schools of Penzance and Truro; he was apprenticed in 1794 to a Penzance surgeon-apothecary. He met Davies Giddy, Gregory Watt, and Thomas Wedgwood, who recommended him to Thomas Beddoes, as superintendent of the laboratory in Beddoes' Pneumatic Medical Institute at Bristol, 1798. In 1801 he was named lecturer in chemistry at the Royal Institution of Great Britain, where he remained as professor of chemistry and director of research until 1813. F.R.S. 1803, Copley medalist 1805, president of the Royal Society, 1820. Knighted 1812, made Baronet 1818. His most important research was in electro-chemistry, where he isolated potassium and sodium; he also proved chlorine was an element and invented a mine safety lamp. Of the many biographies, the most recent is that of Sir Harold Hartley, *Humphry Davy* (London: Thomas Nelson and Sons, 1966).

[24] See, for example, David M. Knight, "The Scientist as Sage," *Studies in Romanticism*, 6 (1967), 65-88, for Davy as a Romantic hero.

[25] This is clearly a reference to the suggestion of L. Pearce Williams, *e.g., Michael Faraday, A Biography* (London: Chapman and Hall, 1965), pp. 62-68, that the poet Samuel Taylor Coleridge introduced Davy to ideas of Kant and of German *naturphilosophie* which were significantly to influence his scientific thought. I wish neither to confirm nor deny that suggestion, but if Davy demonstrably held critical dynamical concepts earlier than Coleridge's return from Germany (1799), it seems clear that any influence of German ideas was supplementary to that of an indigenous neo-mechanism.

[26] The "Essay" was reprinted in John Davy, ed., *Collected Works of Sir Humphry Davy, Bart.* (London: Smith, Elder, and Co., 1839), II, "Early Miscellaneous Papers," 5-86.

an indication of his preconceptions. The particles of bodies, Davy declares, are acted upon by two opposing forces—an approximating power, or attraction, and a repulsive power which the "greater part of chemical philosophers" ascribe to a peculiar elastic fluid called caloric [9, 15]. This repulsive power can, however, be shown by experiment to be not a substance but a peculiar motion, probably a vibration of the corpuscles of bodies. It may, therefore, with propriety, be called repulsive motion [10]. Attraction is compounded of cohesion, gravitation, and superincumbent pressure. The different states in which bodies may exist depend upon the differences in action of attraction and repulsion; solidity being a predominance of attraction, fluidity an equilibrium of the two, and the gaseous state a predominance of repulsion [15-16]. There is another state, that of repulsive projection, where repulsive motion so predominates over attraction that the corpuscles indefinitely separate with great velocity. Light is the only substance which can exist in this fourth state, while light in a condensed state is probably electric fluid [16, 28]. Bodies may have their repulsive motion increased by a transmutation of mechanical motion (as in friction or percussion), by the motions of chemical action, or by the communicated repulsive motions of bodies in apparent contact [17]. "No more sublime idea can be formed of the motions of matter, than to conceive that the different species are continually changing into each other. The gravitative, the mechanical, and the repulsive motions, appear to be continually mutually producing each other, and from these changes all the phaenomena of the mutation of matter probably arise [28n]."

The ingenuity and sophistication of these speculations would be remarkable as the original work of a young, provincial apothecary's apprentice, but Davy reveals their origin in a casual reference to "Macquer's and Hutton's theories [39]." Nearly every part of the "Essay" relating to the nature of matter, of heat, or to the production of heat and light in chemical combination is reminiscent of something in James Hutton's *Dissertations on different Subjects in Natural Philosophy* (1792). Now Davy wrote of his experiments to Thomas Beddoes as early as April 1798, and he wrote the "Essay" under Beddoes' guidance, shortly after October of the same year, to be published in the *Contributions to Physical and Medical Knowledge, principally from the West of England,* edited by Beddoes in 1799.[27] It is appropriate,

[27] Hartley, *Humphry Davy*, pp. 19-20.

therefore, to note that the review of Hutton's *Dissertations*, published in the *Monthly Review* for 1795, was written by Beddoes, who praised the work except for its tedious and obscure style. He noted Hutton's discussion of the release of heat and light in chemical combination (from which Davy may have derived his concept of phosoxygen); he quoted, in detail, from Hutton's distinction between body and active matter, in its two forms of attracting and repelling powers; he particularly noted Hutton's concept of solar substance in its manifestations of light, heat, and electricity; and, finally, he observed, of Hutton's theory of matter, that "it is much more a kin to that of Boscovich than to the common hypothesis of hard, indestructible particles, corpuscles, or atoms, separated by pores."[28]

Davy was shortly to repudiate this early paper as containing "infant speculations," "dreams of misemployed genius which the light of experiment and observation had never conducted to truth." But Davy did not repudiate Huttonian ideas. Years later, in the sixth dialogue of his final work, *Consolations in Travel, or the Last Days of a Philosopher* (1830), he described a world in which apparent rest was an equilibrium of forces, where gravity tended to level everything and, when added to the attractive force of chemical change, would produce an eternal sleep in nature if unopposed. In heat there was an antagonist and vivifying power, but this too, through the changes in temperature it caused, was destructive of material forms.[29] For all its romantic melancholy, this is a description of Hutton's world.

Nor was this a return to "infant" mechanistic ideas, brought on by a premature senility; Davy's writings throughout his active career are filled with such speculations. In the *Syllabus of a Course of Lectures on Chemistry, Delivered at the Royal Institution of Great Britain* (in 1802), he declares that, "chemical phaenomena result from the different arrangements of the particles of bodies, and the powers that produce these arrangements are repulsion, or the agency of heat, and attraction [330]." Solids become fluids, and fluids, gases, on a certain increase in the repulsive powers of their particles; decreasing the repulsion inverts the process. The powers of attraction and repulsion are parts of an arbitrary system, but the laws by which they act are a simple expression of facts which guide the chemist in imitating the opera-

[28] [Thomas Beddoes], Review: "Dissertations on different Subjects in Natural Philosophy. By James Hutton," *Monthly Review; or Literary Journal Enlarged*, 16 (1795), 246-54, esp. 252. I may note, from the *British Museum General Catalogue of Printed Books*, that S. T. Coleridge, whom Davy was to meet in 1799, possessed a copy of Hutton's *Investigation of the Principles of Knowledge* (1795).

[29] Described by David M. Knight, "The Scientist as Sage," p. 81.

tions of nature. "To effect chemical compositions and decompositions, nothing more is required than to bring the particles of the bodies, which are the subjects of experiment, into the sphere of their mutual attractions [332]."[30]

In his *Elements of Chemical Philosophy*, Davy is more detailed concerning his concepts of matter and its action. "Whether matter consists of indivisible corpuscles, or physical points endowed with attraction and repulsion, still the same conclusions may be formed concerning the powers by which they act, and the quantities in which they combine; and the powers seem capable of being measured by their electrical relations, and the quantities on which they act of being expressed by numbers [30]." The forms of matter and their changes depend on the active powers of gravitation, cohesion, calorific repulsion or heat, chemical attraction, and electrical attraction [37]. Davy still does not believe in a substance of heat, preferring the supposition that the particles of a heated solid "are in a constant state of vibratory motion, the particles of the hottest bodies moving with the greatest velocity. . . . Temperature may be conceived to depend upon the velocities of the vibrations . . . [53]." On the other hand, though he does not believe in a universal aether, an unknown matter explaining attraction in general, he does accept the notion of imponderable substances whose motions produce such different effects as radiant heat and light [37, 38], and he finds support for this opinion in the work of John Leslie [42, 49]. It is not clear whether electrical phenomena depend upon one fluid or two different fluids, or whether they may be the exertions of the general attractive powers of matter, but electrical and chemical effects differ only by whether the same primary cause acts upon bodies as masses or upon their individual particles [91-92]. "Matter may ultimately be found to be the same in essence, differing only in the arrangement of its particles; or two or three simple substances may produce all the varieties of compound bodies [101]" —a sublime chemical speculation, sanctioned by the authority of Hooke, Newton, and Boscovich, which must not be confounded with the ideas of the alchemists [279].[31]

This reference to Boscovich is followed by another, in a manuscript journal kept by Davy on his travels in France in 1813 and copied by his assistant, Michael Faraday. "By assuming certain molecules endowed with poles or points of attraction and repulsion as Boscovich has done and giving them gravitation and form,

30 Humphry Davy, *Collected Works*, II, 329-436.

31 Humphry Davy, *Elements of Chemical Philosophy* (Philadelphia: Bradford and Inskeep, 1812).

i.e. weight and measure all the phenomena of chemistry may be accounted for."[32] And a final Boscovichean reference appeared in an unfinished dialogue on atomic theory, found among Davy's papers after his death and presumably intended as a part of his *Consolations in Travel.* "You mistake me, if you suppose," writes Davy in the person of an all-wise "Unknown," "that I suppose the elements to be physical molecules, endowed with the properties of bodies we believe to be indecomposable. On the contrary . . . I consider them with Boscovich, merely as points possessing weight and attractive and repulsive powers. . . ."[33] Explicit association of his views with those of Boscovich came, therefore, comparatively late in Davy's active scientific career, his research having essentially ended shortly after the publication of the *Elements of Chemical Philosophy.* Nor is it necessary to read that particular form of neo-mechanism back into Davy's writings much before he made it explicit. Similar views were available to him, as his mentor Beddoes pointed out, in the writings of Hutton, with which Davy was familiar; they were also to be found in the works of Joseph Priestley—works thought, by Davy, the most likely "to lead a student into the path of discovery."[34] Whether such speculations played a vital role in Davy's most important research is a question to be fought over by historians of nineteenth-century science. Clearly his earliest research is grounded in them, but, equally clearly, Davy was capable of following the lead of experiments without obsessive concern for initial hypotheses. Yet only a Baconian ideologue will suggest that scientists can leave speculations at the doors of their laboratories to pursue experiments at random and interpretations without design. Some indication of the influence of mechanistic ideas on Davy's scientific work and thought may be seen in his failure to adopt the primary materialist parameter of weight in his experiments with gases. Like Priestley, Davy tended to concern himself with volumetric measurements and the determination of densities, to the minimizing of the gravimetric methods being used so effectively by Berzelius and Prout. A clearer, but related, example of Davy's ultimate mechanism is to be seen in his persistent opposition to the atomic chemistry of Dalton, on the grounds that it presumed the indivisibility of the atoms of chemical elements.[35] And

[32] Quoted by L. Pearce Williams, *Michael Faraday*, p. 78.

[33] Davy, *Collected Works*, IX, 388.

[34] Davy, *Collected Works*, VII, 117.

[35] See David M. Knight, "The Atomic Theory and the Elements," *Studies in Romanticism*, 5 (1966), 185-207, and "Steps towards a Dynamical Chemistry," *Ambix*, 14 (1967), 179-97, for a discussion of Davy's role in opposition to the atomic theory of Dalton.

this is also a more important example, for it concerns the influence of Davy's expressed opinion during the formative years of nineteenth-century science. For the young and brilliant lecturer, discoverer, inventor, and science administrator repeatedly to maintain mechanistic ideas in his writings cannot but have made an impact on the thinking of the new generation of British scientists (whatever that influence may have been on Faraday) whether or not those ideas were employed in his research.

Here, then, are four British natural philosophers, representative of those who were educated in the eighteenth century and who, for one reason or for several, were significant figures in the science of the early nineteenth century. As they combined the elements of eighteenth-century mechanism and materialism differently and in different proportions, the direction taken by the work of each of them was different and frequently contradictory. The common possession of a mechanistic view of nature clearly did not insure the common development of scientific theory any more than the common avowal of rigorous experimentalism insured the avoidance of theory in the first place. To argue that eighteenth-century neo-mechanism could uniquely define the future path of nineteenth-century dynamical theory would be patently absurd, when only Young accepted an aetherial wave-theory of light, when Robison and Leslie held a substance theory of heat, and Robison and Young a substance theory of electricity. But the modifications of eighteenth-century concepts, mechanist or materialist, into the dynamical theories of the nineteenth century is a subject for the history of nineteenth-century science. The purpose of this chapter is not to invade that prerogative, but merely to demonstrate that there existed native traditions which permitted British scientists of the nineteenth century to choose a path without necessary external considerations and to show that these traditions were a consequence of the reason as well as the experiments of the previous century. Between the corpuscular dynamics of the early eighteenth century and the mathematical dynamics of the late nineteenth, there is a demonstrable continuity, modified by time, necessity, and foreign influences, but marking British science throughout as essentially and rationally Newtonian. So it is that the history of science, like other histories, does not admit finality, and, as scientific ideas, identified with eighteenth-century Britain, diffused into the nineteenth, this history of those ideas does not so much conclude as fade into a history of nineteenth-century British science.

BIBLIOGRAPHY

Primary Sources

MANUSCRIPT COLLECTIONS

British Museum, London, Department of Manuscripts: Wilson Papers, Add. MSS 3094. Letter from Bryan Robinson to Benjamin Wilson, 25 January 1746.

——— ——— ———: Sloane MSS 4051, vol. xvi, fol. 200-201. Letter of J. T. Desaguliers to Hans Sloane, 4 March 1730/1.

Cambridge University Library. Vigani (Veronens), Joan Francis, MS Dd 12.53 (A), fol. 1-57. "Cours de Chymi."

Cambridge University, Gonville and Caius College Library. MS 619/342. "John Mickleburgh Chemistry Lectures."

Historical Manuscripts Commission Reports. *Portland Manuscripts* (Norwich, HMC Reports, 1899), vol. 5.

Linda Hall Library, Kansas City, Missouri. William Herschel Commonplace Book.

University of Texas, Library, Austin, Texas. Herschel Papers, "Remarks on Dr. Priestley's Disquisitions on Matter and Spirit," c. 1779.

Williams, Wesley C. Private collection, Cleveland, Ohio. William Herschel Papers, read before the Bath Philosophical Society (papers printed in the introduction to Herschel, *Scientific Papers*, vol. 1).

BOOKS AND PAPERS

Alleyne, James. *A New English Dispensatory, in four Parts,* &c. (London: T. Astley & S. Austen, 1733). [Issued in parts, with some parts available in 1732. Discussed in letters published in *Gentleman's Magazine,* 2 (1732), 1099-1100, 1117-18. Thirty pages of Part II's 253 pages are supposedly extracts from Freind's chemistry.]

[Anon.] "Answer to DeDuobus," *Gentleman's Magazine,* 2 (1732), 1117-18.

[Anon.] Review of John Freke's Essay on the cause of electricity, *Gentleman's Magazine,* 16 (1746), 521-22.

[Anon.] Review of Herschel's "Observations on the two lately discovered Celestial Bodies," *Edinburgh Review,* 1 (1802-1803), 430.

[Anon.] Review of Playfair's Illustration, *Edinburgh Review,* 1 (1802), 202-203.

Bacon, Francis. *The Philosophical Works of Francis Bacon . . . methodized and made English, from the Originals* (London: D. Midwinter, A. Bettesworth and C. Hitch, J. and J. Pemberton, C. Ware, C. Rivington, J. & P. Knapton, J. Batby and J. Wood, T. Longman, F. Clay, A. Ward, and R. Hett, 1737), ed. Peter Shaw, 2nd edn.

[Beddoes, Thomas]. Review: "Dissertations on different Subjects in

Natural Philosophy. By James Hutton," *Monthly Review; or Literary Journal Enlarged*, 16 (1795), 246-54.

Bentley, Richard. *Eight Sermons Preach'd at the Honourable Robert Boyle's Lecture, In the First Year, MDCXCII* (Cambridge: for Cornelius Crownfield, James Knapton and Robert Knopstock, 1724), 5th edn.

Black, Joseph. "Joseph Black's Inaugural Dissertation, I; II," translated by Crum Brown, *Journal of Chemical Education*, 12 (1935), 225-28, 268-73.

———. "Experiments upon Magnesia Alba, Quicklime and some other Alcaline Substances," *Essays and Observations, Physical and Literary, Read before a Society in Edinburgh*, 2 (1756), 157-225.

[Black, Joseph]. *An Enquiry into the General Effects of Heat; with Observations on the Theories of Mixture* (London: J. Nourse, 1770).

———. *Lectures on the Elements of Chemistry, delivered in the University of Edinburgh*, now published from his manuscripts by John Robison (Edinburgh: by Mundell and Son, for Longman and Rees, London, and William Creech, Edinburgh, 1803), 2 vols.

Blagden, Charles. "Experiments and Observations in an heated Room," *Philosophical Transactions*, 65 (1775), 111-23.

Boerhaave, Herman. *Elements of Chemistry: being the Annual Lectures*, translated by Timothy Dallowe (London: for J. and J. Pemberton, J. Clarke, A. Millar; and J. Gray, 1735).

———. *A New Method of Chemistry; including the History, Theory, and Practice of the Art: to which are added, Notes and an Appendix, showing the Necessity and Utility of Enlarging the Bounds of Chemistry*. Translated and notes by Peter Shaw (London: T. and T. Longman, 1753), 3rd edn. corr.

Bond, John. "A Letter . . . containing Experiments on the Copper Springs . . . ," *Philosophical Transactions*, 48 (1753-54), 181-90.

Boscovich, Roger Joseph. *A Theory of Natural Philosophy* (Cambridge: The M.I.T. Press, English trans. from the Venetian edn. of 1763, 1966).

Boyle, Robert. *The Works of the Honourable Robert Boyle*, ed. Thomas Birch (London: J. and F. Rivington, L. Davis, W. Johnston, S. Crowder, T. Payne, G. Kearsley, J. Robson, B. White, T. Becket and P. A. DeHondt, T. Davies, T. Cadell, Robinson and Roberts, Richardson and Richardson, J. Knox, W. Woodfall, J. Johnson and T. Evans, 1772), new edition.

Byrom, John. *The Private Journals and Literary Remains of John Byrom*, ed. Richard Parkinson in *Remains Historical and Literary connected with the Palatine Counties of Lancaster and Chester*, vols. 32-33 ([Manchester]: for the Chetham Society, 1854-55).

———. *The Poems of John Byrom, with an Appendix of Unpublished Letters by and to Byrom, Remains, Historical and Literary, connected with the Palatine Counties of Lancaster and Chester*

(Manchester, for the Chetham Society, 1912), Vol. 70 (new series), ed. Adolphus William Ward.

Canton, John. "Electrical Experiments, with an Attempt to account for the several Phaenomena . . . ," *Philosophical Transactions,* 48 (1753-54), pp. 350-58.

Cavallo, Tiberius. *Elements of Natural or Experimental Philosophy* (London: T. Cadell and W. Davies, 1803).

Cavendish, Henry. *The Scientific Papers of the Honourable Henry Cavendish, F.R.S.*; Vol. 1, *Electrical Researches,* ed. James Clerk Maxwell, revised, Sir Joseph Larmor; Vol. 2, *Chemical and Dynamical,* ed. Sir Edward Thorpe (Cambridge: at the University Press, 1921).

Cheyne, George. *An Essay concerning the Improvements of the Theory of Medicine* ([title page missing, 1st edn. Edinburgh, 2nd edn. London], 1702).

———. *Philosophical Principles of Religion: Natural and Reveal'd.* In Two Parts. Part I. Containing the Elements of Natural Philosophy, and the Proofs of Natural Religion arising from them. The Second Edition corrected and Enlarged. Part II. Containing the Nature and Kinds of Infinites, their Arithmetick and Uses: together with the Philosophick Principles of Reveal'd Religion. Now first Published. (London: for George Strahan, 1715). Separate title page for Part II reads "1716."

———. *An Essay of Health and Long Life* (London: for George Strahan and J. Leake, 1724).

———. *An Essay on Regimen. Together with Five Discourses, Medical, Moral, and Philosophical: Serving to illustrate the Principles and Theory of Philosophical Medicin, And point out Some of its Moral Consequences* (London: C. Rivington, and J. Leake, 1740), 2nd edn.

Clarke, John. *An Essay upon Study. Wherein Directions are given for . . . the Collection Of a Library, proper for the Purpose, consisting of the Choicest Books in all the several Parts of Learning* (London: Arthur Battesworth, 1731).

Clarke, Samuel. "Letter occasion'd by the present Controversy among Mathematicians, concerning the Proportion of Velocity and Force in Bodies in Motion," *Philosophical Transactions,* 35 (1728), 381-88.

Cleghorn, William. *de Igne. Disputatio inauguralis theoriam ignis complectens* (Edinburgh, 1779), translated by Douglas McKie and Niels H. de V. Heathcote, with notes, as "William Cleghorn's 'De Igne' (1779)," *Annals of Science,* 14 (1958, pub. 1960), 1-82.

Cotes, Roger. *Hydrostatical and Pneumatical Lectures,* Published with Notes, by Robert Smith, Prof. of Astronomy and Experimental Philosophy at Cambridge (London: for the editor, sold by S. Austen, 1738).

Cullen, William. "Of the Cold produced by evaporating Fluids, and some other means of producing Cold," *Essays and Observations,*

Physical and Literary, Read before a Society in Edinburgh, 2 (1756), 145-56.

———. *First Lines of the Practice of Physic* (Philadelphia: Parry Hall, 1792).

Davy, Humphry. *The Collected Works of Sir Humphry Davy,* ed. John Davy (London: Smith, Elder & Co., 1839). Particularly vol. 2. "Early Miscellaneous Papers," including "An Essay on Heat, Light and the Combinations of Light," first published in Thomas Beddoes, *Contributions to Physical and Medical Knowledge, principally from the West of England* (Bristol, 1798); *A Discourse introductory to a Course of Chemistry* (London, 1802); and *A Syllabus of a Course of Chemistry* (London, 1802).

———. *Elements of Chemical Philosophy* (Philadelphia: Bradford and Inskeep, 1812).

"De Duobus, etc., Tricks of Booksellers," *Gentleman's Magazine,* 2 (1732), 1099-1100.

de Luc, J. A. "First Letter to Dr. James Hutton . . . on his Theory of the Earth," *Monthly Review,* 2 (new series, 1790), 210.

Derham, W[illiam]. *Physico-Theology: or, A Demonstration of the Being and Attributes of God, from his Works of Creation. Being the Substance of Sixteen Sermons Preached in St. Mary-le-Bow Church, London, At the Honourable Mr. Boyle's Lectures, in the Years 1711, and 1712. With large Notes, and many curious Observations* (London: W. Innys and R. Manby, 1737), 9th edn.

———. *Astro-Theology Or, A Demonstration of the Being and Attributes of God, from a Survey of the Heavens* (London: William and John Innys, 1726), 5th edn.

Desaguliers, J. T. Papers in the *Philosophical Transactions,* 31 (1720-21) No. 365, 81-82; 32 (1722-23), No. 375, 269-79; 33 (1724-25), No. 389, 345-47; 35 (1727-28), No. 406, 596-629; 36 (1729-30), No. 407, 6-22; 41 (1739-40), No. 454, 175-85.

———. Review of Stephen Hales' *Vegetable Staticks, Philosophical Transactions,* 34 (1726-27), No. 392, 264-91; 35 (1727-28), No. 399, 323-31.

———. *A Course of Experimental Philosophy* (London: A. Millar [etc.], 1763), 3rd edn., 2 vols.

Doddridge, Philip. *A Course of Lectures on the Principal Subjects in Pneumatology, Ethics, and Divinity* (London: J. Buckland, J. Rivington, R. Baldwin, L. Hawes, W. Clarke and R. Collins, W. Johnston, J. Richardson, S. Crowder and Co., T. Longman, B. Law, T. Field, and H. Payne and W. Cropley, 1763).

Ellicott, John. "Several Essays towards discovering the laws of Electricity," *Philosophical Transactions,* 45 (1748), 195-224.

Euler, Leonhard. *Leonhardi Euleri Opera Omnia,* Ser. III, Vol. 1. "Commentationes Physicae ad Physicam Generalem et ad Theoriam soni pertinentes" (Lipsiae et Berolini, B. G. Teubneri, 1926).

[Fitton, W. T.]. Review of "Elements of Geology" by Charles Lyell, *Monthly Review*, 69 (1839), 443.

Forbes, Duncan. "Letter to a Bishop concerning some important Discoveries in Philosophy and Theology," in *The Works of the Right Honourable Duncan Forbes, Late Lord President of the Court of Session in Scotland* (London: by J. Morton, for T. Hamilton, R. Ogle, and J. Ogle, Edinburgh; and M. Ogle, Glasgow, 1809). "Letter" first published as pamphlet, London, H. Woodfall, 1732.

Franklin, Benjamin. *Experiments and Observations on Electricity* (London: Francis Newbery, 1769), 4th edn.

———. *Benjamin Franklin's Experiments* (Cambridge: Harvard University Press, 1941), ed. I. Bernard Cohen, from the 5th edn. of 1774.

———. *Works* (Philadelphia: William Duane, 1808).

Freind, John. *Chymical Lectures: In which almost all the Operations of Chymistry are Reduced to their True Principles, and the Laws of Nature. Read in the Museum at Oxford, 1704.* Englished by J. M. To which is added, An Appendix, containing the Account given of this Book in the Lipsick Acts, together with the Author's Remarks thereon (London: for Jonah Bower, 1712).

———. *Emmenologia,* translated by Thomas Dale (London: for T. Cox, 1752).

Freke, John. *A Treatise on the Nature and Property of Fire. In three Essays. I. Shewing the Cause of Vitality, and Muscular Motion, with many other Phaenomena. II. On Electricity. III. Shewing the Mechanical Cause of Magnetism; and why the Compass varies in the Manner it does* (London: W. Innys, and J. Richardson, 1752). Includes, as Part II, 3rd edn. of Freke's "Essay to shew the Cause of Electricity."

Geoffroy, E. "Part of a Letter . . . concerning the exact quantity of acid Salts, contained in acid Spirits," *Philosophical Transactions,* 22 (1700-01), 530-34.

———. "Observations upon the Dissolutions and Fermentations which we may call Cold," *Philosophical Transactions,* 22 (1700-01), 951-62.

———. "Des Différents Rapports observés en chimie entre différentes substances," *Mémoires de l'Académie Royale des Sciences* (1718), 202-12.

'sGravesande, William-James. *Mathematical Elements of Natural Philosophy Confirm'd by Experiments; or, an Introduction to Sir Isaac Newton's Philosophy,* translated by J. T. Desaguliers (London: J. Senex, W. and J. Innys; and J. Osborn and T. Longman, 1726), 3rd edn., 2 vols.

Gray, Stephen. Papers in the *Philosophical Transactions*: 37 (1731-32), No. 417, 18-44; No. 422, 227-30; No. 423, 285-91; No. 426, 397-407.

Green[e], Robert. *The Principles of Natural Philosophy, In which is shewn the Insufficiency of the Present Systems, To give us any*

Just Account of that Science: And the Necessity there is of some New Principles, In order to furnish us with a True and Real Knowledge of Nature (Cambridge: at the University-Press, for Edm. Jeffery, and James Knapton, and Benjamin Took, 1712).

———. *The Principles of the Philosophy of the Expansive and Contractive Forces or An Inquiry into the Principles of Modern Philosophy: that is, into the Several Chief Rational Sciences, which are Extant* (Cambridge: Cornelius Crownfield, E. Jefferys, W. Thurlbourne, J. Knapton, R. Knaplock, W. and J. Innys, and B. Motte, 1727).

Hales, Stephen. *Vegetable Staticks: Or, An Account of some Statical Experiments on the Sap in Vegetables: Being an Essay towards a Natural History of Vegetation. Also, a Specimen of An Attempt to Analyse the Air, By a great Variety of Chymio-Statical Experiments; which were read at several Meetings before the Royal Society* (London: W. and J. Innys, and T. Woodward, 1727); 1961 reprint, with foreword by M. A. Hoskin (London: Oldbourne, 1961).

———. *Statical Essays: Containing Haemastaticks; or, An Account of some Hydraulick and Hydrostatical Experiments made on the Blood and Blood-Vessels of Animals. Also An Account of some Experiments on Stones in the Kidneys and Bladder; with an Enquiry into the Nature of those anomalous Concretions. To which is added, An Appendix, Containing Observations and Experiments relating to several Subjects in the first Volume. The greatest Part of which were read at several Meetings before the Royal Society* (London: for W. Innys and R. Manby, and T. Woodward, 1733), 1964 facsimile reprint, with introduction by André Cournand (New York: No. 22, History of Medicine Series, New York Academy of Medicine, by Hafner Publishing Company, 1964).

Haller, Albrecht von. *A Dissertation on the Motion of the Blood and on the Effects of Bleeding. Verified by Experiments made on Living Animals. To which are added, Observations on the Heart, proving that Irritability is the primary Cause of its Motion* (London: J. Whiston and B. White, 1757).

———. "A Dissertation on the Sensible and Irritable Parts of Animals," with an introduction by Owsei Tempkin, *Bulletin of the Institute of Medicine,* 4 (1936), 651-99.

———. *First Lines of Physiology* (New York and London: Johnson Reprint Corporation, from the 1786 Edinburgh edn., Sources of Science No. 32, 1966), with an introduction by Lester S. King.

Harris, John. *Lexicon Technicum, or an Universal Dictionary of Arts and Sciences,* Vol. 2 (1710) (New York and London: Johnson Reprint Corporation, Sources of Science, No. 28, 1966).

Hartley, David. *Various Conjectures on the Perception, Motion, and Generation of Ideas (1746),* translated from the Latin by Robert E. A. Palmer, introduction and notes by Martin Kallich (Los Angeles:

Augustan Reprint Society, No. 77-78, William Andrews Clark Memorial Library, University of California, Los Angeles, 1959).

———. *Observations on Man, his Frame, his Duty, and his Expectations. Part the First: containing Observations on the Frame of the Human Body and Mind, and on their mutual Connexions and Influences* (London: first printed 1749, reprinted by J. Johnson, 1791).

Hauksbee, Francis. Papers in the *Philosophical Transactions*: 25 (1706-07), No. 305, 2223-24, No. 309, 2372-77, No. 311, 2409-11, 2412-13; 26 (1708-09), No. 315, 83-88, No. 318, 217-18, 221-22, No. 319, 267-68, 395-96, No. 320, 306-08; 27 (1710-11), No. 328, 199-203, 204-07, No. 332, 395-96, No. 334, 473-74, No. 336, 539-40; 28 (1712-13), No. 337, 151-52, 153-54, 155-56.

Hauksbee, F[rancis]. *Physico-Mechanical Experiments on Various Subjects* (London: by R. Brugis, for the author, 1709).

Helsham, Richard. *A Course of Lectures in Natural Philosophy* (London: J. Nourse, 1777), 5th edn.

Herschel, Sir William. *The Scientific Papers of Sir William Herschel*, ed. by J. L. E. Dreyer (London: The Royal Society and Royal Astronomical Society, 1912), 2 vols.

Hunter, John. "On the Digestion of the Stomach after Death," *Philosophical Transactions*, 62 (1772), 447-54.

———. "Experiments on Animals and Vegetables with respect to the Power of Producing Heat," *Philosophical Transactions*, 65 (1775), 446-58.

Hutchinson, John. *An Abstract from the Works of John Hutchinson, Esq; being a Summary of his Discoveries in Philosophy and Divinity* (London: for E. Withers, 1755), 2nd edn.

Hutton, James. "Theory of the Earth; or an Investigation of the Laws observeable in the Composition, Dissolution and Restoration of Land upon the Globe," *Transactions of the Royal Society of Edinburgh*, 1 (1788), 209-304.

———. *Dissertations on Different Subjects in Natural Philosophy* (Edinburgh: A. Strahan, and T. Cadell, London, 1792).

———. *An Investigation of the Principles of Knowledge and the Progress of Reason, from Sense to Science and Philosophy* (Edinburgh: A. Strahan, and T. Cadell, London, 1794).

———. *Theory of the Earth* (Edinburgh: Cadell, Junior, and Davies, London; and William Creech, Edinburgh, 1795).

Jones, William. *Philosophical Disquisitions: or, Discourses on the Natural Philosophy of the Elements* (London: J. Rivington and Sons, G. Robinson, D. Prince, Mess. Merrils, W. Keymer, Mrs. Drummons, W. Watson, 1781).

Jurin, James. "An Account of some Experiments . . . with an enquiry into the Cause of the Ascent and Suspension of Water in Capillary Tubes," *Philosophical Transactions*, 30 (1717-19), No. 355, 739-47.

———. "An Account of some new Experiments relating to the Action of Glass Tubes upon Water and Quicksilver," *Philosophical Transactions*, 30 (1717-19), No. 363, 1083-96.

———. "Concerning the Action of Springs," *Philosophical Transactions*, 43 (1744-45), No. 472, 46-71.

Keill, James. *Anatomy of the Human Body Abridg'd: or, A short and full View of all the Parts of the Body. Together with their several Uses drawn from their Compositions and Structures* (London: for John Clarke, 1759), 12th edn., corr.

———. *Essays on Several Parts of the Animal Oeconomy* (London: for George Strahan, 1738), 4th edn.

Keill, John. "In qua Leges Attractionis aliaque Physices Principia traduntur," *Philosophical Transactions*, 26 (1708-09), No. 315, 97-110; translated in Hutton, Shaw, Pearson, *et al.* edn. of *Philosophical Transactions of the Royal Society, Abridged* (London, 1809), vol. 5, pp. 407-24.

———. "Theoremata quaedam infinitam Materiae Divisibilitatem spectantia, quae ejusdem raritatem et tenuem compositionem demonstrant, quorum ope plurimae in Physica tolluntur difficultates," *Philosophical Transactions*, 29 (1714), No. 339, 82-86.

———. *An Examination of Dr. Burnet's Theory of the Earth: with some Remarks on Mr. Whiston's New Theory of the Earth. Also An Examination of the Reflections on the Theory of the Earth; and A Defence of the Remarks on Mr. Whiston's New Theory. To the whole is annexed a Dissertation on the Different Figures of the Caelestial Bodies, &c., with a Summary-Exposition of the Cartesian and Newtonian Systems,* by Mons. de Maupertuis (Oxford: H. Clements; and S. Harding, London, 1734), 2nd edn., corr.

———. *An Introduction to Natural Philosophy; or, Philosophical Lectures read in the University of Oxford, Anno Dom. 1700. To which are Added, The Demonstrations of Monsieur Huygens's theorems, concerning the Centrifugal Force and Circular Motion* (London: for William and John Innys, and John Osborn, 1720).

Knight, Gowin. "A Collection of the magnetical Experiments communicated to the Royal Society," *Philosophical Transactions*, 44 (1746-7), 656-72.

———. *An Attempt to demonstrate, that all the Phaenomena in Nature May be explained by Two simple active Principles, Attraction and Repulsion* (London: J. Nourse, 1754).

Langrish, Browne. "The Croonian Lectures on Muscular Motion," supplement to the *Philosophical Transactions*, 44 (1747-48).

Lavoisier, Antoine. *Elements of Chemistry*, translated by Robert Kerr, with a new introduction by Douglas McKie (New York: Dover Publications, Inc., reprint of the 1790 edn., 1965).

Leslie, John. "On Capillary Action," *Philosophical Magazine*, 14 (1802), 193-205.

———. *An Experimental Inquiry into The Nature and Propagation of Heat* (London: for J. Mawman, and sold by Bell and Bradfute, Edinburgh, 1804).

———. "On Heat and Climate," *Annals of Philosophy*, 14 (1819), 5-27.

———. "Observations on Electrical Theories," *Edinburgh Philosophical Journal*, 11 (1824), 1-39.

———. *Elements of Natural Philosophy* (Edinburgh: Oliver and Boyd, and Geo. B. Whittaker, London, 1829), 2nd edn.

Lovett, Richard. *The Electrical Philosopher. Containing a New System of Physics founded upon the Principle of an universal Plenum of elementary Fire, &c* (Worcester: for the Author, 1774).

Maclaurin, Colin. *An Account of Sir Isaac Newton's Philosophical Discoveries, in Four Books* (London: J. Nourse, W. Strahan, J. and F. Rivington, W. Johnston, D. Wilson, T. Lowndes, T. Cadell, T. Becket, W. Richardson, T. Longman, and W. Otridge, 1775), published from the author's manuscripts by Patrick Murdoch.

Macquer, Pierre Joseph. *Elements of the Theory and Practice of Chymistry*, translated by Andrew Reid (London: A. Millar and J. Nourse, 1764), 2nd edn.

[Macquer, P. J.]. *A Dictionary of Chemistry, containing the Theory and Practice of that Science*, translated by James Keir (London: T. Cadell and P. Elmsly, 1777), 2nd edn.

Martin, Benjamin. *An Essay on Electricity: Being an Enquiry into the Nature, Cause and Properties thereof, On the Principles of Sir Isaac Newton's Theory of Vibrating Motion, Light and Fire; . . . with Some Observations relative to the Uses that may be made of this Wonderful Power of Nature* (Bath: for the Author and Leake, Frederick, Raches, Collins, and Newbury, 1746).

———. *The Philosophical Grammar, Being a View of the Present State of Experimented Physiology or Natural Philosophy* (London: John Noon, 1755), 5th edn.

[Martine, George]. *An Examination of the Newtonian Argument, etc.* (London: T. Cooper, 1740).

Martine, George. *Essays Medical and Philosophical* (London: for A. Millar, 1740).

Maxwell, J. Clerk. "On the Dynamical Evidence of the Molecular Constitution of Bodies," *Journal of the Chemical Society*, 13 (ser. 2, 1875), 493-508.

Mead, Richard. *A Discourse concerning the Action of the Sun and Moon on Animal Bodies; and the Influence which this may have in many Diseases* (London: n.p., 1708).

———. *Mechanical Account of Poisons in several Essays* (London: J. M. for Ralph Smith, 1708), 2nd edn.

———. *The Medical Works of Richard Mead* (Dublin: Thomas Ewing, 1767). Includes: "A Mechanical Account of Poisons, in several Essays" and "A Treatise concerning the Influence of the Sun and Moon upon human Bodies, and the Diseases thereby produced."

Melvill, Thomas. "Discourse concerning the Cause of the different Refrangibility of the Rays of Light," *Philosophical Transactions*, 48 (1753), 261-70.

———. "Observations on Light and Colours," *Essays and Observations, Physical and Literary, Read before a Society in Edinburgh*, 2 (1756), 2-90.

Michell, J[ohn]. *A Treatise of Artificial Magnets; In which is shewn An easy and expeditious Method of making them, Superior to the best Natural Ones; and also, A Way of improving the Natural Ones, and of changing or converting their Poles* (Cambridge: W. and J. Mount, and T. Page, J. & P. Knapton, C. Bathurst, J. Beecroft, London, W. Thurlbourn, T. Merrill, Cambridge, J. Fletcher, Oxford, J. Hildyard, York, 1751) 2nd edn., corr. & improv. (1st edn., 1750).

———. "An Inquiry into the probable Parallax, and Magnitude of the fixed Stars, from the Quantity of Light which they afford us, and the particular Circumstances of their Situation," *Philosophical Transactions*, 57 (1767), 234-64.

———. "On the Means of discovering the Distance, Magnitude, etc., of the Fixed Stars, in consequence of the Diminution of the Velocity of their Light, in case such a Diminution should be found to take place in any of them, and such other Data should be procured from Observations, as would be farther necessary for that Purpose," *Philosophical Transactions*, 74 (1784), 35-57.

Miles, Henry. "Extracts of Two Letters . . . concerning the Effects of a Cane of Black Sealing-Wax, and a Cane of Brimstone in electrical Experiments," *Philosophical Transactions*, 44 (1746-47), 27-32.

———. "Part of a Letter . . . Concerning Electrical Fire," *Philosophical Transactions*, 44 (1746-47), 78-81.

Monroe, Donald. "An Account of some neutral Salts made with vegetable Acids, and with Salt of Amber, which shews that vegetable acids differ from one another . . . ," *Philosophical Transactions*, 57 (1767), 479-516.

Morgan, Tho[mas]. *Philosophical Principles of Medicine, in Three Parts. Containing, I. A Demonstration of the general Laws of Gravity, with their Effects upon Animal Bodys. II. The more particular Laws which obtain in the Motion and Secretion of the vital Fluids, applied to the principal Diseases and Irregularitys of the Animal Machine. III. The primary and chief Intentions of Medicine in the Cure of Diseases, problematically propos'd and mechanically resolv'd* (London: by J. Darby and T. Browne, for J. Osborne, T. Longman and J. Batley, F. Clay, E. Symon, S. Billingsley, and S. Chandler, 1725).

———. *Physico-Theology: or, A Philosophico-Moral Disquisition concerning Human Nature, Free Agency, Moral Government, and Divine Providence* (London: T. Cox, 1741).

Musschenbroek, Pieter van. *The Elements of Natural Philosophy.*

Chiefly intended for the Use of Students in Universities. Translated from the Latin by John Colson (London: J. Nourse, 1744), 2 vols.

Newton, Sir Isaac. *Philosophiae Naturalis Principia Mathematica* (London: Joseph Streator, 1687).

———. *The Mathematical Principles of Natural Philosophy,* translated into English by Andrew Motte (London: Benjamin Motte, 1729).

———. *Opticks: or, A Treatise of the Reflexions, Refractions, Inflexions and Colours of Light. Also Two Treatises of the Species and Magnitude of Curvilinear Figures* (London: Sam. Smith and Benj. Walford, 1704).

———. *Optice: sive de Reflexionibus, Refractionibus, Inflexionibus & Coloribus Lucis libris Tres* (London: Sam. Smith & Benj. Walford, 1706).

———. *Opticks: or, A Treatise of the Reflections, Refractions, Inflections and Colours of Light* (London: W. and J. Innys, 1718), 2nd edn., with additions.

Nicholson, William. *An Introduction to Natural Philosophy* (London: J. Johnson, 1805), 5th edn.

Nieuwentyt, Bernardus. *The Religious Philosopher: Or, the Right Use of Contemplating the Works of the Creator: I. In the wonderful Structure of Animal Bodies, and in particular, Man. II. In the no less wonderful and wise Formation of the Elements and their various Effects upon Animal and Vegetable Bodies. And, III. In the most amazing Structure of the Heavens, with all its Furniture. Designed for the Conviction of Atheists and Infidels.* Translated by John Chamberlayne, with prefixed letter by J. T. Desaguliers (London: by T. Wood, for J. Senex and W. Taylor, 1719), 2nd edn. 3 vols.

Parsons, James. "The Crounian Lectures, on Muscular Motion," supplement to the *Philosophical Transactions,* 43 (1744-45).

———. *Philosophical Observations on the Analogy between the Propagation of Animals and that of Vegetables: In which are answered Some Objections against the Indivisibility of the Soul which have been inadvertently drawn from the late curious and useful Experiments upon the Polypus and other Animals* (London: C. Davis, 1752).

Pemberton, Henry. *A View of Sir Isaac Newton's Philosophy* (London: by S. Palmer, 1728).

———. *A Course of Chymistry,* ed. James Wilson (London: J. Nourse, 1771).

Pike, Samuel. *Philosophia Sacra: or, the Principles of Natural Philosophy. Extracted from Divine Revelation* (London: for the Author, and sold by J. Buckland, 1753).

Pitcairne, Archibald. *The Philosophical and Mathematical Elements of Physick* (London: for W. Innys, T. Longman, and T. Shewell, and Aaron Ward, 1745).

Priestley, Joseph. *History and Present State of Electricity* (London: J. Dodsley, J. Johnson, B. Davenport, and T. Cadell, 1767).
———. *History and Present State of Electricity* (New York: Johnson Reprint Corporation, Sources of Science, No. 18, reprint of 3rd edn. of 1775, 1966).
———. *History and Present State of Discoveries relating to Vision, Light and Colours* (London: J. Johnson, 1772).
———. *Experiments and Observations on different Kinds of Air* (London: J. Johnson, 1775), 2nd edn.
———. *Experiments and Observations on different Kinds of Air*, vol. 2 (London: J. Johnson, 1776), 2nd edn.
———. *Experiments and Observations on different Kinds of Air*, vol. 3 (London: J. Johnson, 1777).
———. *Disquisitions relating to Matter and Spirit* (London: J. Johnson, 1777).
———. *A Free Discussion of the Doctrines of Materialism and Philosophical Necessity, in a Correspondence between Dr. Price and Dr. Priestley* (London: J. Johnson and T. Cadell, 1778).
———. *Experiments and Observations on Various Branches of Natural Philosophy* (London: J. Johnson, 1779).
———. *Experiments and Observations on Various Branches of Natural Philosophy*, vol. 2 (London: Joseph Johnson, 1781).
———. *Experiments and Observations on Various Branches of Natural Philosophy*, vol. 3 (London: J. Johnson, 1786).
———. *Experiments on the Generation of Air from Water* (London: J. Johnson, 1793).
———. *Heads of Lectures on a Course of Experimental Philosophy* (London: J. Johnson, 1794).
———. "Miscellaneous Observations relating to the Doctrine of Air," *New York Medical Repository*, 5 (1802), 264-67.
———. "On the Transmission of Liquid in Vapour, #6 Of the proportion of latent heat in some kinds of Air," *Transactions of the American Philosophical Society*, 5 (1802), 10-11.
———. "Experiments relating to Change of Place in different Kinds of Air, through several interposing Substances," *Transactions of the American Philosophical Society*, 5 (1802), 14-20.
———. *Memoirs of Dr. Joseph Priestley* (London: J. Johnson, 1806).
[Purshall, Conyers]. *An Essay at the Mechanism of the Macrocosm: or the Dependance of Effects upon their Causes, in a New Hypothesis, Accommodated to Our Modern and Experimental Philosophy. In which are solved several Phaenomena, hitherto unaccounted for; as the Cause of Gravitation, Motion, Reflexion, Refractions, & c. With a Method Proposed to found out the Exact Rate that a Ship Runs, and consequently the Longitude at Sea* (London: by F. Collins, for Jeffery Wale, 1705).
Robinson, Bryan. *A Treatise of the Animal Oeconomy* (Dublin: George Grurson, 1732).

———. *Dissertation on the Aether of Sir Isaac Newton* (Dublin: Geo. Ewing and Wil. Smith, 1743).

R[obinson], B[ryan]. *Sir Isaac Newton's Account of the Aether, with some Additions by way of Appendix* (Dublin: G. and A. Ewing, and W. Smith, 1745).

[Robison, John]. Article: "Boscovich," *Supplement, Encyclopaedia Britannica* (Edinburgh: for Thomson Bonar, 1801), vol. 1, pp. 96-110, 3rd edn.

Robison, John. *A System of Mechanical Philosophy*, ed. David Brewster (Edinburgh: for John Murray, 1822).

Rohault, Jacob. *Rohault's System of Natural Philosophy, Illustrated with Dr. Samuel Clarke's Notes Taken mostly out of Sir Isaac Newton's Philosophy. With Additions.* Done into English by John Clarke (London: for James Knapton, 1723), 2 vols.

Rowning, John. *A Compendious System of Natural Philosophy* (London: Sam. Harding, 1737-43); Prefatory material, 1743, 1st edn.; Part I, 1738, 3rd edn.; Part II, 1737, 3rd edn.; Part III, 1743, 2nd edn.; Part IV, 1743, 1st edn.

———. *Compleat Course of Experimental Philosophy and Astronomy* (n.p., n.d.), a syllabus preserved in the Science Museum, Oxford.

Rutherforth, Thomas. *A System of Natural Philosophy, being a Course of Lectures in Mechanics, Optics, Hydrostatics, and Astronomy; which are read in St. John's College Cambridge* (Cambridge: by J. Bentham, for W. Thurlbourn, Cambridge, and sold by J. Beecroft, London, 1748), 2 vols.

Savery, Servington. "Magnetical Observations and Experiments," *Philosophical Transactions*, 36 (1729-30), 295-340.

Shaw, Peter. *Chemical Lectures, Publickly read at London, In the Years 1731, and 1732; And at Scarborough, in 1733; For the Improvement of Arts, Trades, and Natural Philosophy* (London: T. and T. Longman, J. Shuckburgh, and A. Millar, 1755), 2nd edn., corr.

Smith, Robert. *A Compleat System of Opticks* (Cambridge: for the Author, sold by Cornelius Crownfield, and by Stephen Austen and Robert Dodsley, 1738), 2 vols.

Stahl, George Ernest. *Philosophical Principles of Universal Chemistry*, translated from the Collegium Jenese by Peter Shaw (London: J. Osborn and T. Longman, 1730).

Stuart, Alexander. "Explanation of an Essay on the Use of Bile in the Animal Oeconomy," *Philosophical Transactions*, 38 (1733), 5-25.

———. "Three Lectures on Muscular Motion," supplement to the *Philosophical Transactions*, 40 (1739).

Stukeley, William. "The Philosophy of Earthquakes," *Philosophical Transactions*, 46 (1750), 731-750.

———. *The Family Memoirs of the Rev. William Stukeley, M.D. and the Antiquarian and other correspondence of William Stuke-*

ley, Roger and Samuel Gale, etc., ed. W. C. Lukis (Durham: for the Surtees Society, 1882-83).

Symmer, Robert. "New Experiments and Observations concerning Electricity," *Philosophical Transactions*, 51 (1759), 340-89.

Taylor, Brook. "Part of a Letter . . . concerning the Ascent of Water between two Glass Planes," *Philosophical Transactions*, 27 (1710-11), No. 336, 538.

———."Extract of a Letter . . . giving an Account of some Experiments relating to Magnetism," *Philosophical Transactions*, 31 (1720-21), No. 368, 204-08.

Thomson, J. J. *The Corpuscular Theory of Matter* (New York: Charles Scribner's Sons, 1907).

[Thomson, Thomas]. Article: "Chemistry," *Encyclopaedia Britannica, Supplement* to the 3rd edn. (Edinburgh: for Thomson Bonar, 1802).

Tytler, James. Articles: "Electricity," "Fire," and (with John Gleig) "Motion," in *Encyclopaedia Britannica; or, A Dictionary of Arts, Sciences, and Miscellaneous Literature, etc.* (Edinburgh: A. Bell and C. Macfarquhar, 1797), 3rd edn.

Watson, William. Papers on Electricity, in *Philosophical Transactions*: 43 (1744-45), 481-502; 44 (1746-47), 41-50, 704-09; 45 (1747-48), 93-120; 47 (1751-52), 202-11, 362-76; 48 (1753-54), 201-16; 52 (1761-62), 336-43; 54 (1764), 201-27.

Wheler, Granville. "Some electrical experiments, chiefly regarding the repulsive force of electrical bodies," *Philosophical Transactions*, 41 (1739-40), 98-111.

Whiston, William. *Historical Memoirs of the Life of Dr. Samuel Clarke* (London: Fletcher Gyles and J. Roberts, 1730).

Whytt, Robert. *The Works of Robert Whytt* (Edinburgh: T. Becket and P. A. de Hondt, and J. Balfour, 1768).

Wilson, Benjamin. *An Essay towards an Explication of the Phaenomena of Electricity, Deduced from the Æther of Sir Isaac Newton* (London: C. Davis, 1746).

Worster, Benjamin. *A Compendious and Methodical Account of the Principles of Natural Philosophy. As Explained and Illustrated in the Course of Experiments, performed at the Academy in Little-Tower-Street* (London: Stephen Austen, 1730), 2nd edn., revised and corrected.

Young, Thomas. *A Course of Lectures on Natural Philosophy and the Mechanical Arts* (London: J. Johnson, 1807).

Secondary Works

GENERAL REFERENCES I have, of course, made use of the standard reference works, including the *Dictionary of National Biography*, the *Dictionary of American Biography*, (Michaud) *Biographie Universelle*, (Hoefer) *Nouvelle Biographie Générale*, and the catalogues of printed books in the Bibliothèque Nationale, the British Museum, and the Library of Congress.

UNPUBLISHED PAPERS, ESSAYS, AND THESES

Gerstner, Patsy A. "James Hutton's Theory of Matter and his Geological Theory," unpublished M.A. research essay, Case Western Reserve University, 1966.

Gillispie, Charles Coulston. "Devoid of Atoms," unpublished paper, read at a conference on 18th-century chemistry, held in Paris, summer 1959.

[Lindstrom], Carolyn Vetter. "Jean Théophile Desaguliers' theory of elasticity and his theory of matter," unpublished research paper, Case Western Reserve University, 1967.

Massouh, Michael. "Dr. Cadwallader Colden, Iatro-Mechanism to Vitalism in 18th Century Natural Philosophy," unpublished M.A. research essay, Case Western Reserve University, 1967.

McCormmach, Russell. "Electrical Researches of Henry Cavendish," unpublished Ph.D. thesis, Case Western Reserve University, 1967.

Olson, Richard G. "Sir John Leslie: 1766-1832. A Study of the Pursuit of the Exact Sciences in the Scottish Enlightenment," unpublished Ph.D. dissertation, Harvard University, 1967.

Pav, Peter Anton. "Eighteenth-Century Optics: the Age of Unenlightenment," unpublished Ph.D. dissertation, Indiana University, 1964 (University Microfilms 65-3510).

Scott, Wilson L. "Sources of Boerhaave's Medical Lectures on Physics, with Particular Reference to the Significance of these Lectures to Physical Chemistry," unpublished M.A. essay, for The Johns Hopkins University, Faculty of Philosophy, 1955.

Thackray, Arnold W. "The Newtonian Tradition and Eighteenth-Century Chemistry," unpublished Ph.D. dissertation, Cambridge University, 1967.

BOOKS AND PAPERS

Agassi, Joseph. "Towards a Historiography of Science," *History and Theory*, Beiheft 2 (1963).

Alexander, H. G., ed. *Leibniz-Clarke Correspondence* (Manchester: Manchester University Press, 1956).

Armitage, Angus. *William Herschel* (London: Thomas Nelson & Sons Ltd., 1962).

Atkinson, A. D. "William Derham, F.R.S. (1657-1735)," *Annals of Science*, 8 (1952), 368-92.

Bailey, Edward Battersby. *James Hutton—the Founder of Modern Geology* (London: Elsevier Publishing Co. Ltd., 1967).

Balz, Albert G. A. *Cartesian Studies* (New York: Columbia University Press, 1951).

Bastholm, E. *The History of Muscle Physiology* (Copenhagen, Acta Historica Scientiarum Naturalium et Medicinalium, vol. 7, University Library, Copenhagen, 1950).

Becker, Carl. *The Heavenly City of the Eighteenth-Century Philosophers* (New Haven: Yale University Press, 1932).

Berry, A. J. *Henry Cavendish: His Life and Scientific Work* (London: Hutchinson, 1960).

Beth, E. W. "Nieuwentijt's Significance for the Philosophy of Science," *Synthese*, 9 (1955), 447-64.

Brewster, Sir David. *Memoirs of the Life, Writings and Discoveries of Sir Isaac Newton* (New York and London: Johnson Reprint Corporation, Sources of Science No. 14, from the Edinburgh Edn. 1855, 1965).

Brunet, Pierre. *Les Physiciens Hollandais et la Méthode Expérimentale en France au XVIIIe Siècle* (Paris: Librairie Scientifique Albert Blanchard, 1926).

———. *L'introduction des théories de Newton en France au XVIIIe siècle. I: avant 1738* (Paris: Librairie Scientifique Albert Blanchard, 1931).

Burr, Alex C. "Notes on the History of the Concept of Thermal Conductivity," *Isis*, 20 (1933), 246-59.

———. "Notes on the History of Experimental Determinations of the Thermal Conductivity of Gases," *Isis*, 21 (1934), 169-86.

Carmichael, Leonard. "Robert Whytt: A Contribution to the History of Physiological Psychology," *Psychological Review*, 34 (1927), 287-304.

Cassirer, Ernst. *The Platonic Renaissance in England*, translated by James P. Pettegrove (Austin: University of Texas Press, 1953).

Chipman, Robert A. "An Unpublished Letter of Stephen Gray on Electrical Experiments, 1707-1708," *Isis*, 45 (1954), 33-40.

———. "The Manuscript Letters of Stephen Gray, F.R.S. (1666/7-1736)," *Isis*, 49 (1958), 414-33.

Clark-Kennedy, A. E. *Stephen Hales D.D., F.R.S.* (Cambridge: at the University Press, 1929).

Cochrane, Rexmond D. "Francis Bacon and the Rise of the Mechanical Arts in Eighteenth-Century England," *Annals of Science*, 12 (1956, publ. 1957), 137-56.

Cohen, I. Bernard. "Neglected Sources for the Life of Stephen Gray (1666 or 1667-1736)," *Isis*, 45 (1954), 41-50.

———. *Franklin and Newton* (Philadelphia: The American Philosophical Society, 1956).

———, ed. *Isaac Newton's Papers & Letters On Natural Philosophy and related documents* (Cambridge: Harvard University Press, 1958).

———. " 'Quantum In Se Est': Newton's Concept of Inertia in Relation to Descartes and Lucretius," *Notes and Records of the Royal Society of London*, 19 (1964), 131-55.

Coleby, L.J.M. "John Francis Vigani, first professor of chemistry in the University of Cambridge," *Annals of Science*, 8 (1952), 46-60.

———. "John Mickleburgh," *Annals of Science*, 8 (1952), 165-74.

Crosland, Maurice P. "The Use of Diagrams as Chemical 'Equations' in the Lecture Notes of William Cullen and Joseph Black," *Annals of Science*, 15 (1959, publ. 1961), 75-90.

Daumas, Maurice. "Les Conceptions de Lavoisier sur les Affinités Chimiques et la Constitution de la Matière," *Thalès*, 6 (1949-1950, publ. 1951), 69-80.

Davis, Tenney L. "Vicissitudes of Boerhaave's Textbook of Chemistry," *Isis*, 10 (1928), 33-46.

———. "Boyle's conception of element compared with that of Lavoisier," *Isis*, 16 (1931), 82-91.

Dawson, Percy M. "Stephen Hales, the Physiologist," *Bulletin of the Johns Hopkins Hospital*, 15 (1904), 232-37.

De Morgan, Augustus. *A Budget of Paradoxes* (London: Longmans, Green, and Co., 1872).

Dobbin, Leonard. "A Cullen Chemical Manuscript of 1753," *Annals of Science*, 1 (1936), 138-56.

Drake, Stillman. *Discoveries and Opinions of Galileo* (New York: Doubleday Anchor Books, 1959).

Duncan, A. M. "William Keir's *de Attractione Chemica* (1778) and the Concepts of Chemical Saturation, Attraction, and Repulsion," *Annals of Science*, 23 (1967), 149-73.

Dyment, S. A. "Some Eighteenth Century Ideas Concerning Aqueous Vapours and Evaporation," *Annals of Science*, 2 (1937), 465-73.

Farrar, W. V. "Nineteenth-century Speculations on the Complexity of the Chemical Elements," *British Journal for the History of Science*, 2 (1965), 297-323.

Foster, Joseph. *Alumni Oxonienses: The Members of the University of Oxford 1500-1714: Their Parentage, Birthplace, and Year of Birth, with a Record of their Degrees* (Oxford: James Parker & Co., 1891), vol. 3.

Fulton, John F. *A Bibliography of the Honourable Robert Boyle, Fellow of the Royal Society* (Oxford: at the Clarendon Press, 1961), 2nd edn.

Geikie, Sir Archibald. *The Founders of Geology* (New York: Dover Press, reprint of the 2nd edn. of 1905, 1962).

———. *Geology at the Beginning of the Nineteenth Century* (London: Geological Society, 1907).

———. *Memoir of John Michell* (Cambridge: at the University Press, 1918).

Gerstner, Patsy A. "James Hutton's Theory of the Earth and his Theory of Matter," *Isis*, 59 (1968), 26-31.

Gibbs, F. W. "Peter Shaw and the Revival of Chemistry," *Annals of Science*, 7 (1951), 211-37.

———. "William Lewis, M.B., F.R.S. (1708-1781)," *Annals of Science*, 8 (1952), 122-51.

———. *Joseph Priestley, Revolutions of the Eighteenth Century* (Garden City, N.Y.: Doubleday and Co., Inc., 1967).

Gibson, Reginald. *Francis Bacon: A Bibliography of his Works and of Baconiana to the Year 1750* (Oxford: at the Scrivener Press, 1950).

Gillispie, Charles C. *Genesis and Geology* (New York: Harper Torchbook, 1951).

———. *Edge of Objectivity* (Princeton, N.J.: Princeton University Press, 1960).

Gourli, Norah. *The Prince of Botanists. Carl Linnaeus* (London: H. F. and G. Witherby Ltd., 1953).

Guerlac, Henry. "The Continental Reputation of Stephen Hales," *Archives Internationales d'Histoire des Sciences,* 4 (1951), 393-404.

———. "Joseph Black and Fixed Air: A bicentenary Retrospective, with some new or little known material," *Isis,* 48 (1957), 124-51, 433-56.

———. "Newton's Changing Reputation in the Eighteenth Century," in Raymond O. Rockwood, ed., *Carl Becker's Heavenly City Revisited* (Ithaca, N.Y.: Cornell University Press, 1958).

———. *Lavoisier—The Crucial Year* (Ithaca, N.Y.: Cornell University Press, 1961).

———. "Francis Hauksbee: expérimentateur au profit de Newton," *Archives Internationales d'Histoire des Sciences,* 16 (1963), 113-28.

———. "Newton et Epicure," Conférence donnée au Palais de la Découverte le 2 Mars 1963, D 91, Histoire des Sciences (Paris: Université de Paris, Palais de la Découverte, [1963]).

———. "Sir Isaac and the Ingenious Mr. Hauksbee," in I. Bernard Cohen and René Taton, eds., *Mélanges Alexandre Koyré,* vol. 1, "L' Aventure de la Science" (Paris: Hermann, 1964), pp. 228-53.

———. "Newton's Optical Aether," *Notes and Records of the Royal Society of London,* 22 (1967), 45-57.

Gunther, Robert T. *Early Science in Oxford,* vol. 1, "Chemistry, Mathematics, Physics and Surveying" (Oxford: for the subscribers, 1923).

———. *Early Science in Cambridge* (Oxford: for the Author, 1937).

Hall, A. Rupert. *From Galileo to Newton, 1630-1720* (New York and Evanston: Harper & Row, 1963).

Hall, A. Rupert, and Hall, Marie Boas. *Unpublished Scientific Papers of Isaac Newton. A Selection from the Portsmouth Collection in the University Library, Cambridge* (Cambridge: at the University Press, 1962).

Hall, Marie Boas. "Matter in Seventeenth Century Science," in Ernan McMullin, *The Concept of Matter* (Notre Dame: University of Notre Dame Press, 1963), pp. 344-67.

———. *Robert Boyle on Natural Philosophy, An Essay with Selections from His Writings* (Bloomington: Indiana University Press, 1965).

[Hall,] Marie Boas, and Hall, Rupert. "Newton's 'Mechanical Principles,'" *Journal of the History of Ideas,* 20 (1959), 167-78.

Hardin, Clyde L. "The Scientific Work of the Reverend John Michell," *Annals of Science,* 22 (1966), 27-47.

Hardy, W. B. "Historical Notes upon Surface Energy and Forces of Short Range," *Nature*, 109 (1922), 375-78.

Hartley, Sir Harold. *Humphry Davy* (London: Thomas Nelson and Sons, Ltd., 1966).

Hartog, Sir Philip. "Newer Views of Priestley and Lavoisier," *Annals of Science*, 5 (1941), 1-56.

Hesse, Mary B. *Forces and Fields. The concept of Action at a Distance in the history of physics* (London: Thomas Nelson and Sons Ltd., 1961).

Hindle, Brooke. "Cadwallader Colden's Extension of the Newtonian Principles," *William and Mary Quarterly*, 12 (3rd ser., 1956), 459-75.

Hiscock, W. G. *David Gregory, Isaac Newton and their Circle. Extracts from David Gregory's Memoranda 1677-1708* (Oxford: for the editor, 1937).

Hobbs, Wm. H. "James Hutton, the Pioneer of Modern Geology," *Science*, 64 (1926), 265.

Hochdoerfer, Margarete. *The Conflict between the Religious and the Scientific Views of Albrecht von Haller (1708-1777)* (Lincoln, Nebraska: University of Nebraska Studies in Language, Literature, and Criticism, No. 12, 1932).

Holden, Edward S. *Sir William Herschel, His Life and Works* (New York: Charles Scribner's Sons, 1881).

Hoskin, Michael A. *William Herschel. Pioneer of Sidereal Astronomy* (New York: Sheed and Ward, 1959).

———. " 'Mining All Within': Clarke's Notes To Rohault's *Traité de Physique*," *The Thomist*, 24 (1961), 353-63.

———. *William Herschel and the Construction of the Heavens* (London: Oldbourne Press, 1963).

Hughes, Arthur. "Science in English Encyclopaedias, 1704-1875—I," *Annals of Science*, 7 (1951), 340-70.

———. "Science in English Encyclopaedias, 1704-1875—II. Theories of the Elementary Composition of Matter," *Annals of Science*, 8 (1952), 323-67.

Huxley, George. "Roger Cotes and Natural Philosophy," *Scripta Mathematica*, 26 (1961), 231-38.

Jourdain, Philip E. B. "The Principles of Mechanics with Newton from 1666 to 1669," *Monist*, 24 (1914), 188-224.

———. "The Principles of Mechanics with Newton from 1679 to 1687," *Monist*, 24 (1914), 515-64.

———. "Newton's Hypotheses of Ether and of Gravitation from 1672 to 1679," *Monist*, 25 (1915), 79-106.

———. "Newton's Hypotheses of Ether and of Gravitation from 1679 to 1693," *Monist*, 25 (1915), 234-54.

———. "Newton's Hypotheses of Ether and Gravitation from 1693 to 1726," *Monist*, 25 (1915), 418-40.

Kargon, Robert H. "The Decline of the Caloric Theory of Heat: A Case Study," *Centaurus*, 10 (1964), 35-39.

———. *Atomism in England from Hariot to Newton* (Oxford: Clarendon Press, Oxford University Press, 1966).

Kerker, Milton. "Herman Boerhaave and the Development of Pneumatic Chemistry," *Isis,* 46 (1955), 36-49.

King, Lester S. *The Medical World of the Eighteenth Century* (Chicago: University of Chicago Press, 1958).

———. *The Growth of Medical Thought* (Chicago: University of Chicago Press, 1963).

———. "Stahl and Hoffmann: A Study in Eighteenth Century Animism," *Journal of the History of Medicine and Allied Sciences,* 19 (1964), 118-30.

Knight, David M. "The Atomic Theory and the Elements," *Studies in Romanticism,* 5 (1966), 185-207.

———. "Steps towards a Dynamical Chemistry," *Ambix,* 14 (1967), 179-97.

———. "The Scientist as Sage," *Studies in Romanticism,* 6 (1967), 65-88.

Koyré, Alexandre. "The Significance of the Newtonian Synthesis," *Archives Internationales d'Histoire des Sciences,* 29 (1950), 291-311.

———. *Newtonian Studies* (London: Chapman & Hall, 1965).

Kuhn, Thomas S. "The Caloric Theory of Adiabatic Compression," *Isis,* 49 (1958), 132-40.

Lilley, S. "Attitudes to the Nature of Heat about the beginning of the nineteenth century," *Archives Internationales d'Histoire des Sciences,* 27 (1947-48), 630-39.

Lindeboom, G. A. "Pitcairne's Leyden Interlude described from the Documents," *Annals of Science,* 19 (1963), 273-84.

Lovejoy, Arthur O. *Essays in the History of Ideas* (Baltimore: Johns Hopkins Press, 1948).

McCormmach, Russell. "Henry Cavendish: Rational Natural Philosopher," *Isis,* in press.

———. "John Michell and Henry Cavendish—Weighing the Stars," *British Journal of the History of Science,* in press.

McKendrick, John G. "Abstract of Lectures on Physiological Discovery. Lecture I. The Circulation of the Blood, A Problem in Hydrodynamics," *British Medical Journal* 1 (1883), 654-55.

McKie, Douglas. "On Thos. Cochrane's MS notes of Black's Chemical Lectures, 1767/8," *Annals of Science,* 1 (1936), 101-10.

———. "Some Notes on Newton's Chemical Philosophy written upon the Occasion of the Tercentenary of his Birth," *Philosophical Magazine,* 33 (ser. 7, 1942), 847-70.

———. "On some MS. Copies of Black's Chemical Lectures," *Annals of Science,* 21 (1965, publ. 1966), 209-55.

McKie, Douglas, and Heathcote, Niels H. de V. *The Discovery of Specific and Latent Heats* (London: Edward Arnold and Co., 1935).

McKie, Douglas, and Kennedy, David. "On Some Letters of Joseph Black and Others," *Annals of Science*, 16 (1960), 129-70.

McMullin, Ernan, ed. *The Concept of Matter* (Notre Dame: University of Notre Dame Press, 1963).

Manuel, Frank E. "Newton as Autocrat of Science," *Daedalus, the Proceedings of the American Academy of Arts and Sciences*, 97 (1968), 969-1001.

Meldrum, Andrew Norman. "The Development of the Atomic Theory: (3) Newton's Theory and its Influence in the Eighteenth Century," *Memoirs of the Manchester Literary and Philosophical Society*, 55, No. 4 (1910), 1-15.

———. *The Eighteenth Century Revolution in Science—First Phase* (Calcutta: Longmans, Green, 1930).

Mendelsohn, Everett. *Heat and Life, The Development of the Theory of Animal Heat* (Cambridge: Harvard University Press, 1964).

Merz, John Theodore. *History of European Thought in the Nineteenth Century* (New York: Dover Publications Inc., reprint of 1904-12 edn., 1965).

Metzger, Hélène. "La philosophie de la matière chez Stahl et ses disciples," *Isis*, 8 (1926), 427-64.

———. "La théorie de la composition des sels et la théorie de la combustion d'après Stahl et ses disciples," *Isis*, 9 (1927), 294-325.

———. "Newton: La Théorie de l'Émission de la Lumière et la Doctrine Chimique au XVIIIème Siècle," *Archeion*, 11 (1929), 13-25.

———. *Newton, Stahl, Boerhaave et la Doctrine Chimique* (Paris: F. Alcan, 1930).

———. *Attraction Universelle et Religion Naturelle chez quelques Commentateurs Anglais de Newton* (Paris: Hermann & Cie, 1938).

Millington, E. C. "Studies in Capillarity and Cohesion in the Eighteenth Century," *Annals of Science*, 5 (1947), 352-69.

Mischel, T. "Emotion and Motivation in the Development of English Psychology: D. Hartley, James Mill, A. Bain," *Journal of the History of the Behavioral Sciences*, 2 (1966), 123-44.

Nichols, John. *Literary Anecdotes of the Eighteenth Century, &c.* (London: for the Author, 1812).

———. *Illustrations of the Literary History of the Eighteenth Century* (London: for the Author, 1817).

Nicolson, Marjorie Hope. *Newton Demands the Muse* (Princeton: Princeton University Press, 1946).

Pannekoek, A. *A History of Astronomy* (New York: Interscience Publishers, Inc., 1961).

Partington, James R. "The Origins of the Atomic Theory," *Annals of Science*, 4 (1939), 245-82.

———. *History of Chemistry*, vol. 3 (London: Macmillan and Co., Ltd., 1962).

Peacock, George. *Life of Thomas Young, M.D., F.R.S., etc.* (London: John Murray, 1855).

Peck, E. Saville. "John Francis Vigani, first Professor of Chemistry in the University of Cambridge (1703-12), and his Materia Medica Cabinet in the Library of Queens' College," *Communications of the Cambridge Antiquarian Society,* 34 (1933), 34-49.

Piggott, Stuart. *William Stukeley, An Eighteenth-Century Antiquary* (Oxford: at the Clarendon Press, 1950).

Playfair, John. "Biography of James Hutton," *Transactions of the Royal Society of Edinburgh,* 5 (1805), 74.

Ramsay, Sir William. *The Life and Letters of Joseph Black, M.D.* (London: Constable and Company Ltd., 1918).

Rand, Benjamin. "Early Development of Hartley's Doctrine of Association," *Psychological Review,* 30 (1923), 306-20.

Randolph, Herbert, ed. *Life of General Sir Robert Wilson* (London: John Murray, 1862), vol. 1.

Reilly, J., and O'Flynn, N. "Richard Kirwan, an Irish Chemist of the XVIIIth Century," *Isis,* 13 (1929-30), 298-319.

Ritterbush, Philip C. *Overtures to Biology: The Speculations of Eighteenth-Century Naturalists* (New Haven: Yale University Press, 1964).

Rutt, J. T. *Life and Correspondence of Joseph Priestley* (London: R. Hunter, 1831-32), vol. 1.

Sachs, Julius von. *History of Botany* (Oxford: at the Clarendon Press, 1890).

Schenk, H. G. *The Mind of the European Romantics* (London: Constable & Co., Ltd., 1966).

Schofield, Robert E. "John Wesley and Science in Eighteenth-Century England," *Isis,* 44 (1953), 331-40.

———, ed. *A Scientific Autobiography of Joseph Priestley* (1733-1804) (Cambridge: The M.I.T. Press, 1966).

———. "Joseph Priestley Natural Philosopher," *Ambix,* 14 (1967), 1-15.

Seller, William. "Memoir of the Life and Writings of Robert Whytt, M.D., Professor of Medicine in the University of Edinburgh from 1747-1766," *Transactions of the Royal Society of Edinburgh,* 23 (1864), 99-131.

Sinclair, H. M., and Robb-Smith, A.H.T. *A Short History of Anatomical Teaching in Oxford* (Oxford, Oxford University Press 1950).

Singer, Dorothea Waley. "Sir John Pringle and his Circle.—Part II. Public Health," *Annals of Science,* 6 (1950), 229-61.

Smeaton, William A. "Guyton de Morveau and Chemical Affinity," *Ambix,* 11 (1963), 55-64.

Smith, Joe William Ashley. *Birth of Modern Education* (London: Independent Press, 1954).

Smith, Preserved. *The Enlightenment 1687-1776,* vol. II of *A History of Modern Culture* (New York: Collier Books, 1962), first published 1934.

Snow, A. J. *Matter & Gravity in Newton's Physical Philosophy. A*

Study in the Natural Philosophy of Newton's Time (London: Humphry Milford for Oxford University Press, 1926).
Stewart, Agnes Grainger. *The Academic Gregories* (Edinburgh and London: Oliphant Anderson & Ferrier, Famous Scots Series, 1901).
Stones, G. B. "The Atomic View of Matter in the XV, XVI and XVIIth Centuries," *Isis*, 10 (1928), 445-65.
Taylor, E.G.R. *Mathematical Practitioners of Hanoverian England 1714-1840* (Cambridge: for the Institute of Navigation, at the University Press, 1966).
Thomson, John. *An Account of the Life, Lectures, and Writings of William Cullen, M.D.*, first published in 1832, now re-issued along with the second volume and having prefixed to it a biographical notice of the author (Edinburgh and London: William Blackwood & Sons, 1859).
Thorpe, Jocelyn. "Stephen Hales, D.D.F.R.S. 1677-1761," *Notes & Records of the Royal Society*, 3 (1940), 53-63.
Trengrove, L. "Chemistry at the Royal Society of London in the Eighteenth Century—I," *Annals of Science*, 19 (1963), 183-237.
Truesdell, Clifford. "Rational Fluid Mechanics, 1687-1765," Introduction to: *Leonhard Euleri Opera Omnia*, ser. 2, vol. 12, "Commentations mechanicae ad theoream Corporum Fluidorum Pertinentes" (Lausanne: Society of Natural Sciences of Helvetia, 1954).
Vavilov, S. I. "Newton and the Atomic Theory" in *Newton Tercentenary Celebrations 15-19 July 1946* (Cambridge, at the University Press 1947), 43-55.
Venn, John, and Venn, J. A. *Alumni Cantabrigienses* (Cambridge: at the University Press, 1927).
Walker, W. C. "The Detection and Estimation of Electric Charges in the Eighteenth Century," *Annals of Science*, 1 (1936), 66-100.
Watson, Richard A. *The Downfall of Cartesianism, 1673-1712. A Study of Epistemological Issues in Late 17th Century Cartesianism* (The Hague: Martinus Nijhoff, No. II. International Archives of the History of Ideas, 1966).
Weld, D. R. *History of the Royal Society* (London: John W. Parker, 1848).
Whyte, Lancelot Law, ed. *Roger Joseph Boscovich, S.J., F.R.S., 1711-1787* (London: George Allen and Unwin Ltd., 1961).
Wightman, William P. D. "Gregory's Notae in Isaaci Newtoni Principia Philosophiae," *Nature*, 172 (1953), 690.
———. "David Gregory's Commentary on Newton's Principia," *Nature*, 179 (1957), 393-94.
———. "William Cullen and the Teaching of Chemistry—II," *Annals of Science*, 12 (1956, publ. 1957), 192-205.
Williams, L. Pearce. *Michael Faraday, A Biography* (London: Chapman and Hall, 1965).
Wilson, George. *Life of the Honble. Henry Cavendish* (London: for the Cavendish Society, 1851).

Wolf, Abraham. *History of Science, Technology and Philosophy in the 18th Century*, ed. by Douglas McKie (London: Allen and Unwin [1952]), 2nd edn.

Wood, Alex. *Thomas Young, Natural Philosopher 1773-1829* (Cambridge: at the University Press, 1954), completed by Frank Oldham.

Wordsworth, Christopher. *Scholae Academicae: Some Account of the Studies at English Universities in the Eighteenth Century* (Cambridge: at the University Press, 1877).

INDEX